The Philosophy of Science

A Collection of Essays

Series Editor

Lawrence Sklar
University of Michigan

A Garland Series
Readings in Philosophy

Robert Nozick, *Advisor*
Harvard University

Series Contents

1. Explanation, Law and Cause
2. The Nature of Scientific Theory
3. Theory Reduction and Theory Change
4. Probability and Confirmation
5. Bayesian and Non-Inductive Methods
6. The Philosophy of Physics

The Nature of Scientific Theory

Edited with an introduction by

Lawrence Sklar
University of Michigan

GARLAND PUBLISHING, INC.
A MEMBER OF THE TAYLOR & FRANCIS GROUP
New York & London
2000

Library of Congress Cataloging-in-Publication Data

Philosophy of science : a collection of essays / edited with an introduction by Lawrence Sklar.
p. cm.
"A Garland series"—Ser. t.p.
Includes bibliographical references.
Contents: 1. Explanation, law, and cause — 2. The nature of scientific theory — 3. Theory reduction and theory change — 4. Probability and confirmation — 5. Bayesian and non-inductive methods — 6. The philosophy of physics.
ISBN 0-8153-2700-5 (v.1 : alk. paper) — ISBN 0-8153-2701-3 (v. 2 : alk. paper) — ISBN 0-8153-2702-1 (v. 3 : alk. paper) — ISBN 0-8153-2703-X (v. 4 : alk. paper) — ISBN 0-8153-3492-3 (v. 5 : alk. paper) — ISBN 0-8153-3493-1 (v. 6 : alk. paper)
1. Science—Philosophy. 2. Science—Methodology. 3. Physics—Philosophy. I. Sklar, Lawrence.

Q175.P51227 1999
501—dc21 99-40012
CIP

Printed on acid-free, 250-year-life paper
Manufactured in the United States of America

Contents

Introduction

From the earliest days of science, we find scientists proposing to explain the observable phenomena of the world by postulating the existence of entities and properties of entities not known to our common-sense picture of the world. In early Greek astronomy, for example, we have the explanations of the motions of the heavenly bodies that postulated these bodies to be lights attached to "crystalline spheres." Contemporary science, of course, is replete with the postulation of vast numbers of such so-called "theoretical" entities and properties. We find scientists telling us about the existence of genes, of molecules, of atoms and subatomic particles, for example. And even more exotic entities and properties are posited. For example, we are told that there are such things as electromagnetic fields that pervade all of space. We are even told that space-time itself exists as a needed explanatory entity in our scientific account of the world.

But again from the earliest days of science, we find skeptical arguments against the legitimacy of postulating these theoretical entities and properties. The skepticism is most likely to be directed against the reasonableness of postulating things and their features that are taken by science to be beyond the range of direct observation. And the skepticism is at its most intense when there are grounds for claiming that the postulated things and their properties are "in principle" forever outside of the range of possible observation.

Many Greek astronomers claimed, for example, that the aim of astronomy was not to discover the "real mechanisms" behind the motion of the heavenly bodies, but only to "save the phenomena" of the observed positions of the visible lights in the sky, that is to correctly predict the positions of the visible heavenly lights over time. This view of astronomy was appealed to much later, when some Copernicans, wary of opposition to their views by the religious, once again claimed for astronomy only the instrumental role of saving the observable phenomena. One can find similar skepticism with regard to taking the postulated unobservable things and properties of the atomic theory of matter in a straight-forward manner in the proposals of the Energeticists in the nineteenth century to view theories as merely simplest, most "economical," or most elegant ways of summarizing the lawlike relations among the observable features of the world.

Beyond skeptical doubts about the legitimacy of postulating such entities, there are deep questions about how we even understand what we are talking about when we make assertions about the realm of the unobservable. Here it is not so much the truth of claims about the realm of being beyond the possibility of direct observation that is in question, but, rather, questions about how we could ever come to attribute meaning of any kind to the expressions we intend to use to refer to the unobservable things and features of the world.

Many accounts of the roles of theories in science assume that there is a coherent and principled distinction between what is observable and what is not observable. This distinction is generally supposed to be one that is not highly context dependent, and it is supposed to be a kind of distinction "in principle" that allows for philosophically interesting distinctions between the way in which assertions about the observable get their meaning from the way in which meaning accrues to assertions about the unobservable. But it is not easy to characterize what is supposed to count as the genuinely observable features of the world.

This problem is taken up in the readings in Section A of this volume.

Even given some presupposed legitimacy to the distinction of observables from unobservables, many questions remain about how we are to understand the theories of science that make reference to allegedly unobservable entities and properties.

How, for example, assuming that we know the meanings of terms and sentences referring to the observable, are we to determine whether or not assertions about the unobservable are meaningful enough to count as legitimate scientific claims? Early positivists and operationalists thought that for a theoretical term to be meaningful, it would have to be provided with an explicit definition in terms of concepts referring to the observables. From that point of view, there would not be any theoretical terms, properly so-called.

But must there be some sort of explicit definability of the expressions relating to the unobservable in terms of expressions directed at the observable if the terms playing roles in theories are to be intelligible? Might not language referring to the unobservable obtain its legitimacy in some "holistic" manner, by that language playing a significant role in a wide structure of assertions that is tied down to observation only as a whole theoretical network of assertions?

And if we are to understand our theoretical discourse in this more holistic manner, exactly how should we offer a philosophically formal analysis of the ways in which theoretical networks introduce reference to the unobservable into our scientific accounts of the world? There are a number of distinct proposals as to how to use our formal apparatus of syntax and semantics of languages along with formal logic, to try and do justice to the structure of theories in science. The hope is that such formal reconstructions of scientific theories will help us to understand their asserted content and will provide resources useful in addressing philosophical questions about the roles of theories in science.

A number of efforts have been made to come to grips with an understanding of how terms can acquire their meaning from the role that they play in theories. One initial move was to try and understand the meaning of so-called dispositional terms (such as 'soluble' or 'fragile') in terms of a sophisticated generalization of the notion of definition by observation terms (reduction sentences).

Another important contribution to the discussion was the introduction of what are called Ramsey sentences. Here a move is made that will, for theories whose formulation is given in terms of an appropriate logical format, give clear and exact sense to the idea of the meaning of a theoretical term. A term that has been introduced into science by a theory in which that term plays a part, can be construed holistically by specifying the total role played in the theory in question by that term. From this perspective theoretical terms are "placeholders" in the network of general hypotheses that go to make up the total theory.

Several different approaches exist to the idea of giving theories of science formal reconstructions. Some of these approaches look upon theories as sets of sentences organized into deductive structures. Other approaches to formalizing theories in science rely, rather, upon the devices of formal semantics. Here the notions of models, structures used to interpret sentences and to supply reference to their singular and general terms, are invoked in order to give reconstructions of informal theories which will regiment those theories in ways that make them suitable for deeper conceptual analysis.

It is issues of these sorts that are taken up in the papers of Section B.

A long-standing philosophical debate concerns the degree to which we ought to take the apparent asserted content of our theories in a completely straight-forward manner. When we posit theoretical entities and their properties, we seem to be positing a realm of being not present to us in observation. But when these theories are put forth, are we truly claiming the existence of things and features of the world that had previously gone unsuspected in their reality? Or, rather, are we simply constructing some sort of linguistic device that will prove useful to us in our goal of being able to predict what will happen in the observable world and control what will happen in that world?

One source of skepticism about taking the claims of science to be claims that ought to be understood as intended to offer naively true utterances rests on the fact that scientific assertions are often (always?) qualified by conditions under which they are to be taken as accurate. These conditions are often not fully expressed in science, and are, perhaps, not fully and explicitly expressible at all. This has sometimes led to claims to the effect that the assertions ought to be taken not as being intended to give true descriptions of the world, but as serving as "instruments" for scientific prediction and control. Given the applicability of our scientific assertions only against an inexplicit background of qualifications and conditions, how are we to understand the meaning and purpose of the assertions themselves?

A more common ground for denying simple assertion as the intended goal of the scientific theorist is the claim that insofar as the theories reach beyong the realm of the observable, we could never have good reason for asserting their theoretical content to be truly the case. In a stronger version of this thesis, it is alleged that we could never have good reason for thinking that we even understand the content of our theoretical discourse, if that discourse is to be thought of as trying to give naively true descriptions of how things are in the world.

Numerous philosophers of an "empiricist" or "positivist" tendency have offered over the years accounts of theories that argue, sometimes, that the theories ought to be understood not as making asserted claims about how the world of the unobservable is really constructed, but, rather, as providing systematic linguistic instruments to summarize the content of scientifically established relationships within the realm of the observable phenomena. Such claims date back, once again, to Greek astronomy, where, as we have noted, it was often argued that the aim of theory was to "save the phenomena" and not to say how unobservable things really were.

Some versions of such views suggest translatability of theoretical discourse into observational discourse. Some speak of theoretical postulation as the positing of "fictions." Others characterize theories as "formal instruments" that license inferences from observable facts to observable predictions, but which theories fail to make explicit asserted claims that go beyond the realm of the observable.

Another version of such an "empiricism" is to argue that science is indifferent to the truth or falsity of its theoretical claims, but only concerned with the "empirical adequacy" of its postulated theories. Theories might, then, make assertive claims about how things stand with the unobservables, but scientists will only care, when they are trying to decide whether or not to accept a theory, whether the theory does justice to the possible observational facts. According to this view, scientists will be indifferent as to the full truth of the theories they accept, demanding of them only that they be empirically adequate.

Much of the debate about how to understand the meaning and role of theories in science assumes that there are deep differences between the way in which we became committed to the existence of the ordinary things of our everyday environment, and the way in which the scientist becomes committed to the existence of those unobservable entities and properties postulated in scientific theories. But there are also questions about whether such a distinction between everyday discourse and scientific theoretical discourse really exists. That is, there are questions about just why some philosophers think that the issue of the "interpretation" of theories is so important and so difficult, and whether their concerns are well grounded.

Suppose one accepts the view that theories often seem to make assertions about realms of being that are, in some sense, "in principle" immune from observational determination. The possibility then arises that there will be alternative theories, alternatives that at least appear to be incompatible with one another, that will have exactly the same observational consequences. This may be the case if the theories have their differences from one another confined to the realm of the unobservable, and if these differences "cancel out" in such a way as to make the empirical facts predicted by the theories exactly the same. Such pairs of theories are said to be "empirically equivalent" to one another.

The notion of empirical equivalence gives rise to many recurrent philosophical questions: Ought we to think of any two empirically equivalent theories as being fully equivalent to one another, that is as saying exactly the same things about the world despite their apparent differences from one another at the theoretical level? If we think of the theories as not fully equivalent, despite their empirical equivalence, how might we decide which theory to accept or which theory to think of as "true" of the world? In the final analysis on what could the decision rest as to whether to think of the theories as fully equivalent or, instead, to think of them as genuine alternatives between which observational and experimental facts could not decide?

These are the issues taken up in the readings in Section C of this volume.

GROVER MAXWELL

The Ontological Status of Theoretical Entities

That anyone today should seriously contend that the entities referred to by scientific theories are only convenient fictions, or that talk about such entities is translatable without remainder into talk about sense contents or everyday physical objects, or that such talk should be regarded as belonging to a mere calculating device and, thus, without cognitive content—such contentions strike me as so incongruous with the scientific and rational attitude and practice that I feel this paper *should* turn out to be a demolition of straw men. But the instrumentalist views of outstanding physicists such as Bohr and Heisenberg are too well known to be cited, and in a recent book of great competence, Professor Ernest Nagel concludes that "the opposition between [the realist and the instrumentalist] views [of theories] is a conflict over preferred modes of speech" and "the question as to which of them is the 'correct position' has only terminological interest."[1] The phoenix, it seems, will not be laid to rest.

The literature on the subject is, of course, voluminous, and a comprehensive treatment of the problem is far beyond the scope of one essay. I shall limit myself to a small number of constructive arguments (for a radically realistic interpretation of theories) and to a critical examination of some of the more crucial assumptions (sometimes tacit, sometimes explicit) that seem to have generated most of the problems in this area.[2]

[1] E. Nagel, *The Structure of Science* (New York: Harcourt, Brace, and World, 1961), Ch. 6.

[2] For the genesis and part of the content of some of the ideas expressed herein, I am indebted to a number of sources; some of the more influential are H. Feigl, "Existential Hypotheses," *Philosophy of Science*, 17:35–62 (1950); P. K. Feyerabend, "An Attempt at a Realistic Interpretation of Experience," *Proceedings of the Aristotelian Society*, 58:144–170 (1958); N. R. Hanson, *Patterns of Discovery* (Cambridge: Cambridge University Press, 1958); E. Nagel, *loc. cit.*; Karl Popper, *The Logic of Scientific Discovery* (London: Hutchinson, 1959); M. Scriven, "Definitions, Explanations, and Theories," in *Minnesota Studies in the Philosophy of Science*,

Grover Maxwell

The Problem

Although this essay is not comprehensive, it aspires to be fairly self-contained. Let me, therefore, give a pseudohistorical introduction to the problem with a piece of science fiction (or fictional science).

In the days before the advent of microscopes, there lived a Pasteur-like scientist whom, following the usual custom, I shall call Jones. Reflecting on the fact that certain diseases seemed to be transmitted from one person to another by means of bodily contact or by contact with articles handled previously by an afflicted person, Jones began to speculate about the mechanism of the transmission. As a "heuristic crutch," he recalled that there is an obvious *observable* mechanism for transmission of certain afflictions (such as body lice), and he postulated that all, or most, infectious diseases were spread in a similar manner but that in most cases the corresponding "bugs" were too small to be seen and, possibly, that some of them lived inside the bodies of their hosts. Jones proceeded to develop his theory and to examine its testable consequences. Some of these seemed to be of great importance for preventing the spread of disease.

After years of struggle with incredulous recalcitrance, Jones managed to get some of his preventative measures adopted. Contact with or proximity to diseased persons was avoided when possible, and articles which they handled were "disinfected" (a word coined by Jones) either by means of high temperatures or by treating them with certain toxic preparations which Jones termed "disinfectants." The results were spectacular: within ten years the death rate had declined 40 per cent. Jones and his theory received their well-deserved recognition.

However, the "crobes" (the theoretical term coined by Jones to refer to the disease-producing organisms) aroused considerable anxiety among many of the philosophers and philosophically inclined scientists of the day. The expression of this anxiety usually began something like this: "In order to account for the facts, Jones must assume that his crobes are too small to be seen. Thus the very postulates of his theory preclude

Vol. II, H. Feigl, M. Scriven, and G. Maxwell, eds. (Minneapolis: University of Minnesota Press, 1958); Wilfrid Sellars, "Empiricism and the Philosophy of Mind," in *Minnesota Studies in the Philosophy of Science*, Vol. I, H. Feigl and M. Scriven, eds. (Minneapolis: University of Minnesota Press, 1956), and "The Language of Theories," in *Current Issues in the Philosophy of Science*, H. Feigl and G. Maxwell, eds. (New York: Holt, Rinehart, and Winston, 1961).

their being observed; they are *unobservable in principle*." (Recall that no one had envisaged such a thing as a microscope.) This common prefatory remark was then followed by a number of different "analyses" and "interpretations" of Jones' theory. According to one of these, the tiny organisms were merely convenient fictions—*façons de parler*—extremely useful as heuristic devices for facilitating (in the "context of discovery") the thinking of scientists but not to be taken seriously in the sphere of cognitive knowledge (in the "context of justification"). A closely related view was that Jones' theory was merely an instrument, useful for organizing observation statements and (thus) for producing desired results, and that, therefore, it made no more sense to ask what was the nature of the entities to which it referred than it did to ask what was the nature of the entities to which a hammer or any other tool referred.[3] "Yes," a philosopher might have said, "Jones' theoretical expressions are just meaningless sounds or marks on paper which, when correlated with observation sentences by appropriate syntactical rules, enable us to predict successfully and otherwise organize data in a convenient fashion." These philosophers called themselves "instrumentalists."

According to another view (which, however, soon became unfashionable), although expressions containing Jones' theoretical terms were genuine sentences, they were translatable without remainder into a set (perhaps infinite) of observation sentences. For example, 'There are crobes of disease X on this article' was said to translate into something like this: 'If a person handles this article without taking certain precautions, he will (probably) contract disease X; and if this article is first raised to a high temperature, then if a person handles it at any time afterward, before it comes into contact with another person with disease X, he will (probably) not contract disease X; and . . .'

Now virtually all who held any of the views so far noted granted, even insisted, that theories played a useful and legitimate role in the scientific enterprise. Their concern was the elimination of "pseudo problems" which might arise, say, when one began wondering about the "reality of supraempirical entities," etc. However, there was also a school of thought, founded by a psychologist named Pelter, which differed in an

[3] I have borrowed the hammer analogy from E. Nagel, "Science and [Feigl's] Semantic Realism," *Philosophy of Science*, 17:174–181 (1950), but it should be pointed out that Professor Nagel makes it clear that he does not necessarily subscribe to the view which he is explaining.

interesting manner from such positions as these. Its members held that while Jones' crobes might very well exist and enjoy "full-blown reality," they should not be the concern of medical research at all. They insisted that if Jones had employed the correct methodology, he would have discovered, even sooner and with much less effort, all of the observation laws relating to disease contraction, transmission, etc. without introducing superfluous links (the crobes) into the causal chain.

Now, lest any reader find himself waxing impatient, let me hasten to emphasize that this crude parody is not intended to convince anyone, or even to cast serious doubt upon sophisticated varieties of any of the reductionistic positions caricatured (some of them not too severely, I would contend) above. I am well aware that there are theoretical entities and theoretical entities, some of whose conceptual and theoretical statuses differ in important respects from Jones' crobes. (I shall discuss some of these later.) Allow me, then, to bring the Jonesean prelude to our examination of observability to a hasty conclusion.

Now Jones had the good fortune to live to see the invention of the compound microscope. His crobes were "observed" in great detail, and it became possible to identify the specific kind of *microbe* (for so they began to be called) which was responsible for each different disease. Some philosophers freely admitted error and were converted to realist positions concerning theories. Others resorted to subjective idealism or to a thoroughgoing phenomenalism, of which there were two principal varieties. According to one, the one "legitimate" observation language had for its descriptive terms only those which referred to sense data. The other maintained the stronger thesis that *all* "factual" statements were *translatable* without remainder into the sense-datum language. In either case, any two non-sense data (e.g., a theoretical entity and what would ordinarily be called an "observable physical object") had virtually the same status. Others contrived means of modifying their views much less drastically. One group maintained that Jones' crobes actually never had been unobservable in principle, for, they said, the theory did not imply the impossibility of finding a means (e.g., the microscope) of observing them. A more radical contention was that the crobes were not observed at all; it was argued that what was seen by means of the microscope was just a shadow or an image rather than a corporeal organism.

The Observational-Theoretical Dichotomy

Let us turn from these fictional philosophical positions and consider some of the actual ones to which they roughly correspond. Taking the last one first, it is interesting to note the following passage from Bergmann: "But it is only fair to point out that if this . . . methodological and terminological analysis [for the thesis that there are no atoms] . . . is strictly adhered to, even stars and microscopic objects are not physical things in a literal sense, but merely by courtesy of language and pictorial imagination. This might seem awkward. But when I look through a microscope, all I see is a patch of color which creeps through the field like a shadow over a wall. And a shadow, though real, is certainly not a physical thing."[4]

I should like to point out that it is also the case that if this analysis is strictly adhered to, we cannot observe physical things through opera glasses, or even through ordinary spectacles, and one begins to wonder about the status of what we see through an ordinary windowpane. And what about distortions due to temperature gradients—however small and, thus, always present—in the ambient air? It really *does* "seem awkward" to say that when people who wear glasses describe what they see they are talking about shadows, while those who employ unaided vision talk about physical things—or that when we look through a windowpane, we can only *infer* that it is raining, while if we raise the window, we may "observe directly" that it is. The point I am making is that there is, in principle, a continuous series beginning with looking through a vacuum and containing these as members: looking through a windowpane, looking through glasses, looking through binoculars, looking through a low-power microscope, looking through a high-power microscope, etc., in the order given. The important consequence is that, so far, we are left without criteria which would enable us to draw a non-arbitrary line between "observation" and "theory." Certainly, we will often find it convenient to draw such a to-some-extent-arbitrary line; but its position will vary widely from context to context. (For example, if we are determining the resolving characteristics of a certain microscope, we would certainly draw the line beyond ordinary spectacles, probably

[4] G. Bergmann, "Outline of an Empiricist Philosophy of Physics," *American Journal of Physics*, 11: 248–258; 335–342 (1943), reprinted in *Readings in the Philosophy of Science*, H. Feigl and M. Brodbeck, eds. (New York: Appleton-Century-Crofts, 1953), pp. 262–287.

beyond simple magnifying glasses, and possibly beyond another microscope with a lower power of resolution.) But what ontological ice does a mere methodologically convenient observational-theoretical dichotomy cut? Does an entity attain physical thinghood and/or "real existence" in one context only to lose it in another? Or, we may ask, recalling the continuity from observable to unobservable, is what is seen through spectacles a "little bit less real" or does it "exist to a slightly less extent" than what is observed by unaided vision?[5]

However, it might be argued that things seen through spectacles and binoculars look like ordinary physical objects, while those seen through microscopes and telescopes look like shadows and patches of light. I can only reply that this does not seem to me to be the case, particularly when looking at the moon, or even Saturn, through a telescope or when looking at a small, though "directly observable," physical object through a low-power microscope. Thus, again, a continuity appears.

"But," it might be objected, "theory tells us that what we see by means of a microscope is a real image, which is certainly distinct from the object on the stage." Now first of all, it should be remarked that it seems odd that one who is espousing an austere empiricism which requires a sharp observational-language/theoretical-language distinction (and one in which the former language has a privileged status) should need a theory in order to tell him what is observable. But, letting this pass, what is to prevent us from saying that we still observe the object on the stage, even though a "real image" may be involved? Otherwise, we shall be strongly tempted by phenomenalistic demons, and at this point we are considering a physical-object observation language rather than a sense-datum one. (Compare the traditional puzzles: Do I see one physical object or two when I punch my eyeball? Does one object split into two? Or do I see one object and one image? Etc.)

Another argument for the continuous transition from the observable to the unobservable (theoretical) may be adduced from theoretical con-

[5] I am not attributing to Professor Bergmann the absurd views suggested by these questions. He seems to take a sense-datum language as his observation language (the base of what he called "the empirical hierarchy"), and, in some ways, such a position is more difficult to refute than one which purports to take an "observable-physical-object" view. However, I believe that demolishing the straw men with which I am now dealing amounts to desirable preliminary "therapy." Some nonrealist interpretations of theories which embody the presupposition that the observable-theoretical distinction is sharp and ontologically crucial seem to me to entail positions which correspond to such straw men rather closely.

siderations themselves. For example, contemporary valency theory tells us that there is a virtually continuous transition from very small molecules (such as those of hydrogen) through "medium-sized" ones (such as those of the fatty acids, polypeptides, proteins, and viruses) to extremely large ones (such as crystals of the salts, diamonds, and lumps of polymeric plastic). The molecules in the last-mentioned group are macro, "directly observable" physical objects but are, nevertheless, genuine, single molecules; on the other hand, those in the first mentioned group have the same perplexing properties as subatomic particles (de Broglie waves, Heisenberg indeterminacy, etc.). Are we to say that a large protein molecule (e.g., a virus) which can be "seen" only with an electron microscope is a little less real or exists to somewhat less an extent than does a molecule of a polymer which can be seen with an optical microscope? And does a hydrogen molecule partake of only an infinitesimal portion of existence or reality? Although there certainly *is* a continuous transition from observability to unobservability, any talk of such a continuity from full-blown existence to nonexistence is, clearly, nonsense.

Let us now consider the next to last modified position which was adopted by our fictional philosophers. According to them, it is only those entities which are *in principle* impossible to observe that present special problems. What kind of impossibility is meant here? Without going into a detailed discussion of the various types of impossibility, about which there is abundant literature with which the reader is no doubt familiar, I shall assume what usually seems to be granted by most philosophers who talk of entities which are unobservable in principle—i.e., that the theory(s) itself (coupled with a physiological theory of perception, I would add) entails that such entities are unobservable.

We should immediately note that if this analysis of the notion of unobservability (and, hence, of observability) is accepted, then its use as a means of delimiting the observation language seems to be precluded for those philosophers who regard theoretical expressions as elements of a calculating device—as meaningless strings of symbols. For suppose they wished to determine whether or not 'electron' was a theoretical term. First, they must see whether the theory entails the sentence 'Electrons are unobservable.' So far, so good, for their calculating devices are said to be able to select genuine sentences, provided they contain no theoretical terms. But what about the selected "sentence" itself? Suppose that 'electron' is an observation term. It follows that the expression is a genuine sentence and asserts that electrons are unobservable. But this entails that 'electron' is *not* an observation term. Thus if 'electron' is an observation term, then it is *not* an observation term. Therefore it is not an observation term. But then it follows that 'Electrons are unobservable' is not a genuine sentence and does not assert that electrons are unobservable, since it is a meaningless string of marks and does not assert anything whatever. Of course, it could be stipulated that when a theory "selects" a meaningless expression of the form 'Xs are unobservable,' then 'X' is to be taken as a theoretical term. But this seems rather arbitrary.

But, assuming that well-formed theoretical expressions are genuine sentences, what shall we say about unobservability in principle? I shall begin by putting my head on the block and argue that the present-day status of, say, electrons is in many ways similar to that of Jones' crobes before microscopes were invented. I am well aware of the numerous theoretical arguments for the impossibility of observing electrons. But suppose new entities are discovered which interact with electrons in such a mild manner that if an electron is, say, in an eigenstate of position, then, in certain circumstances, the interaction does not disturb it. Suppose also that a drug is discovered which vastly alters the human perceptual apparatus—perhaps even activates latent capacities so that a new sense modality emerges. Finally, suppose that in our altered state we are able to perceive (not necessarily visually) by means of these new entities in a manner roughly analogous to that by which we now see by means of photons. To make this a little more plausible, suppose that the energy eigenstates of the electrons in some of the compounds present in the relevant perceptual organ are such that even the weak interaction with the new entities alters them and also that the cross sections, relative to the new entities, of the electrons and other particles of the gases of the air are so small that the chance of any interaction here is negligible. Then we might be able to "observe directly" the position and possibly the approximate diameter and other properties of some electrons. It would follow, of course, that quantum theory would have to be altered in some respects, since the new entities do not conform to all its principles. But however improbable this may be, it does not, I maintain, involve any logical or conceptual absurdity. Furthermore, the modification necessary for the inclusion of the new entities would not necessarily change the meaning of the term 'electron.'[6]

Consider a somewhat less fantastic example, and one which does not involve any change in physical theory. Suppose a human mutant is born who is able to "observe" ultraviolet radiation, or even X rays, in the same way we "observe" visible light.

Now I think that it is extremely improbable that we will ever observe electrons directly (i.e., that it will ever be reasonable to assert that we have so observed them). But this is neither here nor there; it is not the purpose of this essay to predict the future development of scientific theories, and, hence, it is not its business to decide what actually is observable or what will become observable (in the more or less intuitive sense of 'observable' with which we are now working). After all, we are operating, here, under the assumption that it is theory, and thus science itself, which tells us what is or is not, in this sense, observable (the 'in principle' seems to have become superfluous). And this is the heart of the matter; for it follows that, at least for this sense of 'observable,' there are no a priori or philosophical criteria for separating the observable from the unobservable. By trying to show that we can talk about the *possibility* of observing electrons without committing logical or conceptual blunders, I have been trying to support the thesis that any (nonlogical) term is a *possible* candidate for an observation term.

There is another line which may be taken in regard to delimitation of the observation language. According to it, the proper term with which to work is not 'observable' but, rather 'observed.' There immediately comes to mind the tradition beginning with Locke and Hume (No idea without a preceding impression!), running through Logical Atomism and the Principle of Acquaintance, and ending (perhaps) in contemporary positivism. Since the numerous facets of this tradition have been extensively examined and criticized in the literature, I shall limit myself here to a few summary remarks.

Again, let us consider at this point only observation languages which contain ordinary physical-object terms (along with observation predicates, etc., of course). Now, according to this view, all descriptive terms of the observation language must refer to that which has been observed.

[6] For arguments that it is possible to alter a theory without altering the meanings of its terms, see my "Meaning Postulates in Scientific Theories," in *Current Issues in the Philosophy of Science*, Feigl and Maxwell, eds.

How is this to be interpreted? Not too narrowly, presumably, otherwise each language user would have a different observation language. The name of my Aunt Mamie, of California, whom I have never seen, would not be in my observation language, nor would 'snow' be an observation term for many Floridians. One could, of course, set off the observation language by means of this awkward restriction, but then, obviously, not being the referent of an observation term would have no bearing on the ontological status of Aunt Mamie or that of snow.

Perhaps it is intended that the referents of observation terms must be members of a *kind* some of whose members have been observed or instances of a *property* some of whose instances have been observed. But there are familiar difficulties here. For example, given any entity, we can always find a kind whose only member is the entity in question; and surely expressions such as 'men over 14 feet tall' should be counted as observational even though no instances of the "property" of being a man over 14 feet tall have been observed. It would seem that this approach must soon fall back upon some notion of simples or determinables vs. determinates. But is it thereby saved? If it is held that only those terms which refer to observed simples or observed determinates are observation terms, we need only remind ourselves of such instances as Hume's notorious missing shade of blue. And if it is contended that in order to be an observation term an expression must at least refer to an observed determinable, then we can always find such a determinable which is broad enough in scope to embrace any entity whatever. But even if these difficulties can be circumvented, we see (as we knew all along) that this approach leads inevitably into phenomenalism, which is a view with which we have not been concerning ourselves.

Now it is not the purpose of this essay to give a detailed critique of phenomenalism. For the most part, I simply assume that it is untenable, at least in any of its translatability varieties.[7] However, if there are any unreconstructed phenomenalists among the readers, my purpose, insofar as they are concerned, will have been largely achieved if they will grant what I suppose most of them would stoutly maintain anyway, i.e., that theoretical entities are no worse off than so-called observable physical objects.

[7] The reader is no doubt familiar with the abundant literature concerned with this issue. See, for example, Sellars' "Empiricism and the Philosophy of Mind," which also contains references to other pertinent works.

Nevertheless, a few considerations concerning phenomenalism and related matters may cast some light upon the observational-theoretical dichotomy and, perhaps, upon the nature of the "observation language." As a preface, allow me some overdue remarks on the latter. Although I have contended that the line between the observable and the unobservable is diffuse, that it shifts from one scientific problem to another, and that it is constantly being pushed toward the "unobservable" end of the spectrum as we develop better means of observation—better instruments—it would, nevertheless, be fatuous to minimize the importance of the observation base, for it is absolutely necessary as a confirmation base for statements which do refer to entities which are unobservable at a given time. But we should take as its basis and its unit not the "observational term" but, rather, the quickly decidable sentence. (I am indebted to Feyerabend, *loc. cit.*, for this terminology.) A quickly decidable sentence (in the technical sense employed here) may be defined as a singular, nonanalytic sentence such that a reliable, reasonably sophisticated language user can very quickly decide[8] whether to assert it or deny it when he is reporting on an occurrent situation. 'Observation term' may now be defined as a 'descriptive (nonlogical) term which may occur in a quickly decidable sentence,' and 'observation sentence' as a 'sentence whose only descriptive terms are observation terms.'

Returning to phenomenalism, let me emphasize that I am not among those philosophers who hold that there are no such things as sense contents (even sense data), nor do I believe that they play no important role in our perception of "reality." But the fact remains that the referents of most (not all) of the statements of the linguistic framework used in everyday life and in science are *not* sense contents but, rather, physical objects and other publicly observable entities. Except for pains, odors, "inner states," etc., *we do not usually observe sense contents*; and although there is good reason to believe that they play an indispensable role in observation, *we are usually not aware of them when we* visually or tactilely) *observe physical objects*. For example, when I observe a distorted, obliquely reflected image in a mirror, I may seem to be seeing a baby elephant standing on its head; later I discover it is an image of Uncle Charles taking a nap with his mouth open and his hand in a peculiar position. Or, passing my neighbor's home at a high rate of

[8] We may say "noninferentially" decide, provided this is interpreted liberally enough to avoid starting the entire controversy about observability all over again.

speed, I observe that he is washing a car. If asked to report these observations I could quickly and easily report a baby elephant and a washing of a car; I probably would not, without subsequent observations, be able to report what colors, shapes, etc. (i.e., what sense data) were involved.

Two questions naturally arise at this point. How is it that we can (sometimes) quickly decide the truth or falsity of a pertinent observation sentence? and, What role do sense contents play in the appropriate tokening of such sentences? The heart of the matter is that these are primarily scientific-theoretical questions rather than "purely logical," "purely conceptual," or "purely epistemological." If theoretical physics, psychology, neurophysiology, etc., were sufficiently advanced, we could give satisfactory answers to these questions, using, in all likelihood, the physical-thing language as our observation language and *treating sensations, sense contents, sense data, and "inner states"* as *theoretical* (yes, theoretical!) *entities.*[9]

It is interesting and important to note that, even before we give completely satisfactory answers to the two questions considered above, we can, with due effort and reflection, train ourselves to "observe directly" what were once theoretical entities—the sense contents (color sensations, etc.)—involved in our perception of physical things. As has been pointed out before, we can also come to observe other kinds of entities which were once theoretical. Those which most readily come to mind involve the use of instruments as aids to observation. Indeed, using our painfully acquired theoretical knowledge of the world, we come to see that we "directly observe" many kinds of so-called theoretical things. After listening to a dull speech while sitting on a hard bench, we begin to become poignantly aware of the presence of a considerably strong gravitational field, and as Professor Feyerabend is fond of pointing out, if we were carrying a heavy suitcase in a changing gravitational field, we could observe the changes of the $G_{\mu\nu}$ of the metric tensor.

I conclude that our drawing of the observational-theoretical line at any given point is an accident and a function of our physiological make-

[9] Cf. Sellars, "Empiricism and the Philosophy of Mind." As Professor Sellars points out, this is the crux of the "other-minds" problem. Sensations and inner states (relative to an intersubjective observation language, I would add) are theoretical entities (and they "really exist") and *not* merely actual and/or possible behavior. Surely it is the unwillingness to countenance theoretical entities—the hope that every sentence is translatable not only into some observation language but into the physical-thing language—which is responsible for the "logical behaviorism" of the neo-Wittgensteinians.

up, our current state of knowledge, and the instruments we happen to have available and, therefore, that it has no ontological significance whatever.

What If We COULD Eliminate Theoretical Terms?

Among the candidates for methods of eliminating theoretical terms, three have received the lion's share of current attention: explicit definability, the Ramsey sentence,[10] and implications of Craig's theorem.[11] Today there is almost (not quite) universal agreement that not all theoretical terms can be eliminated by explicitly defining them in terms of observation terms. It seems to have been overlooked that even if this could be accomplished it would not necessarily avoid reference to unobservable (theoretical) entities. One example should make this evident. Within the elementary kinetic theory of gases we could define 'molecules' as 'particles of matter (or stuff), not large enough to be seen even with a microscope, which are in rapid motion, frequently colliding with each other, and are the constituents of all gases.' All the (nonlogical) terms in the definiens are observation terms, and still the definition itself, as well as kinetic theory (and other theoretical considerations), implies that molecules of gases are unobservable (at least for the present).

It seems to me that a large number—certainly not all, however; for example, 'photon,' 'electromagnetic field,' 'ψ-function'—of theoretical terms could be explicitly defined wholly in terms of observation terms, but this would in no way avoid a reference to unobservable entities. This important fact seems to have been quite generally overlooked. It is an important oversight because philosophers today are devoting so much attention to the meaning of theoretical terms (a crucially important problem, to be sure), while the ontological stomach-aches (ultimately unjustifiable, of course) concerning theories seem to have arisen from the fact that the *entities* rather than the *terms* were nonobservational. Implicit, of course, is the mistaken assumption that terms referring to unobservable entities cannot be among those which occur in the observation language (and also, perhaps, the assumption that the referent of a defined term always consists of a mere "bundle" of the entities which are referents of the terms of the definiens).

[10] Frank P. Ramsey, *The Foundations of Mathematics* (New York: Humanities, 1931).

[11] William Craig, "Replacement of Auxiliary Expressions," *Philosophical Review*, 65:38–55 (1956).

Surprisingly enough, both the Ramsey sentence and Craig's theorem provide us with genuine (in principle) methods for eliminating theoretical terms provided we are interested only in the deductive "observational" consequences of an axiomatized theory. That neither can provide a viable method for avoiding reference to theoretical entities has been pointed out clearly by both Hempel and Nagel.[12] I shall discuss these two devices only briefly.[13]

The first step in forming the Ramsey sentence of a theory is to take the conjunction of the axioms of the theory and conjoin it with the so-called correspondence rules (sentences containing both theoretical and observational terms—the "links" between the "purely theoretical" and the observational). This conjunction can be represented as follows:

$$\text{---P---Q--- . . .}$$

where the dashes represent the sentential matrixes (the axioms and C-rules) containing the theoretical terms (which are, of course, almost always predicates or class terms) 'P,' 'Q,' '. . .'; the theoretical terms are then "eliminated" by replacing them with existentially quantified variables. The resulting "Ramsey sentence" is represented, then, by

$$(\exists f)(\exists g)\ .\ .\ .\ (\text{---f---g--- . . .}).$$

Or, consider an informal illustration. Let us represent schematically an oversimplified axiomatization of kinetic theory by

> All gases are composed entirely of molecules. The molecules are in rapid motion and are in frequent collision, etc., etc.

And for simplicity's sake, suppose that 'molecules' is the only theoretical term. The Ramsey sentence would be something like the following:

> There is a kind of entity such that all gases are composed entirely of these entities. They are in rapid motion and are in frequent collision, etc., etc.

Now it is a simple matter to demonstrate that any sentence containing only observation (and logical) terms which is a deductive consequence of the original theory is also a deductive consequence of its Ramsey sentence (see, for example, Rozeboom's article in this volume); thus, as far as any deductive systemization is concerned, any theory may be

[12] Carl G. Hempel, "The Theoretician's Dilemma," in *Minnesota Studies in Philosophy of Science*, Vol. II, Feigl, Scriven, and Maxwell, eds. Nagel, *loc. cit.*

[13] For an extended consideration of the Ramsey sentence see Professor William Rozeboom's essay in this volume.

eliminated and its Ramsey sentence used instead. However, it is also easy to prove (if indeed it is not obvious) that if a given theory (or a theory together with other considerations, theoretical or observational) entails that there exist certain kinds of unobservable entities, then the appropriate Ramsey sentence will also entail that there exist the same number of kinds of *unobservable* entities.[14] Although, insofar as deductive systemization is concerned, the Ramsey sentence can avoid the use of theoretical terms; it cannot, even in letter, much less in spirit (Hempel, *loc. cit.*, was too charitable), eliminate reference to unobservable (theoretical) entities.

The Craig result, like the Ramsey sentence, provides a "method" of reaxiomatizing a postulate set so that any arbitrarily selected class of terms may be eliminated, provided one is interested only in those theorems which contain none of these terms. Its "advantages" over the Ramsey sentence are that it does not quantify over predicates and class terms and that its final reaxiomatization eliminates reference both in spirit and in letter to unobservable entities. However, its shortcomings (for the purposes at hand) render it useless as an instrument of actual scientific practice and also preclude its having, even in principle, any implications for ontology. The resulting number of axioms will, in general, and particularly in the case of the empirical sciences, be infinite in number and practicably unmanageable.

But if the practical objections to the use of Craig's method as a means for elimination of theoretical terms are all but insurmountable, there are objections of principle which are even more formidable. Both Craig's method and the Ramsey device must operate upon theories (containing, of course, theoretical terms) which are "already there." They eliminate theoretical terms only after these terms have already been used in inter-

[14] The proof may be sketched as follows: Let 'T' designate the theory (conjoined, if necessary, with other statements in the accepted body of knowledge) which entails that the kinds of entities C, D, . . . are not observable, i.e., T entails that

$(\exists x)(\exists y) \ldots (Cx \cdot Dy \ldots x \text{ is not observable} \cdot y \text{ is not observable} \ldots)$ which in turn entails

$(\exists f)(\exists g) \ldots (\exists x)(\exists y) \ldots (fx \cdot gy \ldots x \text{ is not observable} \cdot y \text{ is not observable} \ldots)$.

Now the Ramsey result holds for any arbitrary division of nonlogical terms into two classes, so we may put 'observable' into the class with the observation terms, so that the latter formalized statement may be treated as an "observational" consequence of T (transitivity of entailment). But then it is also a consequence of the Ramsey sentence of T. Q.E.D.

mediary steps. Neither provides a method for axiomatization *ab initio* or a recipe or guide for invention of new theories. Consequently neither provides a method for the elimination of theoretical terms in the all-important "context of discovery."[15] It might be argued that this objection is not so telling, after all, for we also lack any recipe for the invention of theories themselves, and it is logically possible that we should discover, without the use of theories as intermediaries, Ramsey sentences or Craig end products which are just as useful for explaining and predicting observations as the theories which we happen to have (accidently) adduced. It might be added that it is also logically possible that we should discover just those observation statements (including predictions, etc.) which happen to be true without the use of any instrumental intermediaries.

We must reply that the accomplished fact that it *is* theories, referring to unobservables, which have been invented for this purpose and that many of them serve it so admirably—this fact, itself, cries out for explanation. To say that theories are *designed* to accomplish this task is no reply unless at least a schema of an instrumentalist recipe for such designing is provided. As far as I know this has not been done. The thesis that theoretical entities are "really" just "bundles" of observable objects or of sense data would, if true, provide an explanation; but it is not taken very seriously by most philosophers today—for the very good reason that it seems to be false. The only reasonable explanation for the success of theories of which I am aware is that well-confirmed theories are conjunctions of well-confirmed, genuine statements and that the entities to which they refer, in all probability, exist. That it is psychologically possible for us to invent such theories is explained by the fact that many of the entities to which they refer resemble in many respects (although

[15] The Ramsey sentence is intuitively tractable enough so that very simple "theories" might be invented as full-blown Ramsey sentences without the use of intermediary terms. However, Craig's theorem provides no means of operating *ab initio.* Craig points out (*loc. cit.*) that once the original theory is "there," reference, in letter, to theoretical entities in the application of his method may be avoided by using the names of theoretical terms rather than using the terms themselves (i.e. by mentioning theoretical terms rather than using them). But surely only a diehard instrumentalist can take more than very scant comfort from this. The question would still remain: Where did the theory come from in the first place, and why are the names of these particular terms arranged in this particular manner such admirable "instruments" for explanation and prediction of observations? Whatever ontological implications this modification of the Craig method may have, they seem to be exactly the same as those of instrumentalism proper.

they may differ radically from them in others) the entities which we have already observed.

It should also be remembered, at this point, that theories, even as instruments, are important not only for deductive systemization but also for inductive systemization (see Hempel, *loc. cit.*). We often reason theoretically using induction, and the conclusions may be either observational or theoretical. Thus we might infer from the facts that a certain substance was paramagnetic, that it catalyzed the recombination of free radicals, and that it *probably* contained a "one-electron" bond; and we might go on to infer, again inductively, that it would probably catalyze the conversion of orthohydrogen to parahydrogen. The Craig result applies only to deductive systemization and, thus, cannot, even in its Pickwickian fashion, eliminate theoretical terms where inductive theoretical reasoning is involved. Although Craig's theorem is of great interest for formal logic, we must conclude, to use Craig's (*loc. cit.*) own words, "[as far as] the meaning [and, I would add, the referents] of such expressions [auxiliary expressions (theoretical terms)] . . . [is concerned] the method . . . fails to provide any . . . clarification."

We have seen that the elimination of theoretical *terms*, even by explicit definition, would not necessarily eliminate reference to theoretical (unobservable) *entities*. We have also seen that, even if reference to theoretical entities could be eliminated after the theories themselves have been used in such an elimination (for example, by a device such as Craig's), the reality (existence) of the theoretical entities is not thereby militated against. But the most crucial point follows. Even if we do come up with a gimmick—a prediction machine or "black box"—into which we can feed data and grind out all the completely veridical observational predictions which we may desire, the possibility—I should say the likelihood—of the existence of unobserved causes for the observed events would still remain. For unless an *explanation* of why any prediction machine or "calculating device" in terms of the established rules of explanation, confirmation, etc., were forthcoming, the task of science would still be incomplete.

This brings us to another mistaken assumption that has been responsible for much mischief in considerations concerning the cognitive status of theories—the assumption that science is concerned solely with the "fruitful" organization of observational data or, more specifically, with successful prediction. Surely the main concerns of, say, a theoretical physicist involve such things as the actual properties and varieties of subatomic particles rather than the mere predictions about where and how intense a certain spectral line will be. The instrumentalist has the picture entirely reversed; as far as pure science is concerned, most observational data—most predictions—are mere instruments and are of value only for their roles in confirming theoretical principles. Even if we obtain the prediction machine, many of the theories extant today are well confirmed enough to argue strongly for the reality of theoretical entities. And they are much more intellectually satisfactory, for they provide an explanation of the occurrence of the observational events which they predict. And—equally important—an explanation for the fact that theories "work" as well as they do is, as already noted, also forthcoming; it is simply that the entities to which they refer exist.

"Criteria" of Reality and Instrumentalism

It was pointed out in the beginning of this article that Professor Ernest Nagel considers the dispute between realists and instrumentalists to be merely a verbal one.[16] There follows here a brief and what I hope is a not too inaccurate summary of his argument. Various criteria of 'real' or 'exist' (runs the argument) are employed by scientists, philosophers, etc., in their considerations of the "reality problem." (Among these criteria—some of them competing, some compatible with each other—are public perceivability, being mentioned in a generally accepted law, being mentioned in more than one law, being mentioned in a "causal" law, and being invariant "under some stipulated set of transformation, projections, or perspectives."[17]) Since, then (it continues) any two disputants will, in all probability, be using 'real' or 'exist' in two different senses, such disputes are merely verbal. Now someone might anticipate the forthcoming objections to this argument by pointing out that the word 'criteria' is a troublesome one and that perhaps, for Nagel, the connection between criteria and reality or existence is a contingent one rather than one based on meaning. But a moment's reflection makes it obvious that for Nagel's argument to have force, 'criteria' must be taken in the latter sense; and, indeed, Nagel explicitly speaks for the connection between criteria and the "senses [sic!] of 'real' or 'exist.' "[18]

[16] *Op. cit.*, pp. 141–152.
[17] Nagel, *op. cit.*, pp. 145–150.
[18] *Op. cit.*, p. 151.

Before proceeding to a criticism of these arguments, let me point out that Professor Gustav Bergmann, completely independently, treats ontological questions in a similar manner. Rather than criteria, he speaks of "patterns," although he does say that he "could instead have spoken of criteria," and he makes explicit reference to various "uses" of 'exist.'[19]

There are two main points that I wish to make regarding this kind of approach to ontological issues. First, it seems to me that it commits the old mistake of confusing meaning with evidence. To be sure, the fact that a kind of entity is mentioned in well-confirmed laws or that such entities are publicly perceptible, etc.—such facts are evidence (very good evidence!) for the existence or "reality" of the entities in question. But I cannot see how a prima-facie—or any other kind of—case can be made for taking such conditions as *defining characteristics* of *existence*.

The second point is even more serious. One would hope that (Professor Norman Malcolm notwithstanding) over nine hundred years of debate and analysis have made it clear that existence is *not* a property. Now surely the characteristics of being mentioned in well-confirmed laws, being publicly perceptible, etc., are properties of sorts; and if these comprised part of the meaning of 'exists,' then 'existence' would be a predicate (and existence a property).

Thus, it is seen that the issue between instrumentalism and realism can be made into a merely verbal one *only* by twisting the meanings of 'existence' and 'reality,' not only beyond their "ordinary" meaning but, also, far beyond any *reasonable* meanings which these terms might be given. In fact, it seems not too much to say that such an interpretation of the "reality problem" commits a fallacy closely akin to that of the Ontological Argument.

What can be said about the meanings of 'real' and 'exists'? I submit that in "ordinary language," the most usual uses of these terms are such that

$$\Phi_s \text{ are real} =_{df} \Phi_s \text{ exist}$$

and that

$$\Phi_s \text{ exist} =_{df} \text{there are } \Phi_s$$

and that the meanings of these definiens are clear enough so that no further explication is seriously needed. (In most "constructed languages," 'There are Φ_s' would, of course, be expressed by '$(\exists x)(\Phi x)$.') Thus, if

[19] "Physics and Ontology," *Philosophy of Science*, 28: 1–14 (1961).

we have a well-confirmed set of statements (laws or theories plus initial conditions) which entail the statement 'There are Φ_s' (or '$(\exists x)(\Phi x)$'), then it is well confirmed that Φ_s are real—full stop!

In summary, let us recall three points concerning instrumentalism. First, as is shown above, it cannot be excused on the grounds that it differs from realism only in terminology. Second, it cannot provide an explanation as to why its "calculating devices" (theories) are so successful. Realism provides the very simple and cogent explanation that the entities referred to by well-confirmed theories exist. Third, it must be acutely embarrassing to instrumentalists when what was once a "purely" theoretical entity becomes, due to better instruments, etc., an observable one.[20]

The Ontological Status of Entities—Theoretical and Otherwise

As I have stated elsewhere (see the second reference in footnote 22), the key to the solution of all significant problems in ontology can be found in Carnap's classic article, "Empiricism, Semantics, and Ontology."[21] Taking this essay as our point of departure, we may say that in order to speak at all about any kind of entities whatever and thus, a fortiori, to consider their existence or nonexistence, one must first accept the "linguistic framework" which "introduces the entities."[22] This simply means that in order to understand considerations concerning the existence of any kind of entities one must understand the meanings of the linguistic expressions (sentences and terms) referring to them—and that such expressions have no meaning unless they are given a place in a linguistic framework which "talks about the world" and which has at least a minimum of comprehensiveness. (Since I am interested, here, primarily in empirical science, I neglect universes of discourse containing only "purely mathematical" or "purely logical" entities.)

Although wide latitude in choosing and constructing frameworks is permissible, any satisfactory framework will embody, at the very least,

[20] Although I cannot agree with all the conclusions of Professor Feyerabend's essay in this volume, the reader is referred to it for an interesting critique of instrumentalism.

[21] R. Carnap, *Meaning and Necessity*, 2nd ed. (Chicago: University of Chicago Press, 1959).

[22] For a more detailed discussion of linguistic frameworks as well as their relevance for ontological problems, see Carnap, *ibid.*; and G. Maxwell, "Theories, Frameworks, and Ontology," *Philosophy of Science*, vol. 28 (1961). For an elaboration of the linguistic theses presupposed by the latter article and, to some extent, by this essay,

the following features: (1) the usual L(ogical)-formation and L-transformation rules and the corresponding set of L-true sentences which they generate; (2) a set of confirmation rules, whose nature I shall not discuss here but which I shall assume are quite similar to those actually used in the sciences; (3) a set of sentences whose truth value is quickly decidable on other than purely linguistic grounds—these correspond to "singular observation statements," but, of course, as we have seen, it is neither necessary nor desirable that such statements be incorrigible or indubitable or that a sharp distinction between observation and theory be drawn; and (4) a set of lawlike sentences, which, among other things, provide that component of meaning which is nonostensive for every descriptive (nonlogical) term of the framework. (I have argued in the references given in footnote 22 that every descriptive term has a meaning component which is nonostensive.[23] Even a term such as 'red' has part of its meaning provided by, for example, the lawlike sentence 'No surface can be both red and green all over at the same time.' Such a view is sometimes stigmatized by the epithet 'holism.' But if there is any holism involved in the view I am advocating, it is completely conceptual or epistemological and not ontological. Just what relations are present, or absent, between the actual entities of the "real world" is an empirical question and must be decided by considerations *within* a descriptive linguistic framework rather than by consideration *about* such frameworks.)

At this point, two views may be mentioned. I will omit consideration of explicitly defined terms, since they are, in principle, always eliminable. According to one view, it is always a proper subset of the lawlike sentence containing a given term which contributes to the term's meaning. The sentences in this subset are A-true[24] (analytic in a broad sense) and are totally devoid of any factual content—their only function is to provide part of the meaning of the term in question. The situation is immensely complicated by the fact that when actual usage is considered,

see G. Maxwell and H. Feigl, "Why Ordinary Language Needs Reforming," *Journal of Philosophy*, 58:488–498 (1961); G. Maxwell, "Meaning Postulates in Scientific Theories," in *Current Issues in the Philosophy of Science*, Feigl and Maxwell, eds.; and my brief article, "The Necessary and the Contingent," in this volume.

[23] Cf. also the writings of Wilfrid Sellars, for example in "Some Reflections on Language Games," *Philosophy of Science*, 21:204–228 (1954).

[24] See R. Carnap, "Beobachtungsprache und theoretisch Sprache," *Dialectica*, 12:236–248 (1957); as well as the references in fn. 22.

a sentence which is A-true in one context may be contingent in another and that even in a given context it is, more often than not, not clear, unless the context is a rational reformation, whether a given sentence is being used as A-true or as contingent. This confusion can be avoided by engaging in rational reformation, i.e., by stipulating (subject to certain broad and very liberal limitations) which sentences are to be taken as A-true and which as contingent. Needless to say, this is the viewpoint which I prefer.

The complication just mentioned, however, has led many philosophers, including Professor Putnam[25]—to say nothing of W. V. Quine—to the other viewpoint. According to it, no segregation of the relevant lawlike sentences into A-true and contingent should be attempted; each lawlike sentence plays a dual role: (1) it contributes to the meanings of its descriptive terms and (2) it provides empirical information. Fortunately, we do not have to choose between these two viewpoints here, for the thesis of realism which I am advocating is (almost) equally well accommodated by either one.

Now when we engage in any considerations about *any* kinds of entities and, a fortiori, considerations about the existence of theoretical entities, it is to the lawlike sentences mentioning the entities—for theoretical entities, the theoretical postulates and the so-called correspondence rules—to which we turn. These sentences tell us, for example, how theoretical entities of a given kind resemble, on the one hand, and differ from, on the other, the entities with which we happen to be more familiar. And the fact that many theoretical entities, for example those of quantum theory, differ a great deal from our ordinary everyday physical objects is no reason whatever to ascribe a questionable ontological status to them or to contend that they are merely "calculating devices." After all, the very air we breathe as well as such things as shadows and mirror images are entities of quite different kinds from chairs and tables, but this provides no grounds for impugning their ontological status. The fact that molecules, atoms, etc., cannot be said in any non-Pickwickian sense to have a color has given some philosophers ontological qualms. But, of course, the air has no color (unless we invoke the color of the sky); and a transparent object whose refractive index was the same as that of air would be completely invisible, although it would have all

[25] See his essay in this volume.

the other properties of ordinary physical objects. Molecules, for example, are in about the same category; they are physical things which possess some but not all of the properties of everyday physical things.

A: Do molecules exist?

B: Certainly. We have an extremely well-confirmed theory, which when conjoined with other true sentences such as 'There are gases' entails that there are molecules.

A: But are they real?

B: What do you mean?

A: Well, I'm not sure. As a starter: Are they physical objects?

B: Certainly the large ones are. Take, for example, that diamond in your ring. As for those which are submicroscopic but still large enough to have large quantum numbers, it seems that in almost any reasonable reformation they would be classified as physical objects. It would seem unjustifiable to withhold from them this status simply because they cannot be said to have a color in any straightforward fashion. In fact, I would even be inclined to call the smallest, the molecule of hydrogen, a physical object. It has mass, a reasonably determinate diameter, and, usually, something which approximates simple location, etc.

A: How about electrons?

B: The decision here is more difficult. We might find it necessary to try several reformations, taking into account many facets of contemporary physical theory, before we arrived at the most satisfactory one. It would also be helpful to have a more specific problem in view than the one which we are now considering. At any rate, we might begin by pointing out that electrons do have mass, even rest mass. They can be simply located at the expense of refraining from ascribing to them a determinate momentum. They can be said to causally interact with "bona fide" physical objects, even by those who have a billiard-ball notion of causality. The important point is that the question 'Are electrons physical objects?' is a request for a rational reformation of a very thoroughgoing variety. For most purposes, a rational reformation would not need to answer it. For your purposes, why not be content to learn in what ways electrons are similar to, and in what ways they differ from, what you would call "ordinary physical objects"? This will enable you to avoid conceptual blunders.

A: Perhaps you are right. However, I am genuinely puzzled about fields, and even photons.

B: Take the last first. We would probably never call them physical objects. For example, they have no rest mass and it would be a conceptual mistake to ask, except in a Pickwickian sense, What is their color? However, it would be reasonable to say that they are a sort of physical continuant; and they can even interact with electrons in a billiard-ball manner. At any rate, we must agree, speaking loosely, that they are "every bit as real" as electrons. The concepts of field theories are so open textured that it is difficult to decide what kinds of reformations one should adopt here. And it is virtually impossible to find similar kinds of entities with which one is prescientifically familiar. Perhaps these theories will someday be enriched until decisions concerning the most appropriate rational reformations are easier to make—perhaps not. But even here, the meanings of the terms involved are usually sufficiently clear to avoid conceptual blunders and ontological anxieties. You might like to consider the "lines of force," which are often spoken of in connection with fields. These are often used as a paradigm of the "convenient fiction" by those who hold such a view of theories.[26] But though convenient, lines of force are *not* fictions. They "really exist." Let me try to make this a little more plausible. Consider the isobars of meteorology, or the isograms which connect points of equal elevation above sea level. Now at this very moment, the 1017 millibar isobar, i.e., the line along which the barometric pressure is 1017 millibars, exists right here in the United States. Its location can even be determined "operationally." And all of this is true whether anyone ever draws, or ever has drawn, a weather map. Since a well-confirmed theory (plus, perhaps, other

[26] Cf. B. Mayo, "The Existence of Theoretical Entities," *Science News*, 32:7–18 (1954), and "More about Theoretical Entities," *ibid.*, 39:42–55 (1956). For a critique of these articles and for excellent constructive remarks concerning theoretical entities, see J. J. C. Smart, "The Reality of Theoretical Entities," *Australasian Journal of Philosophy*, 34:1–12 (1956).

In connection with convenient fictions, we might consider such entities as ideal gases and bodies uninfluenced by external forces. These actually are fictions. But no theory (or theory plus true sentences) entails that there are such things. To understand their function, we need only recourse to the notion of a limit, often used in mathematics. Roughly speaking, what we actually do when we use theories involving such "fictions" is to assume, for example, that the influence of external forces on the body in question is very, very small, or that the behavior of the gas with which we are concerned is approximately given by '$PV = nRT$,' or, in early kinetic theory, that the diameter of a molecule is very, very small compared to the distance between molecules. Note that had van der Waals taken the calculating-device or convenient-fiction view, he probably would not have developed his equation which embodies a correction for the effect due to the finite (greater than zero) diameter of molecules.

> well-confirmed sentences) entails that there are lines of force, lines of force exist. To be sure, they are very different from everyday physical objects. But as long as we are clear about this, what metaphysical—what ontological—problems remain?

One of the exciting aspects of the development of science has been the emergence of reference to strikingly new kinds of entities. This is particularly true in field theories and quantum theory. The great difference between these and the old, familiar categories seems to have caused many philosophers and philosophically inclined scientists to despair of effecting a satisfactory conceptual analysis of these powerful new conceptual tools. The attitude too often has been, "Let us proceed to use those new devices and, if necessary for heuristic reasons, even to behave *as if* they consisted of genuine statements about real entities. But let us remember that, in the last analysis, they are only meaningless calculating devices, or, at best, they talk only of convenient fictions, etc. The only *real* entities are the good old familiar ones which we sense directly everyday." To turn the purpose of a saying of Bertrand Russell's almost completely about-face: such a view has advantages—they are the same as those of theft over honest toil. The compulsion toward metaphysical asepsis which appears to have been the motivation for the espousal of many of these reductionistic philosophies seems, itself, to have arisen from a preoccupation with metaphysical pseudo problems, e.g., the conviction that there are very few ontologically legitimate kinds of entities, perhaps only one.

OBSERVATION

Were the eye not attuned to the Sun,
The Sun could never be seen by it.
GOETHE[1]

A

Consider two microbiologists. They look at a prepared slide; when asked what they see, they may give different answers. One sees in the cell before him a cluster of foreign matter: it is an artefact, a coagulum resulting from inadequate staining techniques. This clot has no more to do with the cell, *in vivo*, than the scars left on it by the archaeologists spade have to do with the original shape of some Grecian urn. The other biologist identifies the clot as a cell organ, a 'Golgi body'. As for techniques, he argues: 'The standard way of detecting a cell organ is by fixing and staining. Why single out this one technique as producing artefacts, while others disclose genuine organs?'

The controversy continues.[2] It involves the whole theory of microscopical technique; nor is it an obviously experimental issue. Yet it affects what scientists say they see. Perhaps there is a sense in which two such observers do not see the same thing, do not begin from the same data, though their eyesight is normal and they are visually aware of the same object.

Imagine these two observing a Protozoon—*Amoeba*. One sees a one-celled animal, the other a non-celled animal. The first sees *Amoeba* in all its analogies with different types of single cells: liver cells, nerve cells, epithelium cells. These have a wall, nucleus, cytoplasm, etc. Within this class *Amoeba* is distinguished only by its independence. The other, however, sees *Amoeba's* homology not with single cells, but with whole animals. Like all animals *Amoeba* ingests its food, digests and assimilates it. It excretes, reproduces and is mobile—more like a complete animal than an individual tissue cell.

This is not an experimental issue, yet it can affect experiment. What either man regards as significant questions or relevant data can be determined by whether he stresses the first or the last term in 'unicellular animal'.[1]

Some philosophers have a formula ready for such situations: 'Of course they see the same thing. They make the same observation since they begin from the same visual data. But they interpret what they see differently. They construe the evidence in different ways.'[2] The task is then to show how these data are moulded by different theories or interpretations or intellectual constructions.

Considerable philosophers have wrestled with this task. But in fact the formula they start from is too simple to allow a grasp of the nature of observation within physics. Perhaps the scientists cited above do not begin their inquiries from the same data, do not make the same observations, do not even see the same thing? Here many concepts run together. We must proceed carefully, for wherever it makes sense to say that two scientists looking at *x* do not see the same thing, there must always be a prior sense in which they do see the same thing. The issue is, then, 'Which of these senses is most illuminating for the understanding of observational physics?'

These biological examples are too complex. Let us consider Johannes Kepler: imagine him on a hill watching the dawn. With him is Tycho Brahe. Kepler regarded the sun as fixed: it was the earth that moved. But Tycho followed Ptolemy and Aristotle in this much at least: the earth was fixed and all other celestial bodies moved around it. *Do Kepler and Tycho see the same thing in the east at dawn?*

We might think this an experimental or observational question, unlike the questions 'Are there Golgi bodies?' and 'Are Protozoa one-celled or non-celled?'. Not so in the sixteenth and seventeenth centuries. Thus Galileo said to the Ptolemaist '...neither Aristotle nor you can prove that the earth is *de facto* the centre of the universe...'.[3] 'Do Kepler and Tycho see the same thing in the east at dawn?' is perhaps not a *de facto* question either, but rather the beginning of an examination of the concepts of seeing and observation.

The resultant discussion might run:

'Yes, they do.'

'No, they don't.'

'Yes, they do!'

'No, they don't!'...

B

That this is possible suggests that there may be reasons for both contentions.[1] Let us consider some points in support of the affirmative answer.

The physical processes involved when Kepler and Tycho watch the dawn are worth noting. Identical photons are emitted from the sun; these traverse solar space, and our atmosphere. The two astronomers have normal vision; hence these photons pass through the cornea, aqueous humour, iris, lens and vitreous body of their eyes in the same way. Finally their retinas are affected. Similar electro-chemical changes occur in their selenium cells. The same configuration is etched on Kepler's retina as on Tycho's. So they see the same thing.

Locke sometimes spoke of seeing in this way: a man sees the sun if his is a normally-formed retinal picture of the sun. Dr Sir W. Russell Brain speaks of our retinal sensations as indicators and signals. Everything taking place behind the retina is, as he says, 'an intellectual operation based largely on non-visual experience...'.[2] What we *see* are the changes in the *tunica retina*. Dr Ida Mann regards the macula of the eye as itself 'seeing details in bright light', and the rods as 'seeing approaching motor-cars'. Dr Agnes Arber speaks of the eye as itself seeing.[3] Often, talk of seeing can direct attention to the retina. Normal people are distinguished from those for whom no retinal pictures can form: we may say of the former that they can see whilst the latter cannot see. Reporting when a certain red dot can be seen may supply the occulist with direct information about the condition of one's retina.[4]

This need not be pursued, however. These writers speak carelessly: seeing the sun is not seeing retinal pictures of the sun. The retinal images which Kepler and Tycho have are four in number, inverted and quite tiny.[5] Astronomers cannot be referring to these when they say they see the sun. If they are hypnotized, drugged, drunk or distracted they may not see the sun, even though their retinas register its image in exactly the same way as usual.

Seeing is an experience. A retinal reaction is only a physical state—a photochemical excitation. Physiologists have not always appreciated the differences between experiences and physical states.[6] People, not their eyes, see. Cameras, and eye-balls, are blind. Attempts to locate within the organs of sight (or within the neurological reticulum behind the eyes) some nameable called 'seeing' may be dismissed. That Kepler and Tycho do, or do not, see the same thing cannot be supported by reference to the physical states of their retinas, optic nerves or visual cortices: there is more to seeing than meets the eyeball.

Naturally, Tycho and Kepler see the same physical object. They are both visually aware of the sun. If they are put into a dark room and asked to report when they see something—anything at all—they may both report the same object at the same time. Suppose that the only object to be seen is a certain lead cylinder. Both men see the same thing: namely this object—whatever it is. It is just here, however, that the difficulty arises, for while Tycho sees a mere pipe, Kepler will see a telescope, the instrument about which Galileo has written to him.

Unless both are visually aware of the same object there can be nothing of philosophical interest in the question whether or not they see the same thing. Unless they both see the sun in this prior sense our question cannot even strike a spark.

Nonetheless, both Tycho and Kepler have a common visual experience of some sort. This experience perhaps constitutes their seeing the same thing. Indeed, this may be a seeing logically more basic than anything expressed in the pronouncement 'I see the sun' (where each means something different by 'sun'). If what they meant by the word 'sun' were the only clue, then Tycho and Kepler could not be seeing the same thing, even though they were gazing at the same object.

If, however, we ask, not 'Do they see the same thing?' but rather 'What is it that they both see?', an unambiguous answer may be forthcoming. Tycho and Kepler are both aware of a brilliant yellow-white disc in a blue expanse over a green one. Such a 'sense-datum' picture is single and uninverted. To be unaware of it is not to have it. Either it dominates one's visual attention completely or it does not exist.

If Tycho and Kepler are aware of anything visual, it must be of some pattern of colours. What else could it be? We do not touch or hear with our eyes, we only take in light.[1] This private pattern is the same for both observers. Surely if asked to sketch the contents of their visual fields they would both draw a kind of semicircle on a horizon-line.[2] They say they see the sun. But they do not see every side of the sun at once; so what they really see is discoid to begin with. It is but a visual aspect of the sun. In any single observation the sun is a brilliantly luminescent disc, a penny painted with radium.

So something about their visual experiences at dawn is the same for both: a brilliant yellow-white disc centred between green and blue colour patches. Sketches of what they both see could be identical—congruent. In this sense Tycho and Kepler see the same thing at dawn. The sun appears to them in the same way. The same view, or scene, is presented to them both.

In fact, we often speak in this way. Thus the account of a recent solar eclipse:[1] 'Only a thin crescent remains; white light is now completely obscured; the sky appears a deep blue, almost purple, and the landscape is a monochromatic green...there are the flashes of light on the disc's circumference and now the brilliant crescent to the left....' Newton writes in a similar way in the *Opticks*: 'These Arcs at their first appearance were of a violet and blue Colour, and between them were white Arcs of Circles, which...became a little tinged in their inward Limbs with red and yellow....'[2] Every physicist employs the language of lines, colour patches, appearances, shadows. In so far as two normal observers use this language of the same event, they begin from the same data: they are making the same observation. Differences between them must arise in the interpretations they put on these data.

Thus, to summarize, saying that Kepler and Tycho see the same thing at dawn just because their eyes are similarly affected is an elementary mistake. There is a difference between a physical state and a visual experience. Suppose, however, that it is argued as above—that they see the same thing because they have the same sense-datum experience. Disparities in their accounts arise in *ex post facto* interpretations of what is seen, not in the fundamental visual data. If this is argued, further difficulties soon obtrude.

B

Normal retinas and cameras are impressed similarly by fig. 1.[3] Our visual sense-data will be the same too. If asked to draw what we see, most of us will set out a configuration like fig. 1.

Do we all see the same thing?[4] Some will see a perspex cube viewed from below. Others will see it from above. Still others will see it as a kind of polygonally-cut gem. Some people see only criss-crossed lines in a plane. It may be seen as a block of ice, an aquarium, a wire frame for a kite—or any of a number of other things.

Do we, then, all see the same thing? If we do, how can these differences be accounted for?

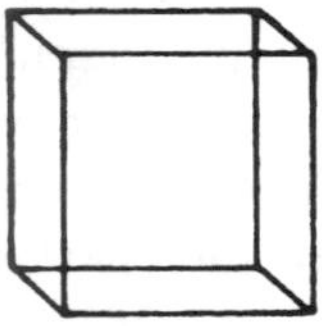

Fig. 1

Here the 'formula' re-enters: 'These are different *interpretations* of what all observers see in common. Retinal reactions to fig. 1 are virtually identical; so too are our visual sense-data, since our drawings of what we see will have the same content. There is no place in the seeing for these differences, so they must lie in the interpretations put on what we see.'

This sounds as if I do two things, not one, when I see boxes and bicycles. Do I put different interpretations on fig. 1 when I see it now as a box from below, and now as a cube from above? I am aware of no such thing. I mean no such thing when I report that the box's perspective has snapped back into the page.[1] If I do not mean this, then the concept of seeing which is natural in this connexion does not designate two diaphanous components, one optical, the other interpretative. Fig. 1 is simply seen now as a box from below, now as a cube from above; one does not first soak up an optical pattern and then clamp an interpretation on it. Kepler and Tycho just see the sun. That is all. That is the way the concept of seeing works in this connexion.

'But', you say, 'seeing fig. 1 first as a box from below, then as a cube from above, involves interpreting the lines differently in each case.' Then for you and me to have a different interpretation of fig. 1 just *is* for us to see something different. This does not mean we see the same thing and then interpret it differently. When I suddenly exclaim 'Eureka—a box from above', I do not refer simply to a different interpretation. (Again, there is a logically prior sense in which seeing fig. 1 as from above and then as from below is seeing the same thing differently, i.e. being aware of the same diagram in different ways. We can refer just to this, but we need not. In this case we do not.)

Besides, the word 'interpretation' is occasionally useful. We know where it applies and where it does not. Thucydides presented the facts objectively; Herodotus put an interpretation on them.

The word does not apply to everything—it has a meaning. Can interpreting always be going on when we see? Sometimes, perhaps, as when the hazy outline of an agricultural machine looms up on a foggy morning and, with effort, we finally identify it. Is this the 'interpretation' which is active when bicycles and boxes are clearly seen? Is it active when the perspective of fig. 1 snaps into reverse? There was a time when Herodotus was half-through with his interpretation of the Graeco-Persian wars. Could there be a time when one is half-through interpreting fig. 1 as a box from above, or as anything else?

'But the interpretation takes very little time—it is instantaneous.' Instantaneous interpretation hails from the Limbo that produced unsensed sensibilia, unconscious inference, incorrigible statements, negative facts and *Objektive*. These are ideas which philosophers force on the world to preserve some pet epistemological or metaphysical theory.

Only in contrast to 'Eureka' situations (like perspective reversals, where one cannot interpret the data) is it clear what is meant by saying that though Thucydides could have put an interpretation on history, he did not. Moreover, whether or not an historian is advancing an interpretation is an empirical question: we know what would count as evidence one way or the other. But whether we are employing an interpretation when we see fig. 1 in a certain way is not empirical. What could count as evidence? In no ordinary sense of 'interpret' do I interpret fig. 1 differently when its perspective reverses for me. If there is some extraordinary sense of word it is not clear, either in ordinary language, or in extraordinary (philosophical) language. To insist that different reactions to fig. 1 *must* lie in the interpretations put on a common visual experience is just to reiterate (without reasons) that the seeing of *x must* be the same for all observers looking at *x*.

'But "I see the figure as a box" means: I am having a particular visual experience which I always have when I interpret the figure as a box, or when I look at a box....' '...if I meant this, I ought to know it. I ought to be able to refer to the experience directly and not only indirectly....'[1]

Ordinary accounts of the experiences appropriate to fig. 1 do not require visual grist going into an intellectual mill: theories and interpretations are 'there' in the seeing from the outset. How can interpretations 'be there' in the seeing? How is it possible to see an object according to an interpretation? 'The question represents it as a queer fact; as if something were being forced into a form it did not really fit. But no squeezing, no forcing took place here.'[1]

Consider now the reversible perspective figures which appear in textbooks on Gestalt psychology: the tea-tray, the shifting (Schröder) staircase, the tunnel. Each of these can be seen as concave, as convex, or as a flat drawing.[2] Do I really see something different each time, or do I only interpret what I see in a different way? To interpret is to think, to do something; seeing is an experiential state.[3] The different ways in which these figures are seen are not due to different thoughts lying behind the visual reactions. What could 'spontaneous' mean if these reactions are not spontaneous? When the staircase 'goes into reverse' it does so spontaneously. One does not think of anything special; one does not think at all. Nor does one interpret. One just sees, now a staircase as from above, now a staircase as from below.

The sun, however, is not an entity with such variable perspective. What has all this to do with suggesting that Tycho and Kepler may see different things in the east at dawn? Certainly the cases are different. But these reversible perspective figures are examples of different things being seen in the same configuration, where this difference is due neither to differing visual pictures, nor to any 'interpretation' superimposed on the sensation.

Some will see in fig. 2 an old Parisienne, others a young woman (à la Toulouse-Lautrec).[4] All normal retinas 'take' the same picture; and our sense-datum pictures must be the same, for even if you see an old lady and I a young lady, the pictures we draw of what we see may turn out to be geometrically indistinguishable. (Some can see this *only* in one way, not both. This is like the difficulty we have after finding a face in a tree-puzzle; we cannot thereafter see the tree without the face.)

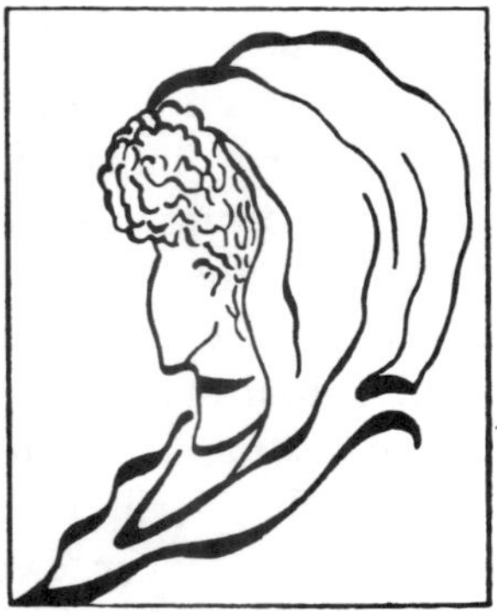

Fig. 2

When what is observed is characterized so differently as 'young woman' or 'old woman', is it not natural to say that the observers see different things? Or must 'see different things' mean only 'see different objects'? This is a primary sense of the expression, to be sure. But is there not also a sense in which one who cannot see the young lady in fig. 2 sees something different from me, who sees the young lady? Of course there is.

Similarly, in Köhler's famous drawing of the Goblet-and-Faces[1] we 'take' the same retinal/cortical/sense-datum picture of the configuration; our drawings might be indistinguishable. I see a goblet, however, and you see two men staring at one another. Do we see the same thing? Of course we do. But then again we do not. (The sense in which we *do* see the same thing begins to lose its philosophical interest.)

I draw my goblet. You say 'That's just what I saw, two men in a staring contest'. What steps must be taken to get you to see what I see? When attention shifts from the cup to the faces does one's visual picture change? How? What is it that changes? What could change? Nothing optical or sensational is modified. Yet one sees different things. The organization of what one sees changes.[2]

How does one describe the difference between the *jeune fille* and the *vieille femme* in fig. 2? Perhaps the difference is not describable: it may just show itself.[3] That two observers have not seen the same things in fig. 2 could show itself in their behaviour. What is the difference between us when you see the zebra as black with white stripes and I see it as white with black stripes? Nothing optical. Yet there might be a context (for instance, in the genetics of animal pigmentation), where such a difference could be important.

A third group of figures will stress further this organizational element of seeing and observing. They will hint at how much more is involved when Tycho and Kepler witness the dawn than 'the formula' suggests.

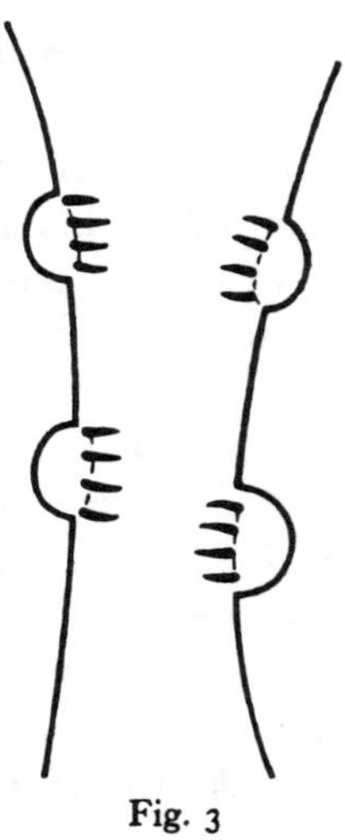

Fig. 3

What is portrayed in fig. 3? Your retinas and visual cortices are affected much as mine are; our sense-datum pictures would not differ. Surely we could all produce an accurate sketch of fig. 3. Do we see the same thing?

I see a bear climbing up the other side of a tree. Did the elements 'pull together'/cohere/organize, when you learned this?[1] You might even say with Wittgenstein 'it has not changed, and yet I see it differently...'.[2] Now, does it not have '...a quite particular "organization"'?

Organization is not itself seen as are the lines and colours of a drawing. It is not itself a line, shape, or a colour. It is not an element in the visual field, but rather the way in which elements are appreciated. Again, the plot is not another detail in the story. Nor is the tune just one more note. Yet without plots and tunes details and notes would not hang together. Similarly the organization of fig. 3 is nothing that registers on the retina along with other details. Yet it gives the lines and shapes a pattern. Were this lacking we would be left with nothing but an unintelligible configuration of lines.

How do visual experiences become organized? How is seeing possible?

Consider fig. 4 in the context of fig. 5:

Fig. 4

Fig. 5

The context gives us the clue. Here, some people could not see the figure as an antelope. Could people who had never seen an antelope, but only birds, see an antelope in fig. 4?

In the context of fig. 6 the figure may indeed stand out as an antelope. It might even be urged that the figure seen in fig. 5 has no similarity to the one in fig. 6 although the two are congruent. Could anything be more opposed to a sense-datum account of seeing?

Of a figure similar to the Necker cube (fig. 1) Wittgenstein writes,

'You could imagine [this] appearing in several places in a text-book. In the relevant text something different is in question every time: here a glass cube, there an inverted open box, there a wire frame of that shape, there three boards forming a solid angle. Each time the text supplies the interpretation of the illustration. But we can also see the illustration now as one thing, now as another. So we interpret it, and see it as we interpret it.'[1]

Fig. 6

Fig. 7

Consider now the head-and-shoulders in fig. 7:

The upper margin of the picture cuts the brow, thus the top of the head is not shown. The point of the jaw, clean shaven and brightly illuminated, is just above the geometric center of the picture. A white mantle...covers the right shoulder. The right upper sleeve is exposed as the rather black area at the lower left. The hair and beard are after the manner of a late mediaeval representation of Christ.[2]

The appropriate aspect of the illustration is brought out by the verbal context in which it appears. It is not an illustration of anything determinate unless it appears in some such context. In the same way, I must talk and gesture around fig. 4 to get you to see the antelope when only the bird has revealed itself. I must provide a context. The context is part of the illustration itself.

Such a context, however, need not be set out explicitly. Often it is 'built into' thinking, imagining and picturing. We are set[1] to appreciate the visual aspect of things in certain ways. Elements in our experience do not cluster at random.

Fig. 8

A trained physicist could see one thing in fig. 8: an X-ray tube viewed from the cathode. Would Sir Lawrence Bragg and an Eskimo baby see the same thing when looking at an X-ray tube? Yes, and no. Yes—they are visually aware of the same object. No—the *ways* in which they are visually aware are profoundly different. Seeing is not only the having of a visual experience; it is also the way in which the visual experience is had.

At school the physicist had gazed at this glass-and-metal instrument. Returning now, after years in University and research, his eye lights upon the same object once again. Does he see the same thing now as he did then? Now he sees the instrument in terms of electrical circuit theory, thermodynamic theory, the theories of metal and glass structure, thermionic emission, optical transmission, refraction, diffraction, atomic theory, quantum theory and special relativity.

Contrast the freshman's view of college with that of his ancient tutor. Compare a man's first glance at the motor of his car with a similar glance ten exasperating years later.

'Granted, one learns all these things', it may be countered, 'but it all figures in the interpretation the physicist puts on what he sees. Though the layman sees exactly what the physicist sees, he cannot interpret it in the same way because he has not learned so much.'

Is the physicist doing more than just seeing? No; he does nothing over and above what the layman does when he sees an X-ray tube. What are you doing over and above reading these words? Are you interpreting marks on a page? When would this ever be a natural way of speaking? Would an infant see what you see here, when you see words and sentences and he sees but marks and lines? One does nothing beyond looking and seeing when one dodges bicycles, glances at a friend, or notices a cat in the garden.

'The physicist and the layman see the same thing', it is objected, 'but they do not make the same thing of it.' The layman can make nothing of it. Nor is that just a figure of speech. I can make nothing of the Arab word for *cat*, though my purely visual impressions may be indistinguishable from those of the Arab who can. I must learn Arabic before I can see what he sees. The layman must learn physics before he can see what the physicist sees.

If one must find a paradigm case of seeing it would be better to regard as such not the visual apprehension of colour patches but things like seeing what time it is, seeing what key a piece of music is written in, and seeing whether a wound is septic.[1]

Pierre Duhem writes:

> Enter a laboratory; approach the table crowded with an assortment of apparatus, an electric cell, silk-covered copper wire, small cups of mercury, spools, a mirror mounted on an iron bar; the experimenter is inserting into small openings the metal ends of ebony-headed pins; the iron oscillates, and the mirror attached to it throws a luminous band upon a celluloid scale; the forward-backward motion of this spot enables the physicist to observe the minute oscillations of the iron bar. But ask him what he is doing. Will he answer 'I am studying the oscillations of an iron bar which carries a mirror'? No, he will say that he is measuring the electric resistance of the spools. If you are astonished, if you ask

him what his words mean, what relation they have with the phenomena he has been observing and which you have noted at the same time as he, he will answer that your question requires a long explanation and that you should take a course in electricity.[1]

The visitor must learn some physics before he can see what the physicist sees. Only then will the context throw into relief those features of the objects before him which the physicist sees as indicating resistance.

This obtains in all seeing. Attention is rarely directed to the space between the leaves of a tree, save when a Keats brings it to our notice.[2] (Consider also what was involved in Crusoe's seeing a vacant space in the sand as a footprint.) Our attention most naturally rests on objects and events which dominate the visual field. What a blooming, buzzing, undifferentiated confusion visual life would be if we all arose tomorrow without attention capable of dwelling only on what had heretofore been overlooked.[3]

The infant and the layman can see: they are not blind. But they cannot see what the physicist sees; they are blind to what he sees.[4] We may not hear that the oboe is out of tune, though this will be painfully obvious to the trained musician. (Who, incidentally, will not hear the tones and *interpret* them as being out of tune, but will simply hear the oboe to be out of tune.[5] We simply see what time it is; the surgeon simply sees a wound to be septic; the physicist sees the X-ray tube's anode overheating.) The elements of the visitor's visual field, though identical with those of the physicist, are not organized for him as for the physicist; the same lines, colours, shapes are apprehended by both, but not in the same way. There are indefinitely many ways in which a constellation of lines, shapes, patches, may be seen. *Why* a visual pattern is seen differently is a question for psychology, but *that* it may be seen differently is important in any examination of the concepts of seeing and observation. Here, as Wittgenstein might have said, the psychological is a symbol of the logical.

You see a bird, I see an antelope; the physicist sees an X-ray tube, the child a complicated lamp bulb; the microscopist sees coelenterate mesoglea, his new student sees only a gooey, formless stuff. Tycho and Simplicius see a mobile sun, Kepler and Galileo see a static sun.[6]

It may be objected, 'Everyone, whatever his state of knowledge, will see fig. 1 as a box or cube, viewed as from above or as from below'. True; almost everyone, child, layman, physicist, will see the figure as box-like one way or another. But could such observations be made by people ignorant of the construction of box-like objects? No. This objection only shows that most of us—the blind, babies, and dimwits excluded—have learned enough to be able to see this figure as a three-dimensional box. This reveals something about the sense in which Simplicius and Galileo do see the same thing (which I have never denied): they both see a brilliant heavenly body. The schoolboy and the physicist both see that the X-ray tube will smash if dropped. Examining how observers see different things in *x* marks something important about their seeing the same thing when looking at *x*. If seeing different things involves having different knowledge and theories about *x*, then perhaps the sense in which they see the same thing involves their sharing knowledge and theories about *x*. Bragg and the baby share no knowledge of X-ray tubes. They see the same thing only in that if they are looking at *x* they are both having some visual experience of it. Kepler and Tycho agree on more: they see the same thing in a stronger sense. Their visual fields are organized in much the same way. Neither sees the sun about to break out in a grin, or about to crack into ice cubes. (The baby is not 'set' even against these eventualities.) Most people today see the same thing at dawn in an even stronger sense: we share much knowledge of the sun. Hence Tycho and Kepler see different things, and yet they see the same thing. That these things can be said depends on their knowledge, experience, and theories.

Kepler and Tycho are to the sun as we are to fig. 4, when I see the bird and you see only the antelope. The elements of their experiences are identical; but their conceptual organization is vastly different. Can their visual fields have a different organization? Then they can see different things in the east at dawn.

It is the sense in which Tycho and Kepler do not observe the same thing which must be grasped if one is to understand disagreements within microphysics. Fundamental physics is primarily a search for intelligibility—it is philosophy of matter. Only secondarily is it a search for objects and facts (though the two endeavours are as hand and glove). Microphysicists seek new modes of conceptual organization. If that can be done the finding of new entities will follow. Gold is rarely discovered by one who has not got the lay of the land.

To say that Tycho and Kepler, Simplicius and Galileo, Hooke and Newton, Priestley and Lavoisier, Soddy and Einstein, De Broglie and Born, Heisenberg and Bohm all make the same observations but use them differently is too easy.[1] It does not explain controversy in research science. Were there no sense in which they were different observations they could not be used differently. This may perplex some: that researchers sometimes do not appreciate data in the same way is a serious matter. It is important to realize, however, that sorting out differences about data, evidence, observation, may require more than simply gesturing at observable objects. It may require a comprehensive reappraisal of one's subject matter. This may be difficult, but it should not obscure the fact that nothing less than this may do.

C

There is a sense, then, in which seeing is a 'theory-laden' undertaking. Observation of *x* is shaped by prior knowledge of *x*. Another influence on observations rests in the language or notation used to express what we know, and without which there would be little we could recognize as knowledge. This will be examined.[2]

I do not mean to identify seeing with *seeing as*. Seeing an X-ray tube is not seeing a glass-and-metal object as an X-ray tube.[3] However, seeing an antelope and seeing an object as an antelope have much in common. Something of the concept of seeing can be discerned from tracing uses of 'seeing...as...'. Wittgenstein is reluctant[4] to concede this, but his reasons are not clear to me. On the contrary, the logic of 'seeing as' seems to illuminate the general perceptual case.[5] Consider again the footprint in the sand. Here all the organizational features of *seeing as* stand out clearly, in the absence of an '*object*'. One can even imagine cases where 'He sees it as a footprint' would be a way of referring to another's apprehension of what actually is a footprint. So, while I do not identify, for example, Hamlet's seeing of a camel in the clouds with his seeing of Yorick's skull, there is still something to be learned about the latter from noting what is at work in the former.

There is, however, a further element in seeing and observation.

If the label 'seeing as' has drawn out certain features of these concepts, 'seeing that...' may bring out more. Seeing a bear in fig. 3 was to see that were the 'tree' circled we should come up behind the beast. Seeing the dawn was for Tycho and Simplicius to see that the earth's brilliant satellite was beginning its diurnal circuit around us, while for Kepler and Galileo it was to see that the earth was spinning them back into the light of our local star. Let us examine 'seeing that' in these examples. It may be the logical element which connects observing with our knowledge, and with our language.

Of course there are cases where the data are confused and where we may have no clue to guide us. In microscopy one often reports sensations in a phenomenal, lustreless way: 'it is green in this light; darkened areas mark the broad end....' So too the physicist may say: 'the needle oscillates, and there is a faint streak near the neon parabola. Scintillations appear on the periphery of the cathodescope....' To deny that these are genuine cases of seeing, even observing, would be unsound, just as is the suggestion that they are the *only* genuine cases of seeing.

These examples are, however, overstressed. The language of shapes, colour patches, oscillations and pointer-readings is appropriate to the unsettled experimental situation, where confusion and even conceptual muddle may dominate. The observer may not know what he is seeing: he aims only to get his observations to cohere against a background of established knowledge. This seeing is the goal of observation. It is in these terms, and not in terms of 'phenomenal' seeing, that new inquiry proceeds. Every physicist forced to observe his data as in an oculist's office finds himself in a special, unusual situation. He is obliged to forget what he knows and to watch events like a child. These are non-typical cases, however spectacular they may sometimes be.

First registering observations and then casting about for knowledge of them gives a simple model of how the mind and the eye fit together. The relationship between seeing and the corpus of our knowledge, however, is not a simple one.

What is it to see boxes, staircases, birds, antelopes, bears, goblets, X-ray tubes? It is (at least) to have knowledge of certain sorts. (Robots and electric eyes are blind, however efficiently they react to light. Cameras cannot see.) It is to see that, were certain things done to objects before our eyes, other things would result. How should we regard a man's report that he sees *x* if we know him to be ignorant of all *x*-ish things? Precisely as we would regard a four-year-old's report that he sees a meson shower. 'Smith sees *x*' suggests that Smith could specify some things pertinent to *x*. To see an X-ray tube is at least to see that, were it dropped on stone, it would smash. To see a goblet is to see something with concave interior. We may be wrong, but not always—not even usually. Besides, deceptions proceed in terms of what is normal, ordinary. Because the world is not a cluster of conjurer's tricks, conjurers can exist. Because the logic of 'seeing that' is an intimate part of the concept of seeing, we sometimes rub our eyes at illusions.

'Seeing as' and 'seeing that' are not components of seeing, as rods and bearings are parts of motors: seeing is not composite. Still, one *can* ask logical questions. What must have occurred, for instance, for us to describe a man as having found a collar stud, or as having seen a bacillus? Unless he had had a visual sensation and knew what a bacillus was (and looked like) we would not say that he had seen a bacillus, except in the sense in which an infant could see a bacillus. 'Seeing as' and 'seeing that', then, are not psychological components of seeing. They are logically distinguishable elements in seeing-talk, in our concept of seeing.

To see fig. 1 as a transparent box, an ice-cube, or a block of glass is to see that it is six-faced, twelve-edged, eight-cornered. Its corners are solid right angles; if constructed it would be of rigid, or semi-rigid material, not of liquescent or gaseous stuff like oil, vapour or flames. It would be tangible. It would take up space in an exclusive way, being locatable here, there, but at least somewhere. Nor would it cease to exist when we blinked. Seeing it as a cube is just to see that all these things would obtain.

This is knowledge: it is knowing what kind of a thing 'box' or 'cube' denotes and something about what materials can make up such an entity. 'Transparent box' or 'glass cube' would not express what was seen were any of these further considerations denied. Seeing a bird in the sky involves seeing that it will not suddenly do vertical snap rolls; and this is more than marks the retina. We could be wrong. But to see a bird, even momentarily, is to see it in all these connexions. As Wisdom would say, every perception involves an aetiology and a prognosis.[1]

C

Sense-datum theorists stress how we can go wrong in our observations, as when we call aeroplanes 'birds'. Thus they seek what we are right about, even in these cases. Preoccupation with this problem obscures another one, namely, that of describing what is involved when we are right about what we say we see; and after all this happens very often. His preoccupation with mistakes leads the phenomenalist to portray a world in which we are usually deceived; but the world of physics is not like that. Were a physicist in an ordinary laboratory situation to react to his visual environment with purely sense-datum responses—as does the infant or the idiot—we would think him out of his mind. We would think him *not* to be seeing what was around him.

'Seeing that' threads knowledge into our seeing; it saves us from re-identifying everything that meets our eye; it allows physicists to observe new data as physicists, and not as cameras. We do not ask 'What's that?' of every passing bicycle. The knowledge is there in the seeing and not an adjunct of it. (The pattern of threads is there in the cloth and not tacked on to it by ancillary operations.) We rarely catch ourselves tacking knowledge on to what meets the eye. Seeing this page as having an opposite side requires no squeezing or forcing, yet nothing optical guarantees that when you turn the sheet it will not cease to exist. This is but another way of saying that ordinary seeing is corrigible, which everybody would happily concede. The search for incorrigible seeing has sometimes led some philosophers to deny that anything less than the incorrigible is seeing at all.

Seeing an object *x* is to see that it may behave in the ways we know *x*'s do behave: if the object's behaviour does not accord with what we expect of *x*'s we may be blocked from seeing it as a straightforward *x* any longer. Now we rarely see dolphin as fish, the earth as flat, the heavens as an inverted bowl or the sun as our satellite. '...what I perceive as the dawning of an aspect is not a property of the object, but an internal relation between it and other objects.'[1] To see in fig. 8 an X-ray tube is to see that a photo-sensitive plate placed below it will be irradiated. It is to see that the target will get extremely hot, and as it has no water-jacket it must be made of metal with a high melting-point—molybdenum or tungsten. It is to see that at high voltages green fluorescence will appear at the anode. Could a physicist see an X-ray tube without seeing that these other things would obtain? Could one see something as an incandescent light bulb and fail to see that it is the wire filament which 'lights up' to a white heat? The answer may sometimes be 'yes', but this only indicates that different things can be meant by 'X-ray tube' and 'incandescent bulb'. Two people confronted with an *x* may mean different things by *x*. Must their saying 'I see *x*' mean that they see the same thing? A child could parrot 'X-ray tube', or 'Kentucky' or 'Winston', when confronted with the figure above, but he would not see that these other things followed. And this is what the physicist does see.

If in the brilliant disc of which he is visually aware Tycho sees only the sun, then he cannot but see that it is a body which will behave in characteristically 'Tychonic' ways. These serve as the foundation for Tycho's general geocentric-geostatic theories about the sun. They are not imposed on his visual impressions as a tandem interpretation: they are 'there' in the seeing. (So too the interpretation of a piece of music is there in the music. Where else could it be? It is not something superimposed upon pure, unadulterated sound.)

Similarly we see fig. 1 as from underneath, as from above, or as a diagram of a rat maze or a gem-cutting project. However construed, the construing is there in the seeing. One is tempted to say 'the construing *is* the seeing'. The thread and its arrangement *is* the fabric, the sound and its composition *is* the music, the colour and its disposition *is* the painting. There are not two operations involved in my seeing fig. 1 as an ice-cube; I simply see it as an ice-cube. Analogously, the physicist sees an X-ray tube, not by first soaking up reflected light and then clamping on interpretations, but just as you see this page before you.

Tycho sees the sun beginning its journey from horizon to horizon. He sees that from some celestial vantage point the sun (carrying with it the moon and planets) could be watched circling our fixed earth. Watching the sun at dawn through Tychonic spectacles would be to see it in something like this way.

Kepler's visual field, however, has a different conceptual organization. Yet a drawing of what he sees at dawn could be a drawing of exactly what Tycho saw,[1] and could be recognized as such by Tycho. But Kepler will see the horizon dipping, or turning away, from our fixed local star. The shift from sunrise to horizon-turn is analogous to the shift-of-aspect phenomena already considered; it is occasioned by differences between what Tycho and Kepler think they know.

These logical features of the concept of seeing are inextricable and indispensable to observation in research physics. Why indispensable? That men do see in a way that permits analysis into 'seeing as' and 'seeing that' factors is one thing; 'indispensable', however, suggests that the world must be seen thus. This is a stronger claim, requiring a stronger argument. Let us put it differently: that observation in physics is not an encounter with unfamiliar and unconnected flashes, sounds and bumps, but rather a calculated meeting with these as flashes, sounds and bumps of a particular kind—this might figure in an account of what observation is. It would not secure the point that observation could not be otherwise. This latter type of argument is now required: it must establish that an alternative account would be not merely false, but absurd. To this I now turn.

D

Fortunately, we do not see the sun and the moon as we see the points of colour and light in the oculist's office; nor does the physicist see his laboratory equipment, his desk, or his hands in the baffled way that he may view a cloud-chamber photograph or an oscillograph pattern. In most cases we could give further information about what sort of thing we see. This might be expressed in a list: for instance, that *x* would break if dropped, that *x* is hollow, and so on.

To see fig. 3 as a bear on a tree is to see that further observations are possible; we can imagine the bear as viewed from the side or from behind. Indeed, seeing fig. 3 as a bear is just to have seen that these other views could all be simultaneous. It is also to see that certain observations are not possible: for example, the bear cannot be waving one paw in the air, nor be dangling one foot. This too is 'there' in the seeing.

'Is it a question of both seeing and thinking? or an amalgam of the two, as I should almost like to say?'[1] Whatever one would like to say, there is more to seeing fig. 3 as a bear, than optics, photochemistry or phenomenalism can explain.[2]

Notice a logical feature: 'see that' and 'seeing that' are always followed by 'sentential' clauses. The addition of but an initial capital letter and a full stop sets them up as independent sentences. One can see an ice-cube, or see a kite as a bird. One cannot see that an ice-cube, nor see that a bird. Nor is this due to limitations of vision. Rather, one may see that *ice-cubes can melt*; that *birds have 'hollow' bones*. Tycho and Simplicius see that *the universe is geocentric*; Kepler and Galileo see that *it is heliocentric*. The physicist sees that *anode-fluorescence will appear in an X-ray tube at high voltages*. The phrases in italics are complete sentential units.

Pictures and statements differ in logical type, and the steps between visual pictures and the statements of what is seen are many and intricate. Our visual consciousness is dominated by pictures; scientific knowledge, however, is primarily linguistic. Seeing is, as I should almost like to say, an amalgam of the two—pictures and language. At the least, the concept of seeing embraces the concepts of visual sensation and of knowledge.[1]

The gap between pictures and language locates the logical function of 'seeing that'. For vision is essentially pictorial, knowledge fundamentally linguistic. Both vision and knowledge are indispensable elements in seeing; but differences between pictorial and linguistic representation may mark differences between the optical and conceptual features of seeing. This may illuminate what 'seeing that' consists in.

Not all the elements of statement correspond to the elements of pictures: only someone who misunderstood the uses of language would expect otherwise.[2] There is a 'linguistic' factor in seeing, although there is nothing linguistic about what forms in the eye, or in the mind's eye. Unless there were this linguistic element, nothing we ever observed could have relevance for our knowledge. We could not speak of significant observations: nothing seen would make sense, and microscopy would only be a kind of kaleidoscopy. For what is it for things to make sense other than for descriptions of them to be composed of meaningful sentences?

We must explore the gulf between pictures and language, between sketching and describing, drawing and reporting. Only by showing how picturing and speaking are different can one suggest how 'seeing that' may bring them together; and brought together they must be if observations are to be *significant* or *noteworthy*.

Knowledge here is of what there is, as factually expressed in books, reports, and essays. How to do things is not our concern. I know how to whistle; but could I express that knowledge in language? Could I describe the taste of salt, even though I know perfectly well how salt tastes? I know how to control a parachute—much of that knowledge was imparted in lectures and drills, but an essential part of it was not *imparted* at all; it was 'got on the spot'. Physicists rely on 'know-how', on the 'feel' of things and the 'look' of a situation, for these control the direction of research. Such imponderables, however, rarely affect the corpus of physical truths. It is not Galileo's insight, Newton's genius and Einstein's imagination which have *per se* changed our knowledge of what there is: it is the true things they have said. 'Physical knowledge', therefore, will mean 'what is reportable in the texts, reports and discussions of physics.' We are concerned with *savoir*, not *savoir faire*.[1]

The 'foundation' of the language of physics, the part closest to mere sensation, is a series of statements. Statements are true or false. Pictures are not at all like statements: they are neither true nor false; retinal, cortical, or sense-datum pictures are neither true nor false. Yet what we see can determine whether statements like 'The sun is above the horizon' and 'The cube is transparent', are true or false. Our visual sensations may be 'set' by language forms; how else could they be appreciated in terms of what we know? Until they *are* so appreciated they do not constitute observation: they are more like the buzzing confusion of fainting or the vacant vista of aimless staring through a railway window.[2] Knowledge of the world is not a *montage* of sticks, stones, colour patches and noises, but a system of propositions.

Fig. 8, p. 15, asserts nothing. It could be inaccurate, but it could not be a lie. This is the wedge between pictures and language.

Significance, relevance—these notions depend on what we already know. Objects, events, pictures, are not intrinsically significant or relevant. If seeing were just an optical-chemical process, then nothing we saw would ever be relevant to what we know, and nothing known could have significance for what we see. Visual life would be unintelligible; intellectual life would lack a visual aspect. Man would be a blind computer harnessed to a brainless photoplate.[3]

Pictures often copy originals. All the elements of a copy, however, have the same kind of function. The lines depict elements in the original. The arrangement of the copy's elements shows the disposition of elements in the original. Copy and original are of the same logical type; you and your reflexion are of the same type. Similarly, language might copy what it describes.[1]

Consider fig. 3 alongside 'The bear is on the tree'. The picture contains a bear-element and a tree-element. If it is true to life, then in the original there is a bear and a tree. If the sentence is true *of* life, then (just as it contains 'bear' and 'tree') the situation it describes contains a bear and a tree. The picture combines its elements, it mirrors the actual relation of the bear and the tree. The sentence likewise conjoins 'bear' and 'tree' in the schema 'The —— is on the ——'. This verbal relation signifies the actual relation of the real bear and the real tree. Both picture and sentence are true copies: they contain nothing the original lacks, and lack nothing the original contains. The elements of the picture stand for (represent) elements of the original: so do 'bear' and 'tree'. This is more apparent when expressed symbolically as $b\ R\ t$, where b = bear, t = tree and R = the relation of being on.

By the arrangement of their elements these copies show the arrangement in the original situation. Thus fig. 3, 'The bear is on the tree', and '*bRt*' show what obtains with the real bear and the real tree; while 'The tree is on the bear', and '*tRb*', and a certain obvious cluster of lines do not show what actually obtains.

The copy is of the same type as the original. We can sketch the bear's teeth, but not his growl, any more than we could see the growl of the original bear. Leonardo could draw Mona Lisa's smile, but not her laugh. Language, however, is more versatile. Here is a dissimilarity between picturing and asserting which will grow to fracture the account once tendered by Wittgenstein, Russell and Wisdom. Language can encapsulate scenes and sounds, teeth and growls, smiles and laughs; a picture, or a gramophone, can do one or the other, but not both. Pictures and recordings stand for things by possessing certain properties of the original itself. Images, reflections, pictures and maps duplicate the spatial properties of what they image, reflect, picture or map; gramophone recordings duplicate audio-temporal properties. Sentences are not like this. They do not stand for things in virtue of possessing properties of the original; they do not *stand for* anything. They can state what is, or could be, the case. They can be used to make assertions, convey descriptions, supply narratives, accounts, etc., none of which depend on the possession of some property in common with what the statement is about. We need not write 'THE BEAR is bigger than ITS CUB' to show our meaning.

Images, reflexions, pictures and maps in fact copy originals with different degrees of strictness. A reflexion of King's Parade does not copy in the same sense that a charcoal sketch does, and both differ from the representation of 'K.P.' on a map of Cambridge and from a town-planner's drawing. The more like a reflection a map becomes, the less useful it is as a map. Drawings are less like copies of originals than are photographs. Of a roughly sketched ursoid shape one says either 'That's a bear' or 'That's supposed to be a bear'. Similarly with maps. Of a certain dot on the map one says either 'This is Cambridge' or 'This stands for Cambridge'.

Language copies least of all. There are exceptional words like 'buzz', 'tinkle' and 'toot', but they only demonstrate how conventional our languages and notations are. Nothing about 'bear' looks like a bear; nothing in the sound of 'bear' resembles a growl. That b-e-a-r can refer to bears is due to a convention which co-ordinates the word with the object. There is nothing dangerous about a red flag, yet it is a signal for danger. Of fig. 3 we might say 'There is a bear'. We would never say this of the word 'bear'. At the cinema we say 'It's a bear', or 'There's K.P.'—not 'That stands for a bear', or 'That denotes K.P.' It is words that denote; but they are rarely like what they denote.

Sentences do not show, for example, bears climbing trees, but they can state that bears climb trees. Showing the sun climbing into the sky consists in representing sun and sky and arranging them appropriately. Stating that the sun is climbing into the sky consists in referring to the sun and then characterizing it as climbing into the sky. The differences between representing and referring, between arranging and characterizing—these are the differences between picturing and language-using.

These differences exist undiminished between visual sense-data and basic sentences. Early logical constructionists were inattentive to the difficulties in fitting visual sense-data to basic sentences. Had they heeded the differences between pictures and maps, they might have detected greater differences still between pictures and language. One's visual awareness of a brown ursoid patch is logically just as remote from the utterance '(I am aware of a) brown, ursoid patch now', as with any of the pictures and sentences we have considered. The picture is of x; the statement is to the effect that x. The picture shows x; the statement refers to and describes x. The gap between pictures and language is not closed one millimetre by focusing on sense-data and basic sentences.

The prehistory of languages need not detain us. The issue concerns differences between *our* languages and *our* pictures, and not the smallness of those differences at certain historical times. Wittgenstein is misleading about this: '...and from [hieroglyphic writing] came the alphabet without the essence of representation being lost.'[1] This strengthened the picture theory of meaning, a truth-functional account of language and a theory of atomic sentences. But unless the essence of representation had been lost, languages could not be used in speaking the truth, telling lies, referring and characterizing.

Not all elements of a sentence do the same work. All the elements of pictures, however, just represent.[2] A picture of the dawn could be cut into small pictures, but sentences like 'The sun is on the horizon' and 'I perceive a solaroid patch now' cannot be cut into small sentences. All the elements of the picture show something; none of the elements of the sentences state anything. 'Bear!' may serve as a statement, as may 'Tree!' from the woodcutter, or 'Sun!' in eclipse-observations. But 'the', 'is' and 'on' are not likely ever to behave as statements.

Pictures are of the picturable. Recordings are of the recordable. You cannot play a smile or a wink on the gramophone. But language is more versatile: we can describe odours, sounds, feels, looks, smiles and winks. This freedom makes type-mistakes possible: for example, 'They found his pituitary but not his mind', 'We surveyed his retina but could not find his sight'. Only when we are free from the natural limitations of pictures and recordings can such errors occur. They are just possible in maps; of the hammer and sickle which signifies Russia on a school map, for instance, a child might ask 'How many miles long is the sickle?'. Maps with their partially conventional characters must be read (as pictures and photographs need not be); yet they must copy.

There is a corresponding gap between visual pictures and what we know. Seeing bridges this, for while seeing is at the least a 'visual copying' of objects, it is also more than that. It is a certain sort of seeing of objects: seeing that if x were done to them y would follow. This fact got lost in all the talk about knowledge arising from sense experience, memory, association and correlation. Memorizing, associating, correlating and comparing mental pictures may be undertaken *ad indefinitum* without one step having been taken towards scientific knowledge, that is, propositions known to be true. How long must one shuffle photographs, diagrams and sketches of antelopes before the statement 'antelopes are ungulates' springs forth?

When language and notation are ignored in studies of observation, physics is represented as resting on sensation and low-grade experiment. It is described as repetitious, monotonous concatenation of spectacular sensations, and of school-laboratory experiments. But physical science is not just a systematic exposure of the senses to the world; it is also a way of thinking about the world, a way of forming conceptions. The paradigm observer is not the man who sees and reports what all normal observers see and report, but the man who sees in familiar objects what no one else has seen before.[1]

4. EMPIRICIST CRITERIA OF COGNITIVE SIGNIFICANCE: PROBLEMS AND CHANGES

1. THE GENERAL EMPIRICIST CONCEPTION OF COGNITIVE AND EMPIRICAL SIGNIFICANCE

It is a basic principle of contemporary empiricism that a sentence makes a cognitively significant assertion, and thus can be said to be either true or false, if and only if either (1) it is analytic or contradictory—in which case it is said to have purely logical meaning or significance—or else (2) it is capable, at least potentially, of test by experiential evidence—in which case it is said to have empirical meaning or significance. The basic tenet of this principle, and especially of its second part, the so-called testability criterion of empirical meaning (or better: meaningfulness), is not peculiar to empiricism alone: it is characteristic also of contemporary operationism, and in a sense of pragmatism as well; for the pragmatist maxim that a difference must make a difference to be a difference may well be construed as insisting that a verbal difference between two sentences must make a difference in experiential implications if it is to reflect a difference in meaning.

How this general conception of cognitively significant discourse led to the rejection, as devoid of logical and empirical meaning, of various formulations in speculative metaphysics, and even of certain hypotheses offered within

This essay combines, with certain omissions and some other changes, the contents of two articles: "Problems and Changes in the Empiricist Criterion of Meaning," *Revue Internationale de Philosophie* No. 11, pp. 41–63 (January, 1950); and "The Concept of Cognitive Significance: A Reconsideration," *Proceedings of the American Academy of Arts and Sciences* 80, No. 1, pp. 61–77 (1951). This material is reprinted with kind permission of the Director of *Revue Internationale de Philosophie* and of the American Academy of Arts and Sciences.

empirical science, is too well known to require recounting. I think that the general intent of the empiricist criterion of meaning is basically sound, and that notwithstanding much oversimplification in its use, its critical application has been, on the whole, enlightening and salutary. I feel less confident, however, about the possibility of restating the general idea in the form of precise and general criteria which establish sharp dividing lines (a) between statements of purely logical and statements of empirical significance, and (b) between those sentences which do have cognitive significance and those which do not.

In the present paper, I propose to reconsider these distinctions as conceived in recent empiricism, and to point out some of the difficulties they present. The discussion will concern mainly the second of the two distinctions; in regard to the first, I shall limit myself to a few brief remarks.

2. THE EARLIER TESTABILITY CRITERIA OF MEANING AND THEIR SHORTCOMINGS

Let us note first that any general criterion of cognitive significance will have to meet certain requirements if it is to be at all acceptable. Of these, we note one, which we shall consider here as expressing a necessary, though by no means sufficient, *condition of adequacy* for criteria of cognitive significance.

(A) If under a given criterion of cognitive significance, a sentence N is nonsignificant, then so must be all truth-functional compound sentences in which N occurs nonvacuously as a component. For if N cannot be significantly assigned a truth value, then it is impossible to assign truth values to the compound sentences containing N; hence, they should be qualified as nonsignificant as well.

We note two corollaries of requirement (A):

(A1) If under a given criterion of cognitive significance, a sentence S is nonsignificant, then so must be its negation, $\sim S$.

(A2) If under a given criterion of cognitive significance, a sentence N is nonsignificant, then so must be any conjunction $N \cdot S$ and any disjunction $N v S$, no matter whether S is significant under the given criterion or not.

We now turn to the initial attempts made in recent empiricism to establish general criteria of cognitive significance. Those attempts were governed by the consideration that a sentence, to make an empirical assertion must be capable of being borne out by, or conflicting with, phenomena which are potentially capable of being directly observed. Sentences describing such potentially observable phenomena—no matter whether the latter do actually occur or not—may be called observation sentences. More specifically, an *observation sentence* might be construed as a sentence—no matter whether true or false—which asserts or denies that a specified object, or group of objects, of macroscopic size has a particular *observable characteristic,* i.e., a characteristic whose presence or absence can, under favorable circumstances, be ascertained by direct observation.[1]

The task of setting up criteria of empirical significance is thus transformed into the problem of characterizing in a precise manner the relationship which obtains between a hypothesis and one or more observation sentences whenever the phenomena described by the latter either confirm or disconfirm the hypothesis in question. The ability of a given sentence to enter into that relationship to some set of observation sentences would then characterize its testability-in-principle, and thus its empirical significance. Let us now briefly examine the major attempts that have been made to obtain criteria of significance in this manner.

One of the earliest criteria is expressed in the so-called *verifiability requirement.* According to it, a sentence is empirically significant if and only if it is not analytic and is capable, at least in principle, of complete verification by observational evidence; i.e., if observational evidence can be described which, if actually obtained, would conclusively establish the truth of the sentence.[2] With the

1. Observation sentences of this kind belong to what Carnap has called the thing-language, cf., e.g., (1938), pp. 52-53. That they are adequate to formulate the data which serve as the basis for empirical tests is clear in particular for the intersubjective testing procedures used in science as well as in large areas of empirical inquiry on the common-sense level. In epistemological discussions, it is frequently assumed that the ultimate evidence for beliefs about empirical matters consists in perceptions and sensations whose description calls for a phenomenalistic type of language. The specific problems connected with the phenomenalistic approach cannot be discussed here; but it should be mentioned that at any rate all the critical considerations presented in this article in regard to the testability criterion are applicable, *mutatis mutandis,* to the case of a phenomenalistic basis as well.

2. Originally, the permissible evidence was meant to be restricted to what is observable by the speaker and perhaps his fellow beings during their life times. Thus construed, the criterion rules out, as cognitively meaningless, all statements about the distant future or the remote past, as has been pointed out, among others, by Ayer (1946), chapter I; by Pap (1949), chapter 13, esp. pp. 333 ff.; and by Russell (1948), pp. 445-47. This difficulty is avoided, however, if we permit the evidence to consist of any finite set of "logically possible observation data", each of them formulated in an observation sentence. Thus, e.g., the sentence S_1, "The tongue of the largest dinosaur in New York's Museum of Natural History was blue or black" is completely verifiable in our sense; for it is a logical consequence of the sentence S_2, "The tongue of the largest dinosaur in New York's Museum of Natural History was blue"; and this is an observation sentence, in the sense just indicated.

And if the concept of *verifiability in principle* and the more general concept of *confirmability in principle,* which will be considered later, are construed as referring to *logically possible evidence* as expressed by observation sentences, then it follows similarly that the class of statements which are verifiable, or at least confirmable, in principle include such assertions as that the planet Neptune and the Antarctic Continent existed before they were discovered, and that atomic warfare, if not checked, will lead to the extermination of this planet. The objections which Russell (1948), pp. 445 and 447, raises against the verifiability criterion by reference to those examples do not apply therefore if the criterion is understood in the manner here suggested.

(continued overeaf)

help of the concept of observation sentence, we can restate this requirement as follows: A sentence S has empirical meaning if and only if it is possible to indicate a finite set of observation sentences, $O_1, O_2, \ldots, O_n$, such that if these are true, then S is necessarily true, too. As stated, however, this condition is satisfied also if S is an analytic sentence or if the given observation sentences are logically incompatible with each other. By the following formulation, we rule these cases out and at the same time express the intended criterion more precisely:

(2.1) REQUIREMENT OF COMPLETE VERIFIABILITY IN PRINCIPLE. A sentence has empirical meaning if and only if it is not analytic and follows logically from some finite and logically consistent class of observation sentences.[3] These observation sentences need not be true, for what the criterion is to explicate is testability by "potentially observable phenomena," or testability "in principle."

In accordance with the general conception of cognitive significance outlined earlier, a sentence will now be classified as cognitively significant if either it is analytic or contradictory, or it satisfies the verifiability requirement.

This criterion, however, has several serious defects. One of them has been noted by several writers:

a. Let us assume that the properties of being a stork and of being red-legged

3. As has frequently been emphasized in the empiricist literature, the term "verifiability" is to indicate, of course, the conceivability, or better, the logical possibility, of evidence of an observational kind which, if actually encountered, would constitute conclusive evidence for the given sentence; it is not intended to mean the technical possibility of performing the tests needed to obtain such evidence, and even less the possibility of actually finding directly observable phenomena which constitute conclusive evidence for that sentence—which would be tantamount to the actual existence of such evidence and would thus imply the truth of the given sentence. Analogous remarks apply to the terms "falsifiability" and "confirmability". This point has clearly been disregarded in some critical discussions of the verifiability criterion. Thus, e.g., Russell (1948), p. 448 construes verifiability as the actual existence of a set of conclusively verifying occurrences. This conception, which has never been advocated by any logical empiricist, must naturally turn out to be inadequate since according to it the empirical meaningfulness of a sentence could not be established without gathering empirical evidence, and moreover enough of it to permit a conclusive proof of the sentence in question! It is not surprising, therefore, that his extraordinary interpretation of verifiability leads Russell to the conclusion: "In fact, that a proposition is verifiable is itself not verifiable" (*l.c.*). Actually, under the empiricist interpretation of complete verifiability, any statement asserting the verifiability of some sentence S whose text is quoted, is either analytic or contradictory; for the decision whether there exists a class of observation sentences which entail S, i.e., whether such observation sentences can be formulated, no matter whether they are true or false—that decision is a purely logical matter.

Incidentally, statements of the kind mentioned by Russell, which are not actually verifiable by any human being, were explicitly recognized as cognitively significant already by Schlick (1936), Part V, who argued that the impossibility of verifying them was "merely empirical." The characterization of verifiability with the help of the concept of observation sentence as suggested here might serve as a more explicit and rigorous statement of that conception.

are both observable characteristics, and that the former does not logically entail the latter. Then the sentence

($S1$) All storks are red-legged

is neither analytic nor contradictory; and clearly, it is not deducible from a finite set of observation sentences. Hence, under the contemplated criterion, $S1$ is devoid of empirical significance; and so are all other sentences purporting to express universal regularities or general laws. And since sentences of this type constitute an integral part of scientific theories, the verifiability requirement must be regarded as overly restrictive in this respect.

Similarly, the criterion disqualifies all sentences such as 'For any substance there exists some solvent', which contain both universal and existential quantifiers (i.e., occurrences of the terms 'all' and 'some' or their equivalents); for no sentences of this kind can be logically deduced from any finite set of observation sentences.

Two further defects of the verifiability requirement do not seem to have been widely noticed:

b. As is readily seen, the negation of $S1$

($\sim S1$) There exists at least one stork that is not red-legged

is deducible from any two observation sentences of the type 'a is a stork' and 'a is not red-legged'. Hence, $\sim S1$ is cognitively significant under our criterion, but $S1$ is not, and this constitutes a violation of condition (A1).

c. Let S be a sentence which does, and N a sentence which does not satisfy the verifiability requirement. Then S is deducible from some set of observation sentences; hence, by a familiar rule of logic, $S \mathrm{v} N$ is deducible from the same set, and therefore cognitively significant according to our criterion. This violates condition (A2) above.[4]

Strictly analogous considerations apply to an alternative criterion, which

4. The arguments here adduced against the verifiability criterion also prove the inadequacy of a view closely related to it, namely that two sentences have the same cognitive significance if any set of observation sentences which would verify one of them would also verify the other, and conversely. Thus, e.g., under this criterion, any two general laws would have to be assigned the same cognitive significance, for no general law is verified by any set of observation sentences. The view just referred to must be clearly distinguished from a position which Russell examines in his critical discussion of the positivistic meaning criterion. It is "the theory that two propositions whose verified consequences are identical have the same significance" (1948), p. 448. This view is untenable indeed, for what consequences of a statement have actually been verified at a given time is obviously a matter of historical accident which cannot possibly serve to establish identity of cognitive significance. But I am not aware that any logical empiricist ever subscribed to that "theory."

makes complete falsifiability in principle the defining characteristic of empirical significance. Let us formulate this criterion as follows:

(2.2) Requirement of complete falsifiability in principle. A sentence has empirical meaning if and only if its negation is not analytic and follows logically from some finite logically consistent class of observation sentences.

This criterion qualifies a sentence as empirically meaningful if its negation satisfies the requirement of complete verifiability; as it is to be expected, it is therefore inadequate on similar grounds as the latter:

(a) It denies cognitive significance to purely existential hypotheses, such as 'There exists at least one unicorn', and all sentences whose formulation calls for mixed—i.e., universal and existential—quantification, such as 'For every compound there exists some solvent', for none of these can possibly be conclusively falsified by a finite number of observation sentences.

(b) If '*P*' is an observation predicate, then the assertion that all things have the property *P* is qualified as significant, but its negation, being equivalent to a purely existential hypothesis, is disqualified [cf. (*a*)]. Hence, criterion (2.2) gives rise to the same dilemma as (2.1).

(c) If a sentence *S* is completely falsifiable whereas *N* is a sentence which is not, then their conjunction, *S*·*N* (i.e., the expression obtained by connecting the two sentences by the word 'and') is completely falsifiable; for if the negation of *S* is entailed by a class of observation sentences, then the negation of *S*·*N* is, *a fortiori*, entailed by the same class. Thus, the criterion allows empirical significance to many sentences which an adequate empiricist criterion should rule out, such as 'All swans are white and the absolute is perfect.'

In sum, then, interpretations of the testability criterion in terms of complete verifiability or of complete falsifiability are inadequate because they are overly restrictive in one direction and overly inclusive in another, and because both of them violate the fundamental requirement A.

Several attempts have been made to avoid these difficulties by construing the testability criterion as demanding merely a partial and possibly indirect confirmability of empirical hypotheses by observational evidence.

A formulation suggested by Ayer[5] is characteristic of these attempts to set up a clear and sufficiently comprehensive criterion of confirmability. It states, in effect, that a sentence *S* has empirical import if from *S* in conjunction with suitable subsidiary hypotheses it is possible to derive observation sentences which are not derivable from the subsidiary hypotheses alone.

This condition is suggested by a closer consideration of the logical structure of

5. (1936, 1946), Chap. I. The case against the requirements of verifiability and of falsifiability, and in favor of a requirement of partial confirmability and disconfirmability, is very clearly presented also by Pap (1949), chapter 13.

scientific testing; but it is much too liberal as it stands. Indeed, as Ayer himself has pointed out in the second edition of his book, *Language, Truth, and Logic*,[6] his criterion allows empirical import to any sentence whatever. Thus, e.g., if *S* is the sentence 'The absolute is perfect', it suffices to choose as a subsidiary hypothesis the sentence 'If the absolute is perfect then this apple is red' in order to make possible the deduction of the observation sentence 'This apple is red', which clearly does not follow from the subsidiary hypothesis alone.

To meet this objection, Ayer proposed a modified version of his testability criterion. In effect, the modification restricts the subsidiary hypotheses mentioned in the previous version to sentences which either are analytic or can independently be shown to be testable in the sense of the modified criterion.[7]

But it can readily be shown that this new criterion, like the requirement of complete falsifiability, allows empirical significance to any conjunction $S{\cdot}N$, where *S* satisfies Ayer's criterion while *N* is a sentence such as 'The absolute is perfect', which is to be disqualified by that criterion. Indeed, whatever consequences can be deduced from *S* with the help of permissible subsidiary hypotheses can also be deduced from $S{\cdot}N$ by means of the same subsidiary hypotheses; and as Ayer's new criterion is formulated essentially in terms of the deducibility of a certain type of consequence from the given sentence, it countenances $S{\cdot}N$ together with *S*. Another difficulty has been pointed out by Church, who has shown[8] that if there are any three observation sentences none of which alone entails any of the others, then it follows for any sentence *S* whatsoever that either it or its denial has empirical import according to Ayer's revised criterion.

All the criteria considered so far attempt to explicate the concept of empirical significance by specifying certain logical connections which must obtain between a significant sentence and suitable observation sentences. It seems now that this type of approach offers little hope for the attainment of precise criteria of meaningfulness: this conclusion is suggested by the preceding survey of some representative attempts, and it receives additional support from certain further considerations, some of which will be presented in the following sections.

3. CHARACTERIZATION OF SIGNIFICANT SENTENCES BY CRITERIA FOR THEIR CONSTITUENT TERMS

An alternative procedure suggests itself which again seems to reflect well

6. (1946), 2d ed., pp. 11-12.

7. This restriction is expressed in recursive form and involves no vicious circle. For the full statement of Ayer's criterion, see Ayer (1946), p. 13.

8. Church (1949). An alternative criterion recently suggested by O'Connor (1950) as a revision of Ayer's formulation is subject to a slight variant of Church's stricture: It can be shown that if there are three observation sentences none of which entails any of the others, and if *S* is any noncompound sentence, then either *S* or $\sim S$ is significant under O'Connor's criterion.

the general viewpoint of empiricism: It might be possible to characterize cognitively significant sentences by certain conditions which their constituent terms have to satisfy. Specifically, it would seem reasonable to say that all extralogical terms[9] in a significant sentence must have experiential reference, and that therefore their meanings must be capable of explication by reference to observables exclusively.[10] In order to exhibit certain analogies between this approach and the previous one, we adopt the following terminological conventions:

Any term that may occur in a cognitively significant sentence will be called a *cognitively significant term*. Furthermore, we shall understand by an *observation term* any term which either (a) is an *observation predicate*, i.e., signifies some observable characteristic (as do the terms 'blue', 'warm', 'soft', 'coincident with', 'of greater apparent brightness than') or (b) names some physical object of macroscopic size (as do the terms 'the needle of this instrument', 'the Moon', 'Krakatoa Volcano', 'Greenwich, England', 'Julius Caesar').

Now while the testability criteria of meaning aimed at characterizing the cognitively significant sentences by means of certain inferential connections in which they must stand to some observation sentences, the alternative approach under consideration would instead try to specify the vocabulary that may be used in forming significant sentences. This vocabulary, the class of significant terms, would be characterized by the condition that each of its elements is either a logical term or else a term with empirical significance; in the latter case, it has to stand in certain definitional or explicative connections to some observation terms. This approach certainly avoids any violations of our earlier conditions of adequacy. Thus, e.g., if *S* is a significant sentence, i.e., contains cognitively significant terms only, then so is its denial, since the denial sign, and its verbal equivalents, belong to the vocabulary of logic and are thus significant. Again, if *N* is a sentence containing a non-significant term, then so is any compound sentence which contains *N*.

But this is not sufficient, of course. Rather, we shall now have to consider a crucial question analogous to that raised by the previous approach: Precisely how are the logical connections between empirically significant terms and observation terms to be construed if an adequate criterion of cognitive significance is to result? Let us consider some possibilities.

9. An extralogical term is one that does not belong to the specific vocabulary of logic. The following phrases, and those definable by means of them, are typical examples of logical terms: 'not', 'or', 'if... then', 'all', 'some', '... is an element of class...'. Whether it is possible to make a sharp theoretical distinction between logical and extra-logical terms is a controversial issue related to the problem of discriminating between analytic and synthetic sentences. For the purpose at hand, we may simply assume that the logical vocabulary is given by enumeration.

10. For a detailed exposition and critical discussion of this idea, see H. Feigl's stimulating and enlightening article (1950).

(3.1) The simplest criterion that suggests itself might be called the *requirement of definability*. It would demand that any term with empirical significance must be explicitly definable by means of observation terms.

This criterion would seem to accord well with the maxim of operationism that all significant terms of empirical science must be introduced by operational definitions. However, the requirement of definability is vastly too restrictive, for many important terms of scientific and even pre-scientific discourse cannot be explicitly defined by means of observation terms.

In fact, as Carnap[11] has pointed out, an attempt to provide explicit definitions in terms of observables encounters serious difficulties as soon as disposition terms, such as 'soluble', 'malleable', 'electric conductor', etc., have to be accounted for; and many of these occur even on the pre-scientific level of discourse.

Consider, for example, the word 'fragile'. One might try to define it by saying that an object x is fragile if and only if it satisfies the following condition: If at any time t the object is sharply struck, then it breaks at that time. But if the statement connectives in this phrasing are construed truth-functionally, so that the definition can be symbolized by

$$\text{(D)} \qquad Fx \equiv (t)\,(Sxt \supset Bxt)$$

then the predicate 'F' thus defined does not have the intended meaning. For let a be any object which is not fragile (e.g., a raindrop or a rubber band), but which happens not to be sharply struck at any time throughout its existence. Then 'Sat' is false and hence '$Sat \supset Bat$' is true for all values of 't'; consequently, 'Fa' is true though a is not fragile.

To remedy this defect, one might construe the phrase 'if. . . then. . .' in the original definiens as having a more restrictive meaning than the truth-functional conditional. This meaning might be suggested by the subjunctive phrasing 'If x were to be sharply struck at any time t, then x would break at t.' But a satisfactory elaboration of this construal would require a clarification of the meaning and the logic of counterfactual and subjunctive conditionals, which is a thorny problem.[12]

An alternative procedure was suggested by Carnap in his theory of reduction sentences.[13] These are sentences which, unlike definitions, specify the meaning of a term only conditionally or partially. The term 'fragile', for example, might be introduced by the following reduction sentence:

$$\text{(R)} \qquad (x)\,(t)\,[Sxt \supset (Fx \equiv Bxt)]$$

11. Cf. (1936–37), especially section 7.

12. On this subject, see for example Langford (1941); Lewis (1946), pp. 210–30; Chisholm (1946); Goodman (1947); Reichenbach (1947), Chapter VIII; Hempel and Oppenheim (1948), Part III; Popper (1949); and especially Goodman's further analysis (1955).

13. Cf. Carnap, *loc. cit.* note 11. For a brief elementary presentation of the main idea, see Carnap (1938), Part III. The sentence R here formulated for the predicate 'F' illustrates only the simplest type of reduction sentence, the so-called bilateral reduction sentence.

which specifies that if x is sharply struck at any time t, then x is fragile if and only if x breaks at t.

Our earlier difficulty is now avoided, for if a is a nonfragile object that is never sharply struck, then that expression in R which follows the quantifiers is true of a; but this does not imply that 'Fa' is true. But the reduction sentence R specifies the meaning of 'F' only for application to those objects which meet the "test condition" of being sharply struck at some time; for these it states that fragility then amounts to breaking. For objects that fail to meet the test condition, the meaning of 'F' is left undetermined. In this sense, reduction sentences have the character of partial or conditional definitions.

Reduction sentences provide a satisfactory interpretation of the experiential import of a large class of disposition terms and permit a more adequate formulation of so-called operational definitions, which, in general, are not complete definitions at all. These considerations suggest a greatly liberalized alternative to the requirement of definability:

(3.2) *The requirement of reducibility*. Every term with empirical significance must be capable of introduction, on the basis of observation terms, through chains of reduction sentences.

This requirement is characteristic of the liberalized versions of positivism and physicalism which, since about 1936, have superseded the older, overly narrow conception of a full definability of all terms of empirical science by means of observables,[14] and it avoids many of the shortcomings of the latter. Yet, reduction sentences do not seem to offer an adequate means for the introduction of the central terms of advanced scientific theories, often referred to as theoretical constructs. This is indicated by the following considerations: A chain of reduction sentences provides a necessary and a sufficient condition for the applicability of the term it introduces. (When the two conditions coincide, the chain is tantamount to an explicit definition.) But now take, for example, the concept of length as used in classical physical theory. Here, the length in centimeters of the distance between two points may assume any positive real number as its value; yet it is clearly impossible to formulate, by means of observation terms, a sufficient condition for the applicability of such expressions as 'having a length of $\sqrt{2}$ cm' and 'having a length of $\sqrt{2} + 10^{-100}$ cm'; for such conditions would provide a possibility for discrimination, in observational terms, between two lengths which differ by only 10^{-100} cm.[15]

14. Cf. the analysis in Carnap (1936-37), especially section 15; also see the briefer presentation of the liberalized point of view in Carnap (1938).

15. (Added in 1964.) This is not strictly correct. For a more circumspect statement, see note 12 in "A Logical Appraisal of Operationism" and the fuller discussion in section 7 of the essay "The Theoretician's Dilemma." Both of these pieces are reprinted in the present volume.

It would be ill-advised to argue that for this reason, we ought to permit only such values of the magnitude, length, as permit the statement of sufficient conditions in terms of observables. For this would rule out, among others, all irrational numbers and would prevent us from assigning, to the diagonal of a square with sides of length 1, the length $\sqrt{2}$, which is required by Euclidean geometry. Hence, the principles of Euclidean geometry would not be universally applicable in physics. Similarly, the principles of the calculus would become inapplicable, and the system of scientific theory as we know it today would be reduced to a clumsy, unmanageable torso. This, then, is no way of meeting the difficulty. Rather, we shall have to analyze more closely the function of constructs in scientific theories, with a view to obtaining through such an analysis a more adequate characterization of cognitively significant terms.

Theoretical constructs occur in the formulation of scientific theories. These may be conceived of, in their advanced stages, as being stated in the form of deductively developed axiomatized systems. Classical mechanics, or Euclidean or some Non-Euclidean form of geometry in physical interpretation, present examples of such systems. The extralogical terms used in a theory of this kind may be divided, in familiar manner, into primitive or basic terms, which are not defined within the theory, and defined terms, which are explicitly defined by means of the primitives. Thus, e.g., in Hilbert's axiomatization of Euclidean geometry, the terms 'point', 'straight line', 'between' are among the primitives, while 'line segment', 'angle', 'triangle', 'length' are among the defined terms. The basic and the defined terms together with the terms of logic constitute the vocabulary out of which all the sentences of the theory are constructed. The latter are divided, in an axiomatic presentation, into primitive statements (also called postulates or basic statements) which, in the theory, are not derived from any other statements, and derived ones, which are obtained by logical deduction from the primitive statements.

From its primitive terms and sentences, an axiomatized theory can be developed by means of purely formal principles of definition and deduction, without any consideration of the empirical significance of its extralogical terms. Indeed, this is the standard procedure employed in the axiomatic development of uninterpreted mathematical theories such as those of abstract groups or rings or lattices, or any form of pure (i.e., noninterpreted) geometry.

However, a deductively developed system of this sort can constitute a scientific theory only if it has received an empirical interpretation[16] which

16. The interpretation of formal theories has been studied extensively by Reichenbach, especially in his pioneer analyses of space and time in classical and in relativistic physics. He describes such interpretation as the establishment of *coordinating definitions* (Zuordnungsdefinitionen) for certain terms of the formal theory. See, for example, Reichenbach (1928). More recently,

(continued overleaf)

renders it relevant to the phenomena of our experience. Such interpretation is given by assigning a meaning, in terms of observables, to certain terms or sentences of the formalized theory. Frequently, an interpretation is given not for the primitive terms or statements but rather for some of the terms definable by means of the primitives, or for some of the sentences deducible from the postulates.[17] Furthermore, interpretation may amount to only a partial assignment of meaning. Thus, e.g., the rules for the measurement of length by means of a standard rod may be considered as providing a *partial* empirical interpretation for the term 'the length, in centimeters, of interval *i*', or alternatively, for some sentences of the form 'the length of interval *i* is *r* centimeters'. For the method is applicable only to intervals of a certain medium size, and even for the latter it does not constitute a full interpretation since the use of a standard rod does not constitute the only way of determining length: various alternative procedures are available involving the measurement of other magnitudes which are connected, by general laws, with the length that is to be determined.

This last observation, concerning the possibility of an indirect measurement of length by virtue of certain laws, suggests an important reminder. It is not correct to speak, as is often done, of "the experiential meaning" of a term or a sentence in isolation. In the language of science, and for similar reasons even in pre-scientific discourse, a single statement usually has no experiential implications. A single sentence in a scientific theory does not, as a rule, entail any observation sentences; consequences asserting the occurrence of certain observable phenomena can be derived from it only by conjoining it with a set of other, subsidiary, hypotheses. Of the latter, some will usually be observation sentences, others will be previously accepted theoretical statements. Thus, e.g., the relativistic theory of the deflection of light rays in the gravitational field of the sun entails assertions about observable phenomena only if it is conjoined with a considerable body of astronomical and optical theory as well as a large number of specific statements about the instruments used in those observations of solar eclipses which serve to test the hypothesis in question.

Hence, the phrase, 'the experiential meaning of expression *E*' is elliptical: What

17. A somewhat fuller account of this type of interpretation may be found in Carnap (1939), §24. The articles by Spence (1944) and by MacCorquodale and Meehl (1948) provide enlightening illustrations of the use of theoretical constructs in a field outside that of the physical sciences, and of the difficulties encountered in an attempt to analyze in detail their function and interpretation.

Northrop [cf. (1947), Chap. VII, and also the detailed study of the use of deductively formulated theories in science, ibid., Chaps. IV, V, VI] and H. Margenau [cf., for example, (1935)] have discussed certain aspects of this process under the title of *epistemic correlation*.

a given expression "means" in regard to potential empirical data is relative to two factors, namely:

I. *the linguistic framework L* to which the expression belongs. Its rules determine, in particular, what sentences—observational or otherwise—may be inferred from a given statement or class of statements;

II. the theoretical context in which the expression occurs, i.e., the class of those statements in *L* which are available as subsidiary hypotheses.

Thus, the sentence formulating Newton's law of gravitation has no experiential meaning by itself; but when used in a language whose logical apparatus permits the development of the calculus, and when combined with a suitable system of other hypotheses—including sentences which connect some of the theoretical terms with observation terms and thus establish a partial interpretation—then it has a bearing on observable phenomena in a large variety of fields. Analogous considerations are applicable to the term 'gravitational field', for example. It can be considered as having experiential meaning only within the context of a theory, which must be at least partially interpreted; and the experiential meaning of the term—as expressed, say, in the form of operational criteria for its application—will depend again on the theoretical system at hand, and on the logical characteristics of the language within which it is formulated.

4. COGNITIVE SIGNIFICANCE AS A CHARACTERISTIC OF INTERPRETED SYSTEMS

The preceding considerations point to the conclusion that a satisfactory criterion of cognitive significance cannot be reached through the second avenue of approach here considered, namely by means of specific requirements for the terms which make up significant sentences. This result accords with a general characteristic of scientific (and, in principle, even pre-scientific) theorizing: Theory formation and concept formation go hand in hand; neither can be carried on successfully in isolation from the other.

If, therefore, cognitive significance can be attributed to anything, then only to entire theoretical systems formulated in a language with a well-determined structure. And the decisive mark of cognitive significance in such a system appears to be the existence of an interpretation for it in terms of observables. Such an interpretation might be formulated, for example, by means of conditional or biconditional sentences connecting nonobservational terms of the system with observation terms in the given language; the latter as well as the connecting sentences may or may not belong to the theoretical system.

But the requirement of partial interpretation is extremely liberal; it is satisfied, for example, by the system consisting of contemporary physical theory combined with some set of principles of speculative metaphysics, even if the latter have no empirical interpretation at all. Within the total system, these metaphysical principles play the role of what K. Reach and also O. Neurath liked to call *isolated sentences:* They are neither purely formal truths or falsehoods, demonstrable or refutable by means of the logical rules of the given language system; nor do they have any experiential bearing; i.e., their omission from the theoretical system would have no effect on its explanatory and predictive power in regard to potentially observable phenomena (i.e., the kind of phenomena described by observation sentences). Should we not, therefore, require that a cognitively significant system contain no isolated sentences? The following criterion suggests itself:

(4.1) A theoretical system is cognitively significant if and only if it is partially interpreted to at least such an extent that none of its primitive sentences is isolated.

But this requirement may bar from a theoretical system certain sentences which might well be viewed as permissible and indeed desirable. By way of a simple illustration, let us assume that our theoretical system T contains the primitive sentence

$$(S1) \qquad (x)\,[P_1x \supset (Qx \equiv P_2x)]$$

where 'P_1' and 'P_2' are observation predicates in the given language L, while 'Q' functions in T somewhat in the manner of a theoretical construct and occurs in only one primitive sentence of T, namely $S1$. Now $S1$ is not a truth or falsehood of formal logic; and furthermore, if $S1$ is omitted from the set of primitive sentences of T, then the resulting system, T', possesses exactly the same systematic, i.e., explanatory and predictive, power as T. Our contemplated criterion would therefore qualify $S1$ as an isolated sentence which has to be eliminated—excised by means of Occam's razor, as it were—if the theoretical system at hand is to be cognitively significant.

But it is possible to take a much more liberal view of $S1$ by treating it as a partial definition for the theoretical term 'Q'. Thus conceived, $S1$ specifies that in all cases where the observable characteristic P_1 is present, 'Q' is applicable if and only if the observable characteristic P_2 is present as well. In fact, $S1$ is an instance of those partial, or conditional, definitions which Carnap calls bilateral reduction sentences. These sentences are explicitly qualified by Carnap as analytic (though not, of course, as truths of formal logic), essentially on the ground that all their consequences which are expressible by means of observation predicates (and logical terms) alone are truths of formal logic.[18]

Let us pursue this line of thought a little further. This will lead us to some observations on analytic sentences and then back to the question of the adequacy of (4.1).

18. Cf. Carnap (1936-37), especially sections 8 and 10.

Suppose that we add to our system T the further sentence

$$(S2) \qquad (x)[P_3x \supset (Qx \equiv P_4x)]$$

where 'P_3', 'P_4' are additional observation predicates. Then, on the view that "every bilateral reduction sentence is analytic",[19] $S2$ would be analytic as well as $S1$. Yet, the two sentences jointly entail non-analytic consequences which are expressible in terms of observation predicates alone, such as[20]

$$(O) \qquad (x)\,[\sim(P_1x \cdot P_2x \cdot Px_3 \cdot \sim P_4x) \cdot \sim(P_1x \cdot \sim P_2x \cdot P_3x \cdot P_4x)]$$

But one would hardly want to admit the consequence that the conjunction of two analytic sentences may be synthetic. Hence if the concept of analyticity can be applied at all to the sentences of interpreted deductive systems, then it will have to be relativized with respect to the theoretical context at hand. Thus, e.g., $S1$ might be qualified as analytic relative to the system T, whose remaining postulates do not contain the term 'Q', but as synthetic relative to the system T enriched by $S2$. Strictly speaking, the concept of analyticity has to be relativized also in regard to the rules of the language at hand, for the latter determine what observational or other consequences are entailed by a given sentence. This need for at least a twofold relativization of the concept of analyticity was almost to be expected in view of those considerations which required the same twofold relativization for the concept of experiential meaning of a sentence.

If, on the other hand, we decide not to permit $S1$ in the role of a partial definition and instead reject it as an isolated sentence, then we are led to an analogous conclusion: Whether a sentence is isolated or not will depend on the linguistic frame and on the theoretical context at hand: While $S1$ is isolated relative to T (and the language in which both are formulated), it acquires definite experiential implications when T is enlarged by $S2$.

Thus we find, on the level of interpreted theoretical systems, a peculiar rapprochement, and partial fusion, of some of the problems pertaining to the concepts of cognitive significance and of analyticity: Both concepts need to be relativized; and a large class of sentences may be viewed, apparently with equal right, as analytic in a given context, or as isolated, or nonsignificant, in respect to it.

In addition to barring, as isolated in a given context, certain sentences which could just as well be construed as partial definitions, the criterion (4.1) has another serious defect. Of two logically equivalent formulations of a theoretical system it may qualify one as significant while barring the other as containing

19. Carnap (1936–37), p. 452.

20. The sentence O is what Carnap calls the *representative sentence* of the couple consisting of the sentences $S1$ and $S2$; see (1936–37), pp. 450–53.

an isolated sentence among its primitives. For assume that a certain theoretical system *T*1 contains among its primitive sentences *S*,′ *S*″, . . . exactly one, *S*′, which is isolated. Then *T*1 is not significant under (4.1). But now consider the theoretical system *T*2 obtained from *T*1 by replacing the two first primitive sentences, *S*′, *S*″, by one, namely their conjunction. Then, under our assumptions, none of the primitive sentences of *T*2 is isolated, and *T*2, though equivalent to *T*1, is qualified as significant by (4.1). In order to do justice to the intent of (4.1), we would therefore have to lay down the following stricter requirement:

(4.2) A theoretical system is cognitively significant if and only if it is partially interpreted to such an extent that in no system equivalent to it at least one primitive sentence is isolated.

Let us apply this requirement to some theoretical system whose postulates include the two sentences *S*1 and *S*2 considered before, and whose other postulates do not contain '*Q*' at all. Since the sentences *S*1 and *S*2 together entail the sentence *O*, the set consisting of *S*1 and *S*2 is logically equivalent to the set consisting of *S*1, *S*2 and *O*. Hence, if we replace the former set by the latter, we obtain a theoretical system equivalent to the given one. In this new system, both *S*1 and *S*2 are isolated since, as can be shown, their removal does not affect the explanatory and predictive power of the system in reference to observable phenomena. To put it intuitively, the systematic power of *S*1 and *S*2 is the same as that of *O*. Hence, the original system is disqualified by (4.2). From the viewpoint of a strictly sensationalist positivism as perhaps envisaged by Mach, this result might be hailed as a sound repudiation of theories making reference to fictitious entities, and as a strict insistence on theories couched exclusively in terms of observables. But from a contemporary vantage point, we shall have to say that such a procedure overlooks or misjudges the important function of constructs in scientific theory: The history of scientific endeavor shows that if we wish to arrive at precise, comprehensive, and well-confirmed general laws, we have to rise above the level of direct observation. The phenomena directly accessible to our experience are not connected by general laws of great scope and rigor. Theoretical constructs are needed for the formulation of such higher-level laws. One of the most important functions of a well-chosen construct is its potential ability to serve as a constituent in ever new general connections that may be discovered; and to such connections we would blind ourselves if we insisted on banning from scientific theories all those terms and sentences which could be "dispensed with" in the sense indicated in (4.2). In following such a narrowly phenomenalistic or positivistic course, we would deprive ourselves of the tremendous fertility of theoretical constructs, and we would often render the formal structure of the expurgated theory clumsy and inefficient.

Criterion (4.2), then, must be abandoned, and considerations such as those outlined in this paper seem to lend strong support to the conjecture that no adequate alternative to it can be found; i.e., that it is not possible to formulate general and precise criteria which would separate those partially interpreted systems whose isolated sentences might be said to have a significant function from those in which the isolated sentences are, so to speak, mere useless appendages.

We concluded earlier that cognitive significance in the sense intended by recent empiricism and operationism can at best be attributed to sentences forming a theoretical system, and perhaps rather to such systems as wholes. Now, rather than try to replace (4.2) by some alternative, we will have to recognize further that cognitive significance in a system is a matter of degree: Significant systems range from those whose entire extralogical vocabulary consists of observation terms, through theories whose formulation relies heavily on theoretical constructs, on to systems with hardly any bearing on potential empirical findings. Instead of dichotomizing this array into significant and non-significant systems it would seem less arbitrary and more promising to appraise or compare different theoretical systems in regard to such characteristics as these:

a. the clarity and precision with which the theories are formulated, and with which the logical relationships of their elements to each other and to expressions couched in observational terms have been made explicit;
b. the systematic, i.e., explanatory and predictive, power of the systems in regard to observable phenomena;
c. the formal simplicity of the theoretical system with which a certain systematic power is attained;
d. the extent to which the theories have been confirmed by experiential evidence.

Many of the speculative philosophical approaches to cosmology, biology, or history, for example, would make a poor showing on practically all of these counts and would thus prove no matches to available rival theories, or would be recognized as so unpromising as not to warrant further study or development.

If the procedure here suggested is to be carried out in detail, so as to become applicable also in less obvious cases, then it will be necessary, of course, to develop general standards, and theories pertaining to them, for the appraisal and comparison of theoretical systems in the various respects just mentioned. To what extent this can be done with rigor and precision cannot well be judged in advance. In recent years, a considerable amount of work has been done towards a definition and theory of the concept of degree of confirmation, or logical probability, of a theoretical system;[21] and several contributions have been made towards the

21. Cf., for example, Carnap (1945)1 and (1945)2, and especially (1950). Also see Helmer and Oppenheim (1945).

clarification of some of the other ideas referred to above.[22] The continuation of this research represents a challenge for further constructive work in the logical and methodological analysis of scientific knowledge.

22. On simplicity, cf. especially Popper (1935), Chap. V; Reichenbach (1938), § 42; Goodman (1949)1, (1949)2, (1950); on explanatory and predictive power, cf. Hempel and Oppenheim (1948), Part IV.

REFERENCES

Ayer, A. J., *Language, Truth and Logic*, London, 1936; 2nd ed. 1946.

Carnap, R., "Testability and Meaning," *Philosophy of Science*, 3 (1936) and 4 (1937).

Carnap, R., "Logical Foundations of the Unity of Science," in: *International Encyclopedia of Unified Science*, I, 1; Chicago, 1938.

Carnap, R., *Foundations of Logic and Mathematics*, Chicago, 1939.

Carnap, R., "On Inductive Logic," *Philosophy of Science*, 12 (1945). Referred to as (1945)1 in this article.

Carnap, R., "The Two Concepts of Probability," *Philosophy and Phenomenological Research*, 5 (1945). Referred to as (1945)2 in this article.

Carnap, R., *Logical Foundations of Probability*, Chicago, 1950.

Chisholm, R. M., "The Contrary-to-Fact Conditional," *Mind*, 55 (1946).

Church, A., Review of Ayer (1946), *The Journal of Symbolic Logic*, 14 (1949), 52-53.

Feigl, H., "Existential Hypotheses: Realistic vs. Phenomenalistic Interpretations," *Philosophy of Science*, 17 (1950).

Goodman, N., "The Problem of Counterfactual Conditionals," *The Journal of Philosophy*, 44 (1947).

Goodman, N., "The Logical Simplicity of Predicates," *The Journal of Symbolic Logic*, 14 (1949). Referred to as (1949)1 in this article.

Goodman, N., "Some Reflections on the Theory of Systems," *Philosophy and Phenomenological Research*, 9 (1949). Referred to as (1949)2 in this article.

Goodman, N., "An Improvement in the Theory of Simplicity," *The Journal of Symbolic Logic*, 15 (1950).

Goodman, N., *Fact, Fiction, and Forecast*, Cambridge, Massachusetts, 1955.

Helmer, O. and P. Oppenheim, "A Syntactical Definition of Probability and of Degree of Confirmation." *The Journal of Symbolic Logic*, 10 (1945).

Hempel, C. G. and P. Oppenheim, "Studies in the Logic of Explanation," *Philosophy of Science*, 15 (1948). (Reprinted in this volume.)

Langford, C. H., Review in *The Journal of Symbolic Logic*, 6 (1941), 67-68.

Lewis, C. I., *An Analysis of Knowledge and Valuation*, La Salle, Ill., 1946.

MacCorquodale, K. and P. E. Meehl, "On a Distinction Between Hypothetical Constructs and Intervening Variables," *Psychological Review*, 55 (1948).

Margenau, H., "Methodology of Modern Physics," *Philosophy of Science*, 2 (1935).

Northrop, F. S. C., *The Logic of the Sciences and the Humanities*, New York, 1947.

O'Connor, D. J., "Some Consequences of Professor A. J. Ayer's Verification Principle," *Analysis*, 10 (1950).

Pap, A., *Elements of Analytic Philosophy*, New York, 1949.

Popper, K., *Logik der Forschung*, Wien, 1935.

Popper, K., "A Note on Natural Laws and So-Called 'Contrary-to-Fact Conditionals'," *Mind*, 58 (1949).

Reichenbach, H., *Philosophie der Raum-Zeit-Lehre*, Berlin, 1928.

Reichenbach, H., *Elements of Symbolic Logic*, New York, 1947.

Russell, B., *Human Knowledge*, New York, 1948.

Schlick, M., "Meaning and Verification," *Philosophical Review*, 45 (1936). Also reprinted in Feigl, H. and W. Sellars, (eds.) *Readings in Philosophical Analysis*, New York, 1949.

Spence, Kenneth W., "The Nature of Theory Construction in Contemporary Psychology," *Psychological Review*, 51 (1944).

TWO DOGMAS OF EMPIRICISM

Modern empiricism has been conditioned in large part by two dogmas. One is a belief in some fundamental cleavage between truths which are *analytic*, or grounded in meanings independently of matters of fact, and truths which are *synthetic*, or grounded in fact. The other dogma is *reductionism*: the belief that each meaningful statement is equivalent to some logical construct upon terms which refer to immediate experience. Both dogmas, I shall argue, are ill-founded. One effect of abandoning them is, as we shall see, a blurring of the supposed boundary between speculative metaphysics and natural science. Another effect is a shift toward pragmatism.

1. Background for Analyticity

Kant's cleavage between analytic and synthetic truths was foreshadowed in Hume's distinction between relations of ideas and matters of fact, and in Leibniz's distinction between truths of reason and truths of fact. Leibniz spoke of the truths of reason as true in all possible worlds. Picturesqueness aside, this is to say that the truths of reason are those which could not possibly be false. In the same vein we hear analytic statements defined as statements whose denials are self-contradictory. But this definition has small explanatory value; for the notion of self-contradictoriness, in the quite broad sense needed for this definition of analyticity, stands in exactly the same need of clarification as does the notion of analyticity itself. The two notions are the two sides of a single dubious coin.

Kant conceived of an analytic statement as one that attributes to its subject no more than is already conceptually contained in the subject. This formulation has two shortcomings: it limits itself to statements of subject-predicate form, and it appeals to a notion of containment which is left at a metaphorical level. But Kant's intent, evident more from the use he makes of the notion of analyticity than from his definition of it, can be restated thus: a statement is analytic when it is true by virtue of meanings and independently of fact. Pursuing this line, let us examine the concept of *meaning* which is presupposed.

Meaning, let us remember, is not to be identified with naming.[1] Frege's example of 'Evening Star' and 'Morning Star', and Russell's of 'Scott' and 'the author of *Waverley*', illustrate that terms can name the same thing but differ in meaning. The distinction between meaning and naming is no less important at the level of abstract terms. The terms '9' and 'the number of the planets' name one and the same abstract entity but presumably must be regarded as unlike in meaning; for astronomical observation was needed, and not mere reflection on meanings, to determine the sameness of the entity in question.

The above examples consist of singular terms, concrete and abstract. With general terms, or predicates, the situation is somewhat different but parallel. Whereas a singular term purports to name an entity, abstract or concrete, a general term does not; but a general term is *true of* an entity, or of each of many, or of none.[2] The class of all entities of which a general term is true is called the *extension* of the term. Now paralleling the contrast between the meaning of a singular term and the entity named, we must distinguish equally between the meaning of a general term and its extension. The general terms 'creature with a heart' and 'creature with kidneys', for example, are perhaps alike in extension but unlike in meaning.

Confusion of meaning with extension, in the case of general terms, is less common than confusion of meaning with naming in the case of singular terms. It is indeed a commonplace in philosophy to oppose intension (or meaning) to extension, or, in a variant vocabulary, connotation to denotation.

[1] See above, p. 9.

[2] See above, p. 10, and below, pp. 107-115.

The Aristotelian notion of essence was the forerunner, no doubt, of the modern notion of intension or meaning. For Aristotle it was essential in men to be rational, accidental to be two-legged. But there is an important difference between this attitude and the doctrine of meaning. From the latter point of view it may indeed be conceded (if only for the sake of argument) that rationality is involved in the meaning of the word 'man' while two-leggedness is not; but two-leggedness may at the same time be viewed as involved in the meaning of 'biped' while rationality is not. Thus from the point of view of the doctrine of meaning it makes no sense to say of the actual individual, who is at once a man and a biped, that his rationality is essential and his two-leggedness accidental or vice versa. Things had essences, for Aristotle, but only linguistic forms have meanings. Meaning is what essence becomes when it is divorced from the object of reference and wedded to the word.

For the theory of meaning a conspicuous question is the nature of its objects: what sort of things are meanings? A felt need for meant entities may derive from an earlier failure to appreciate that meaning and reference are distinct. Once the theory of meaning is sharply separated from the theory of reference, it is a short step to recognizing as the primary business of the theory of meaning simply the synonymy of linguistic forms and the analyticity of statements; meanings themselves, as obscure intermediary entities, may well be abandoned.[3]

The problem of analyticity then confronts us anew. Statements which are analytic by general philosophical acclaim are not, indeed, far to seek. They fall into two classes. Those of the first class, which may be called *logically true*, are typified by:

(1) No unmarried man is married.

The relevant feature of this example is that it not merely is true as it stands, but remains true under any and all reinterpretations of 'man' and 'married'. If we suppose a prior inventory of *logical* particles, comprising 'no', 'un-', 'not', 'if', 'then', 'and', etc., then in general a logical truth is a statement which is true

[3] See above, pp. 11f, and below, pp. 48f.

and remains true under all reinterpretations of its components other than the logical particles.

But there is also a second class of analytic statements, typified by:

(2) No bachelor is married.

The characteristic of such a statement is that it can be turned into a logical truth by putting synonyms for synonyms; thus (2) can be turned into (1) by putting 'unmarried man' for its synonym 'bachelor'. We still lack a proper characterization of this second class of analytic statements, and therewith of analyticity generally, inasmuch as we have had in the above description to lean on a notion of "synonymy" which is no less in need of clarification than analyticity itself.

In recent years Carnap has tended to explain analyticity by appeal to what he calls state-descriptions.[4] A state-description is any exhaustive assignment of truth values to the atomic, or noncompound, statements of the language. All other statements of the language are, Carnap assumes, built up of their component clauses by means of the familiar logical devices, in such a way that the truth value of any complex statement is fixed for each state-description by specifiable logical laws. A statement is then explained as analytic when it comes out true under every state description. This account is an adaptation of Leibniz's "true in all possible worlds." But note that this version of analyticity serves its purpose only if the atomic statements of the language are, unlike 'John is a bachelor' and 'John is married', mutually independent. Otherwise there would be a state-description which assigned truth to 'John is a bachelor' and to 'John is married', and consequently 'No bachelors are married' would turn out synthetic rather than analytic under the proposed criterion. Thus the criterion of analyticity in terms of state-descriptions serves only for languages devoid of extra-logical synonym-pairs, such as 'bachelor' and 'unmarried man'—synonym-pairs of the type which give rise to the "second class" of analytic statements. The criterion in terms of state-descrip-

[4] Carnap [3], pp. 9ff; [4], pp. 70ff.

tions is a reconstruction at best of logical truth, not of analyticity.

I do not mean to suggest that Carnap is under any illusions on this point. His simplified model language with its state-descriptions is aimed primarily not at the general problem of analyticity but at another purpose, the clarification of probability and induction. Our problem, however, is analyticity; and here the major difficulty lies not in the first class of analytic statements, the logical truths, but rather in the second class, which depends on the notion of synonymy.

2. Definition

There are those who find it soothing to say that the analytic statements of the second class reduce to those of the first class, the logical truths, by *definition*; 'bachelor', for example, is *defined* as 'unmarried man'. But how de we find that 'bachelor' is defined as 'unmarried man'? Who defined it thus, and when? Are we to appeal to the nearest dictionary, and accept the lexicographer's formulation as law? Clearly this would be to put the cart before the horse. The lexicographer is an empirical scientist, whose business is the recording of antecedent facts; and if he glosses 'bachelor' as 'unmarried man' it is because of his belief that there is a relation of synonymy between those forms, implicit in general or preferred usage prior to his own work. The notion of synonymy presupposed here has still to be clarified, presumably in terms relating to linguistic behavior. Certainly the "definition" which is the lexicographer's report of an observed synonymy cannot be taken as the ground of the synonymy.

Definition is not, indeed, an activity exclusively of philologists. Philosophers and scientists frequently have occasion to "define" a recondite term by paraphrasing it into terms of a more familiar vocabulary. But ordinarily such a definition, like the philologist's, is pure lexicography, affirming a relation of synonymy antecedent to the exposition in hand.

Just what it means to affirm synonymy, just what the interconnections may be which are necessary and sufficient in order that two linguistic forms be properly describable as synonymous, is far from clear; but, whatever these interconnections may be, ordinarily they are grounded in usage. Definitions reporting selected instances of synonymy come then as reports upon usage.

There is also, however, a variant type of definitional activity which does not limit itself to the reporting of preëxisting synonymies. I have in mind what Carnap calls *explication*—an activity to which philosophers are given, and scientists also in their more philosophical moments. In explication the purpose is not merely to paraphrase the definiendum into an outright synonym, but actually to improve upon the definiendum by refining or supplementing its meaning. But even explication, though not merely reporting a preëxisting synonymy between definiendum and definiens, does rest nevertheless on *other* preexisting synonymies. The matter may be viewed as follows. Any word worth explicating has some contexts which, as wholes, are clear and precise enough to be useful; and the purpose of explication is to preserve the usage of these favored contexts while sharpening the usage of other contexts. In order that a given definition be suitable for purposes of explication, therefore, what is required is not that the definiendum in its antecedent usage be synonymous with the definiens, but just that each of these favored contexts of the definiendum, taken as a whole in its antecedent usage, be synonymous with the corresponding context of the definiens.

Two alternative definientia may be equally appropriate for the purposes of a given task of explication and yet not be synonymous with each other; for they may serve interchangeably within the favored contexts but diverge elsewhere. By cleaving to one of these definientia rather than the other, a definition of explicative kind generates, by fiat, a relation of synonymy between definiendum and definiens which did not hold before. But such a definition still owes its explicative function, as seen, to preexisting synonymies.

There does, however, remain still an extreme sort of definition which does not hark back to prior synonymies at all: namely, the explicitly conventional introduction of novel notations for purposes of sheer abbreviation. Here the definiendum becomes synonymous with the definiens simply because it has been created expressly for the purpose of being synonymous with the definiens. Here we have a really transparent case of synonymy created by definition; would that all species of synonymy were as intelligible. For the rest, definition rests on synonymy rather than explaining it.

The word 'definition' has come to have a dangerously reassuring sound, owing no doubt to its frequent occurrence in logical and mathematical writings. We shall do well to digress now into a brief appraisal of the role of definition in formal work.

In logical and mathematical systems either of two mutually antagonistic types of economy may be striven for, and each has its peculiar practical utility. On the one hand we may seek economy of practical expression—ease and brevity in the statement of multifarious relations. This sort of economy calls usually for distinctive concise notations for a wealth of concepts. Second, however, and oppositely, we may seek economy in grammar and vocabulary; we may try to find a minimum of basic concepts such that, once a distinctive notation has been appropriated to each of them, it becomes possible to express any desired further concept by mere combination and iteration of our basic notations. This second sort of economy is impractical in one way, since a poverty in basic idioms tends to a necessary lengthening of discourse. But it is practical in another way: it greatly simplifies theoretical discourse *about* the language, through minimizing the terms and the forms of construction wherein the language consists.

Both sorts of economy, though prima facie incompatible, are valuable in their separate ways. The custom has consequently arisen of combining both sorts of economy by forging in effect two languages, the one a part of the other. The inclusive language, though redundant in grammar and vocabulary, is economical in message lengths, while the part, called primitive notation, is economical in grammar and vocabulary. Whole and part are correlated by rules of translation whereby each idiom not in primitive notation is equated to some complex built up of primitive notation. These rules of translation are the so-called *definitions* which appear in formalized systems. They are best viewed not as adjuncts to one language but as correlations between two languages, the one a part of the other.

But these correlations are not arbitrary. They are supposed to show how the primitive notations can accomplish all purposes, save brevity and convenience, of the redundant language. Hence the definiendum and its definiens may be expected, in each case, to be related in one or another of the three ways lately noted. The definiens may be a faithful paraphrase of the definiendum into the narrower notation, preserving a direct synonymy[5] as of antecedent usage; or the definiens may, in the spirit of explication, improve upon the antecedent usage of the definiendum; or finally, the definiendum may be a newly created notation, newly endowed with meaning here and now.

In formal and informal work alike, thus, we find that definition—except in the extreme case of the explicitly conventional introduction of new notations—hinges on prior relations of synonymy. Recognizing then that the notion of definition does not hold the key to synonymy and analyticity, let us look further into synonymy and say no more of definition.

3. Interchangeability

A natural suggestion, deserving close examination, is that the synonymy of two linguistic forms consists simply in their interchangeability in all contexts without change of truth value —interchangeability, in Leibniz's phrase, *salva veritate.*[6] Note that synonyms so conceived need not even be free from vagueness, as long as the vaguenesses match.

[5] According to an important variant sense of 'definition', the relation preserved may be the weaker relation of mere agreement in reference; see below, p. 132. But definition in this sense is better ignored in the present connection, being irrelevant to the question of synonymy.

[6] Cf. Lewis [1], p. 373.

But it is not quite true that the synonyms 'bachelor' and 'unmarried man' are everywhere interchangeable *salva veritate*. Truths which become false under substitution of 'unmarried man' for 'bachelor' are easily constructed with the help of 'bachelor of arts' or 'bachelor's buttons'; also with the help of quotation, thus:

'Bachelor' has less than ten letters.

Such counterinstances can, however, perhaps be set aside by treating the phrases 'bachelor of arts' and 'bachelor's buttons' and the quotation ' 'bachelor' ' each as a single indivisible word and then stipulating that the interchangeability *salva veritate* which is to be the touchstone of synonymy is not supposed to apply to fragmentary occurrences inside of a word. This account of synonymy, supposing it acceptable on other counts, has indeed the drawback of appealing to a prior conception of "word" which can be counted on to present difficulties of formulation in its turn. Nevertheless some progress might be claimed in having reduced the problem of synonymy to a problem of wordhood. Let us pursue this line a bit, taking "word" for granted.

The question remains whether interchangeability *salva veritate* (apart from occurrences within words) is a strong enough condition for synonymy, or whether, on the contrary, some heteronymous expressions might be thus interchangeable. Now let us be clear that we are not concerned here with synonymy in the sense of complete identity in psychological associations or poetic quality; indeed no two expressions are synonymous in such a sense. We are concerned only with what may be called *cognitive* synonymy. Just what this is cannot be said without successfully finishing the present study; but we know something about it from the need which arose for it in connection with analyticity in §1. The sort of synonymy needed there was merely such that any analytic statement could be turned into a logical truth by putting synonyms for synonyms. Turning the tables and assuming analyticity, indeed, we could explain cognitive synonymy of terms as follows (keeping to the familiar example): to say that 'bachelor' and 'unmarried man' are cognitively synonymous is to say no more nor less than that the statement:

(3) All and only bachelors are unmarried men

is analytic.[7]

What we need is an account of cognitive synonymy not presupposing analyticity—if we are to explain analyticity conversely with help of cognitive synonymy as undertaken in §1. And indeed such an independent account of cognitive synonymy is at present up for consideration, namely, interchangeability *salva veritate* everywhere except within words. The question before us, to resume the thread at last, is whether such interchangeability is a sufficient condition for cognitive synonymy. We can quickly assure ourselves that it is, by examples of the following sort. The statement:

(4) Necessarily all and only bachelors are bachelors

is evidently true, even supposing 'necessarily' so narrowly construed as to be truly applicable only to analytic statements. Then, if 'bachelor' and 'unmarried man' are interchangeable *salva veritate*, the result:

(5) Necessarily all and only bachelors are unmarried men

of putting 'unmarried man' for an occurrence of 'bachelor' in (4) must, like (4), be true. But to say that (5) is true is to say that (3) is analytic, and hence that 'bachelor' and 'unmarried man' are cognitively synonymous.

Let us see what there is about the above argument that gives it its air of hocus-pocus. The condition of interchangeability *salva veritate* varies in its force with variations in the richness of the language at hand. The above argument supposes we are working with a language rich enough to contain the adverb 'necessarily', this adverb being so construed as to yield truth

[7] This is cognitive synonymy in a primary, broad sense. Carnap ([3], pp. 56ff) and Lewis ([2], pp. 83ff) have suggested how, once this notion is at hand, a narrower sense of cognitive synonymy which is preferable for some purposes can in turn be derived. But this special ramification of concept-building lies aside from the present purposes and must not be confused with the broad sort of cognitive synonymy here concerned.

when and only when applied to an analytic statement. But can we condone a language which contains such an adverb? Does the adverb really make sense? To suppose that it does is to suppose that we have already made satisfactory sense of 'analytic'. Then what are we so hard at work on right now?

Our argument is not flatly circular, but something like it. It has the form, figuratively speaking, of a closed curve in space.

Interchangeability *salva veritate* is meaningless until relativized to a language whose extent is specified in relevant respects. Suppose now we consider a language containing just the following materials. There is an indefinitely large stock of one-place predicates (for example, 'F' where 'Fx' means that x is a man) and many-place predicates (for example, 'G' where 'Gxy' means that x loves y), mostly having to do with extralogical subject matter. The rest of the language is logical. The atomic sentences consist each of a predicate followed by one or more variables 'x', 'y', etc.; and the complex sentences are built up of the atomic ones by truth functions ('not', 'and', 'or', etc.) and quantification.[8] In effect such a language enjoys the benefits also of descriptions and indeed singular terms generally, these being contextually definable in known ways.[9] Even abstract singular terms naming classes, classes of classes, etc., are contextually definable in case the assumed stock of predicates includes the two-place predicate of class membership.[10] Such a language can be adequate to classical mathematics and indeed to scientific discourse generally, except in so far as the latter involves debatable devices such as contrary-to-fact conditionals or modal adverbs like 'necessarily'.[11] Now a language of this type is extensional, in this sense: any two predicates which agree extensionally (that is, are true of the same objects) are interchangeable *salva veritate*.[12]

[8] Pp. 81ff, below, contain a description of just such a language, except that there happens there to be just one predicate, the two-place predicate 'ϵ'.

[9] See above, pp. 5-8; also below, pp. 85f, 166f.

[10] See below, p. 87.

[11] On such devices see also Essay VIII.

[12] This is the substance of Quine [1], *121.

In an extensional language, therefore, interchangeability *salva veritate* is no assurance of cognitive synonymy of the desired type. That 'bachelor' and 'unmarried man' are interchangeable *salva veritate* in an extensional language assures us of no more than that (3) is true. There is no assurance here that the extensional agreement of 'bachelor' and 'unmarried man' rests on meaning rather than merely on accidental matters of fact, as does the extensional agreement of 'creature with a heart' and 'creature with kidneys'.

For most purposes extensional agreement is the nearest approximation to synonymy we need care about. But the fact remains that extensional agreement falls far short of cognitive synonymy of the type required for explaining analyticity in the manner of §1. The type of cognitive synonymy required there is such as to equate the synonymy of 'bachelor' and 'unmarried man' with the analyticity of (3), not merely with the truth of (3).

So we must recognize that interchangeability *salva veritate*, if construed in relation to an extensional language, is not a sufficient condition of cognitive synonymy in the sense needed for deriving analyticity in the manner of §1. If a language contains an intensional adverb 'necessarily' in the sense lately noted, or other particles to the same effect, then interchangeability *salva veritate* in such a language does afford a sufficient condition of cognitive synonymy; but such a language is intelligible only in so far as the notion of analyticity is already understood in advance.

The effort to explain cognitive synonymy first, for the sake of deriving analyticity from it afterward as in §1, is perhaps the wrong approach. Instead we might try explaining analyticity somehow without appeal to cognitive synonymy. Afterward we could doubtless derive cognitive synonymy from analyticity satisfactorily enough if desired. We have seen that cognitive synonymy of 'bachelor' and 'unmarried man' can be explained as analyticity of (3). The same explanation works for any pair of one-place predicates, of course, and it can be extended in obvious fashion to many-place predicates. Other syntactical categories can also be accommodated in fairly parallel fashion.

Singular terms may be said to be cognitively synonymous when the statement of identity formed by putting '=' between them is analytic. Statements may be said simply to be cognitively synonymous when their biconditional (the result of joining them by 'if and only if') is analytic.[13] If we care to lump all categories into a single formulation, at the expense of assuming again the notion of "word" which was appealed to early in this section, we can describe any two linguistic forms as cognitively synonymous when the two forms are interchangeable (apart from occurrences within "words") *salva* (no longer *veritate* but) *analyticitate*. Certain technical questions arise, indeed, over cases of ambiguity or homonymy; let us not pause for them, however, for we are already digressing. Let us rather turn our backs on the problem of synonymy and address ourselves anew to that of analyticity.

4. Semantical Rules

Analyticity at first seemed most naturally definable by appeal to a realm of meanings. On refinement, the appeal to meanings gave way to an appeal to synonymy or definition. But definition turned out to be a will-o'-the-wisp, and synonymy turned out to be best understood only by dint of a prior appeal to analyticity itself. So we are back at the problem of analyticity.

I do not know whether the statement 'Everything green is extended' is analytic. Now does my indecision over this example really betray an incomplete understanding, an incomplete grasp of the "meanings", of 'green' and 'extended'? I think not. The trouble is not with 'green' or 'extended', but with 'analytic'.

It is often hinted that the difficulty in separating analytic statements from synthetic ones in ordinary language is due to the vagueness of ordinary language and that the distinction is clear when we have a precise artificial language with explicit "semantical rules." This, however, as I shall now attempt to show, is a confusion.

[13] The 'if and only if' itself is intended in the truth functional sense. See Carnap [3], p. 14.

The notion of analyticity about which we are worrying is a purported relation between statements and languages: a statement S is said to be *analytic for* a language L, and the problem is to make sense of this relation generally, that is, for variable 'S' and 'L'. The gravity of this problem is not perceptibly less for artificial languages than for natural ones. The problem of making sense of the idiom 'S is analytic for L', with variable 'S' and 'L', retains its stubbornness even if we limit the range of the variable 'L' to artificial languages. Let me now try to make this point evident.

For artificial languages and semantical rules we look naturally to the writings of Carnap. His semantical rules take various forms, and to make my point I shall have to distinguish certain of the forms. Let us suppose, to begin with, an artificial language L_0 whose semantical rules have the form explicitly of a specification, by recursion or otherwise, of all the analytic statements of L_0. The rules tell us that such and such statements, and only those, are the analytic statements of L_0. Now here the difficulty is simply that the rules contain the word 'analytic', which we do not understand! We understand what expressions the rules attribute analyticity to, but we do not understand what the rules attribute to those expressions. In short, before we can understand a rule which begins 'A statement S is analytic for language L_0 if and only if . . .', we must understand the general relative term 'analytic for'; we must understand 'S is analytic for L' where 'S' and 'L' are variables.

Alternatively we may, indeed, view the so-called rule as a conventional definition of a new simple symbol 'analytic-for-L_0', which might better be written untendentiously as 'K' so as not to seem to throw light on the interesting word 'analytic'. Obviously any number of classes K, M, N, etc. of statements of L_0 can be specified for various purposes or for no purpose; what does it mean to say that K, as against M, N, etc., is the class of the "analytic" statements of L_0?

By saying what statements are analytic for L_0 we explain 'analytic-for-L_0' but not 'analytic', not 'analytic for'. We do not begin to explain the idiom 'S is analytic for L' with variable

'*S*' and '*L*', even if we are content to limit the range of '*L*' to the realm of artificial languages.

Actually we do know enough about the intended significance of 'analytic' to know that analytic statements are supposed to be true. Let us then turn to a second form of semantical rule, which says not that such and such statements are analytic but simply that such and such statements are included among the truths. Such a rule is not subject to the criticism of containing the un-understood word 'analytic'; and we may grant for the sake of argument that there is no difficulty over the broader term 'true'. A semantical rule of this second type, a rule of truth, is not supposed to specify all the truths of the language; it merely stipulates, recursively or otherwise, a certain multitude of statements which, along with others unspecified, are to count as true. Such a rule may be conceded to be quite clear. Derivatively, afterward, analyticity can be demarcated thus: a statement is analytic if it is (not merely truė but) true according to the semantical rule.

Still there is really no progress. Instead of appealing to an unexplained word 'analytic', we are now appealing to an unexplained phrase 'semantical rule'. Not every true statement which says that the statements of some class are true can count as a semantical rule—otherwise *all* truths would be "analytic" in the sense of being true according to semantical rules. Semantical rules are distinguishable, apparently, only by the fact of appearing on a page under the heading 'Semantical Rules'; and this heading is itself then meaningless.

We can say indeed that a statement is *analytic-for-*L_0 if and only if it is true according to such and such specifically appended "semantical rules," but then we find ourselves back at essentially the same case which was originally discussed: '*S* is analytic-for-L_0 if and only if. . . .' Once we seek to explain '*S* is analytic for *L*' generally for variable '*L*' (even allowing limitation of '*L*' to artificial languages), the explanation 'true according to the semantical rules of *L*' is unavailing; for the relative term 'semantical rule of' is as much in need of clarification, at least, as 'analytic for'.

It may be instructive to compare the notion of semantical rule with that of postulate. Relative to a given set of postulates, it is easy to say what a postulate is: it is a member of the set. Relative to a given set of semantical rules, it is equally easy to say what a semantical rule is. But given simply a notation, mathematical or otherwise, and indeed as thoroughly understood a notation as you please in point of the translations or truth conditions of its statements, who can say which of its true statements rank as postulates? Obviously the question is meaningless—as meaningless as asking which points in Ohio are starting points. Any finite (or effectively specifiable infinite) selection of statements (preferably true ones, perhaps) is as much *a* set of postulates as any other. The word 'postulate' is significant only relative to an act of inquiry; we apply the word to a set of statements just in so far as we happen, for the year or the moment, to be thinking of those statements in relation to the statements which can be reached from them by some set of transformations to which we have seen fit to direct our attention. Now the notion of semantical rule is as sensible and meaningful as that of postulate, if conceived in a similarly relative spirit—relative, this time, to one or another particular enterprise of schooling unconversant persons in sufficient conditions for truth of statements of some natural or artificial language *L*. But from this point of view no one signalization of a subclass of the truths of *L* is intrinsically more a semantical rule than another; and, if 'analytic' means 'true by semantical rules', no one truth of *L* is analytic to the exclusion of another.[14]

It might conceivably be protested that an artificial language *L* (unlike a natural one) is a language in the ordinary sense *plus* a set of explicit semantical rules—the whole constituting, let us say, an ordered pair; and that the semantical rules of *L* then are specifiable simply as the second component of the pair *L*. But, by the same token and more simply, we might construe an artificial language *L* outright as an ordered pair whose second

[14] The foregoing paragraph was not part of the present essay as originally published. It was prompted by Martin (see Bibliography), as was the end of Essay VII.

component is the class of its analytic statements; and then the analytic statements of L become specifiable simply as the statements in the second component of L. Or better still, we might just stop tugging at our bootstraps altogether.

Not all the explanations of analyticity known to Carnap and his readers have been covered explicitly in the above considerations, but the extension to other forms is not hard to see. Just one additional factor should be mentioned which sometimes enters: sometimes the semantical rules are in effect rules of translation into ordinary language, in which case the analytic statements of the artificial language are in effect recognized as such from the analyticity of their specified translations in ordinary language. Here certainly there can be no thought of an illumination of the problem of analyticity from the side of the artificial language.

From the point of view of the problem of analyticity the notion of an artificial language with semantical rules is a *feu follet par excellence*. Semantical rules determining the analytic statements of an artificial language are of interest only in so far as we already understand the notion of analyticity; they are of no help in gaining this understanding.

Appeal to hypothetical languages of an artificially simple kind could conceivably be useful in clarifying analyticity, if the mental or behavioral or cultural factors relevant to analyticity—whatever they may be—were somehow sketched into the simplified model. But a model which takes analyticity merely as an irreducible character is unlikely to throw light on the problem of explicating analyticity.

It is obvious that truth in general depends on both language and extralinguistic fact. The statement 'Brutus killed Caesar' would be false if the world had been different in certain ways, but it would also be false if the word 'killed' happened rather to have the sense of 'begat'. Thus one is tempted to suppose in general that the truth of a statement is somehow analyzable into a linguistic component and a factual component. Given this supposition, it next seems reasonable that in some statements the factual component should be null; and these are the analytic statements. But, for all its a priori reasonableness, a boundary between analytic and synthetic statements simply has not been drawn. That there is such a distinction to be drawn at all is an unempirical dogma of empiricists, a metaphysical article of faith.

5. The Verification Theory and Reductionism

In the course of these somber reflections we have taken a dim view first of the notion of meaning, then of the notion of cognitive synonymy, and finally of the notion of analyticity. But what, it may be asked, of the verification theory of meaning? This phrase has established itself so firmly as a catchword of empiricism that we should be very unscientific indeed not to look beneath it for a possible key to the problem of meaning and the associated problems.

The verification theory of meaning, which has been conspicuous in the literature from Peirce onward, is that the meaning of a statement is the method of empirically confirming or infirming it. An analytic statement is that limiting case which is confirmed no matter what.

As urged in §1, we can as well pass over the question of meanings as entities and move straight to sameness of meaning, or synonymy. Then what the verification theory says is that statements are synonymous if and only if they are alike in point of method of empirical confirmation or infirmation.

This is an account of cognitive synonymy not of linguistic forms generally, but of statements.[15] However, from the concept of synonymy of statements we could derive the concept of synonymy for other linguistic forms, by considerations somewhat similar to those at the end of §3. Assuming the notion of "word," indeed, we could explain any two forms as synonymous when the

[15] The doctrine can indeed be formulated with terms rather than statements as the units. Thus Lewis describes the meaning of a term as "*a criterion in mind*, by reference to which one is able to apply or refuse to apply the expression in question in the case of presented, or imagined, things or situations" ([2], p. 133).—For an instructive account of the vicissitudes of the verification theory of meaning, centered however on the question of meaning*fulness* rather than synonymy and analyticity, see Hempel.

putting of the one form for an occurrence of the other in any statement (apart from occurrences within "words") yields a synonymous statement. Finally, given the concept of synonymy thus for linguistic forms generally, we could define analyticity in terms of synonymy and logical truth as in §1. For that matter, we could define analyticity more simply in terms of just synonymy of statements together with logical truth; it is not necessary to appeal to synonymy of linguistic forms other than statements. For a statement may be described as analytic simply when it is synonymous with a logically true statement.

So, if the verification theory can be accepted as an adequate account of statement synonymy, the notion of analyticity is saved after all. However, let us reflect. Statement synonymy is said to be likeness of method of empirical confirmation or infirmation. Just what are these methods which are to be compared for likeness? What, in other words, is the nature of the relation between a statement and the experiences which contribute to or detract from its confirmation?

The most naïve view of the relation is that it is one of direct report. This is *radical reductionism*. Every meaningful statement is held to be translatable into a statement (true or false) about immediate experience. Radical reductionism, in one form or another, well antedates the verification theory of meaning explicitly so called. Thus Locke and Hume held that every idea must either originate directly in sense experience or else be compounded of ideas thus originating; and taking a hint from Tooke we might rephrase this doctrine in semantical jargon by saying that a term, to be significant at all, must be either a name of a sense datum or a compound of such names or an abbreviation of such a compound. So stated, the doctrine remains ambiguous as between sense data as sensory events and sense data as sensory qualities; and it remains vague as to the admissible ways of compounding. Moreover, the doctrine is unnecessarily and intolerably restrictive in the term-by-term critique which it imposes. More reasonably, and without yet exceeding the limits of what I have called radical reductionism, we may take full statements as our significant units—thus demanding that our statements as wholes be translatable into sense-datum language, but not that they be translatable term by term.

This emendation would unquestionably have been welcome to Locke and Hume and Tooke, but historically it had to await an important reorientation in semantics—the reorientation whereby the primary vehicle of meaning came to be seen no longer in the term but in the statement. This reorientation, seen in Bentham and Frege, underlies Russell's concept of incomplete symbols defined in use;[16] also it is implicit in the verification theory of meaning, since the objects of verification are statements.

Radical reductionism, conceived now with statements as units, set itself the task of specifying a sense-datum language and showing how to translate the rest of significant discourse, statement by statement, into it. Carnap embarked on this project in the *Aufbau*.

The language which Carnap adopted as his starting point was not a sense-datum language in the narrowest conceivable sense, for it included also the notations of logic, up through higher set theory. In effect it included the whole language of pure mathematics. The ontology implicit in it (that is, the range of values of its variables) embraced not only sensory events but classes, classes of classes, and so on. Empiricists there are who would boggle at such prodigality. Carnap's starting point is very parsimonious, however, in its extralogical or sensory part. In a series of constructions in which he exploits the resources of modern logic with much ingenuity, Carnap succeeds in defining a wide array of important additional sensory concepts which, but for his constructions, one would not have dreamed were definable on so slender a basis. He was the first empiricist who, not content with asserting the reducibility of science to terms of immediate experience, took serious steps toward carrying out the reduction.

If Carnap's starting point is satisfactory, still his construc-

[16] See above, p. 6.

tions were, as he himself stressed, only a fragment of the full program. The construction of even the simplest statements about the physical world was left in a sketchy state. Carnap's suggestions on this subject were, despite their sketchiness, very suggestive. He explained spatio-temporal point-instants as quadruples of real numbers and envisaged assignment of sense qualities to point-instants according to certain canons. Roughly summarized, the plan was that qualities should be assigned to point-instants in such a way as to achieve the laziest world compatible with our experience. The principle of least action was to be our guide in constructing a world from experience.

Carnap did not seem to recognize, however, that his treatment of physical objects fell short of reduction not merely through sketchiness, but in principle. Statements of the form 'Quality *q* is at point-instant *x;y;z;t*' were, according to his canons, to be apportioned truth values in such a way as to maximize and minimize certain over-all features, and with growth of experience the truth values were to be progressively revised in the same spirit. I think this is a good schematization (deliberately oversimplified, to be sure) of what science really does; but it provides no indication, not even the sketchiest, of how a statement of the form 'Quality *q* is at *x;y;z;t*' could ever be translated into Carnap's initial language of sense data and logic. The connective 'is at' remains an added undefined connective; the canons counsel us in its use but not in its elimination.

Carnap seems to have appreciated this point afterward; for in his later writings he abandoned all notion of the translatability of statements about the physical world into statements about immediate experience. Reductionism in its radical form has long since ceased to figure in Carnap's philosophy.

But the dogma of reductionism has, in a subtler and more tenuous form, continued to influence the thought of empiricists. The notion lingers that to each statement, or each synthetic statement, there is associated a unique range of possible sensory events such that the occurrence of any of them would add to the likelihood of truth of the statement, and that there is associated also another unique range of possible sensory events whose occurrence would detract from that likelihood. This notion is of course implicit in the verification theory of meaning.

The dogma of reductionism survives in the supposition that each statement, taken in isolation from its fellows, can admit of confirmation or infirmation at all. My countersuggestion, issuing essentially from Carnap's doctrine of the physical world in the *Aufbau,* is that our statements about the external world face the tribunal of sense experience not individually but only as a corporate body.[17]

The dogma of reductionism, even in its attenuated form, is intimately connected with the other dogma—that there is a cleavage between the analytic and the synthetic. We have found ourselves led, indeed, from the latter problem to the former through the verification theory of meaning. More directly, the one dogma clearly supports the other in this way: as long as it is taken to be significant in general to speak of the confirmation and infirmation of a statement, it seems significant to speak also of a limiting kind of statement which is vacuously confirmed, *ipso facto,* come what may; and such a statement is analytic.

The two dogmas are, indeed, at root identical. We lately reflected that in general the truth of statements does obviously depend both upon language and upon extralinguistic fact; and we noted that this obvious circumstance carries in its train, not logically but all too naturally, a feeling that the truth of a statement is somehow analyzable into a linguistic component and a factual component. The factual component must, if we are empiricists, boil down to a range of confirmatory experiences. In the extreme case where the linguistic component is all that matters, a true statement is analytic. But I hope we are now impressed with how stubbornly the distinction between analytic and synthetic has resisted any straightforward drawing. I am impressed also, apart from prefabricated examples of black and white balls in an urn, with how baffling the problem has always

[17] This doctrine was well argued by Duhem, pp. 303-328. Or see Lowinger, pp. 132-140.

been of arriving at any explicit theory of the empirical confirmation of a synthetic statement. My present suggestion is that it is nonsense, and the root of much nonsense, to speak of a linguistic component and a factual component in the truth of any individual statement. Taken collectively, science has its double dependence upon language and experience; but this duality is not significantly traceable into the statements of science taken one by one.

The idea of defining a symbol in use was, as remarked, an advance over the impossible term-by-term empiricism of Locke and Hume. The statement, rather than the term, came with Bentham to be recognized as the unit accountable to an empiricist critique. But what I am now urging is that even in taking the statement as unit we have drawn our grid too finely. The unit of empirical significance is the whole of science.

6. Empiricism without the Dogmas

The totality of our so-called knowledge or beliefs, from the most casual matters of geography and history to the profoundest laws of atomic physics or even of pure mathematics and logic, is a man-made fabric which impinges on experience only along the edges. Or, to change the figure, total science is like a field of force whose boundary conditions are experience. A conflict with experience at the periphery occasions readjustments in the interior of the field. Truth values have to be redistributed over some of our statements. Reëvaluation of some statements entails reëvaluation of others, because of their logical interconnections —the logical laws being in turn simply certain further statements of the system, certain further elements of the field. Having reëvaluated one statement we must reëvaluate some others, which may be statements logically connected with the first or may be the statements of logical connections themselves. But the total field is so underdetermined by its boundary conditions, experience, that there is much latitude of choice as to what statements to reëvaluate in the light of any single contrary experience. No particular experiences are linked with any particular statements in the interior of the field, except indirectly through considerations of equilibrium affecting the field as a whole.

If this view is right, it is misleading to speak of the empirical content of an individual statement—especially if it is a statement at all remote from the experiential periphery of the field. Furthermore it becomes folly to seek a boundary between synthetic statements, which hold contingently on experience, and analytic statements, which hold come what may. Any statement can be held true come what may, if we make drastic enough adjustments elsewhere in the system. Even a statement very close to the periphery can be held true in the face of recalcitrant experience by pleading hallucination or by amending certain statements of the kind called logical laws. Conversely, by the same token, no statement is immune to revision. Revision even of the logical law of the excluded middle has been proposed as a means of simplifying quantum mechanics; and what difference is there in principle between such a shift and the shift whereby Kepler superseded Ptolemy, or Einstein Newton, or Darwin Aristotle?

For vividness I have been speaking in terms of varying distances from a sensory periphery. Let me try now to clarify this notion without metaphor. Certain statements, though *about* physical objects and not sense experience, seem peculiarly germane to sense experience—and in a selective way: some statements to some experiences, others to others. Such statements, especially germane to particular experiences, I picture as near the periphery. But in this relation of "germaneness" I envisage nothing more than a loose association reflecting the relative likelihood, in practice, of our choosing one statement rather than another for revision in the event of recalcitrant experience. For example, we can imagine recalcitrant experiences to which we would surely be inclined to accommodate our system by reëvaluating just the statement that there are brick houses on Elm Street, together with related statements on the same topic. We can imagine other recalcitrant experiences to which we would be inclined to accommodate our system by reëvaluating just the statement that there are no centaurs, along with kindred statements. A recalcitrant experience can, I have urged, be accommodated by any of various alternative reëvaluations in various alternative quarters of the total system; but, in the cases which we are now imagining, our natural tendency to disturb the total system as little as possible would lead us to focus our revisions upon these specific statements concerning brick houses or centaurs. These statements are felt, therefore, to have a sharper empirical reference than highly theoretical statements of physics or logic or ontology. The latter statements may be thought of as relatively centrally located within the total network, meaning merely that little preferential connection with any particular sense data obtrudes itself.

As an empiricist I continue to think of the conceptual scheme of science as a tool, ultimately, for predicting future experience in the light of past experience. Physical objects are conceptually imported into the situation as convenient intermediaries—not by definition in terms of experience, but simply as irreducible posits[18] comparable, epistemologically, to the gods of Homer. For my part I do, qua lay physicist, believe in physical objects and not in Homer's gods; and I consider it a scientific error to believe otherwise. But in point of epistemological footing the physical objects and the gods differ only in degree and not in kind. Both sorts of entities enter our conception only as cultural posits. The myth of physical objects is epistemologically superior to most in that it has proved more efficacious than other myths as a device for working a manageable structure into the flux of experience.

Positing does not stop with macroscopic physical objects. Objects at the atomic level are posited to make the laws of macroscopic objects, and ultimately the laws of experience, simpler and more manageable; and we need not expect or demand full definition of atomic and subatomic entities in terms of macroscopic ones, any more than definition of macroscopic

[18] Cf. pp. 17f above.

things in terms of sense data. Science is a continuation of common sense, and it continues the common-sense expedient of swelling ontology to simplify theory.

Physical objects, small and large, are not the only posits. Forces are another example; and indeed we are told nowadays that the boundary between energy and matter is obsolete. Moreover, the abstract entities which are the substance of mathematics—ultimately classes and classes of classes and so on up—are another posit in the same spirit. Epistemologically these are myths on the same footing with physical objects and gods, neither better nor worse except for differences in the degree to which they expedite our dealings with sense experiences.

The over-all algebra of rational and irrational numbers is underdetermined by the algebra of rational numbers, but is smoother and more convenient; and it includes the algebra of rational numbers as a jagged or gerrymandered part.[19] Total science, mathematical and natural and human, is similarly but more extremely underdetermined by experience. The edge of the system must be kept squared with experience; the rest, with all its elaborate myths or fictions, has as its objective the simplicity of laws.

Ontological questions, under this view, are on a par with questions of natural science.[20] Consider the question whether to countenance classes as entities. This, as I have argued elsewhere,[21] is the question whether to quantify with respect to variables which take classes as values. Now Carnap [6] has maintained that this is a question not of matters of fact but of choosing a convenient language form, a convenient conceptual scheme or framework for science. With this I agree, but only on the proviso that the same be conceded regarding scientific hypotheses generally. Carnap ([6], p. 32n) has recognized that he is able to preserve a double standard for ontological questions and scientific hypotheses only by assuming an absolute distinc-

[19] Cf. p. 18 above.

[20] "L'ontologie fait corps avec la science elle-même et ne peut en être separée." Meyerson, p. 439.

[21] Above, pp. 12f; below, pp. 102ff.

tion between the analytic and the synthetic; and I need not say again that this is a distinction which I reject.[22]

The issue over there being classes seems more a question of convenient conceptual scheme; the issue over there being centaurs, or brick houses on Elm Street, seems more a question of fact. But I have been urging that this difference is only one of degree, and that it turns upon our vaguely pragmatic inclination to adjust one strand of the fabric of science rather than another in accommodating some particular recalcitrant experience. Conservatism figures in such choices, and so does the quest for simplicity.

Carnap, Lewis, and others take a pragmatic stand on the question of choosing between language forms, scientific frameworks; but their pragmatism leaves off at the imagined boundary between the analytic and the synthetic. In repudiating such a boundary I espouse a more thorough pragmatism. Each man is given a scientific heritage plus a continuing barrage of sensory stimulation; and the considerations which guide him in warping his scientific heritage to fit his continuing sensory promptings are, where rational, pragmatic.

[22] For an effective expression of further misgivings over this distinction, see White [2].

CARL G. HEMPEL

The Theoretician's Dilemma

A STUDY IN THE LOGIC OF THEORY CONSTRUCTION

1. Deductive and Inductive Systematization

Scientific research in its various branches seeks not merely to record particular occurrences in the world of our experience: it tries to discover regularities in the flux of events and thus to establish general laws which may be used for prediction, postdiction,[1] and explanation.

The principles of Newtonian mechanics, for example, make it possible, given the present positions and momenta of the celestial objects that make up the solar system, to predict their positions and momenta for a specified future time or to postdict them for a specified time in the past; similarly, those principles permit an explanation of the present positions and momenta by reference to those at some earlier time. In addition to thus accounting for particular facts such as those just mentioned, the principles of Newtonian mechanics also explain certain "general facts," i.e., empirical uniformities such as Kepler's laws of planetary motion; for the latter can be deduced from the former.[2]

Scientific explanation, prediction, and postdiction all have the same logical character: they show that the fact under consideration can be inferred from certain other facts by means of specified general laws. In the simplest case, this type of argument may be schematized as a deductive inference of the following form:

$$(1.1) \qquad \frac{\begin{array}{c} C_1, C_2 \ldots C_k \\ L_1, L_2 \ldots L_r \end{array}}{E}$$

AUTHOR'S NOTE: I am indebted to the Council of the Humanities at Princeton University for the award of a Senior Fellowship for the academic year 1956–57, which, by reducing my teaching load, gave me additional time for research. The present study is part of the work done during my tenure as a Senior Fellow.

Here, C_1, C_2, C_k are statements of particular occurrences (e.g., of the position and momenta of certain celestial bodies at a specified time), and L_1, L_2 . . . L_r are general laws (e.g., those of Newtonian mechanics); finally, E is a sentence stating whatever is being explained, predicted, or postdicted. And the argument has its intended force only if its conclusion, E, follows deductively from the premises.[3]

While explanation, prediction, and postdiction are alike in their logical structure, they differ in certain other respects. For example, an argument of the form (1.1) will qualify as a prediction only if E refers to an occurrence at a time later than that at which the argument is offered; in the case of a postdiction, the event must occur before the presentation of the argument. These differences, however, require no fuller study here, for the purpose of the preceding discussion was simply to point out the role of general laws in scientific explanation, prediction, and postdiction.

For these three types of scientific procedure, we will use the common term '*(deductive) systematization*'. More precisely, that term will be used to refer, first, to any argument of the form (1.1) that meets the requirements indicated above, no matter whether it serves as an explanation, a prediction, a postdiction, or in still some other capacity; second, to the procedure of establishing arguments of the kind just characterized.

So far, we have considered only those cases of explanation, prediction, and related procedures which can be construed as deductive arguments. There are many instances of scientific explanation and prediction, however, which do not fall into a strictly deductive pattern. For example, when Johnny comes down with the measles, this might be explained by pointing out that he caught the disease from his sister, who is just recovering from it. The particular antecedent facts here invoked are that of Johnny's exposure and, let us assume, the further fact that Johnny had not had the measles previously. But to connect these with the event to be explained, we cannot adduce a general law to the effect that under the specified circumstances, the measles is invariably transmitted to the exposed person: what can be asserted is only a high probability (in the sense of statistical frequency) of transmission. The same type of argument can be used also for predicting or postdicting the occurrence of a case of the measles.

Similarly, in a psychoanalytic explanation of the neurotic behavior of an adult by reference to certain childhood experiences, the generalizations which might be invoked to connect the antecedent events with those to be explained can be construed at best as establishing more or less high probabilities for the connections at hand, but surely not as expressions of unexceptional uniformities.

Explanations, predictions, and postdictions of the kind here illustrated differ from those previously discussed in two important respects: The laws invoked are of a different form, and the statement to be established does not follow deductively from the explanatory statements adduced. We will now consider these differences somewhat more closely.

The laws referred to in connection with the schema (1.1), such as the laws of Newtonian mechanics, are what we will call *statements of strictly universal form, or strictly universal statements.* A statement of this kind is an assertion—which may be true or false—to the effect that all cases which meet certain specified conditions will unexceptionally have such and such further characteristics. For example, the statement 'All crows are black' is a sentence of strictly universal form; and so is Newton's first law of motion, that any material body which is not acted upon by an external force persists in its state of rest or of rectilinear motion at constant speed.

The laws invoked in the second type of explanatory and related arguments, on the other hand, are, as we will say, of *statistical form;* they are *statistical probability statements.* A statement of this kind is an assertion—which may be true or false—to the effect that for cases which meet conditions of a specified kind, the probability of having such and such further characteristics is so-and-so much.[4]

To put the distinction in a nutshell: A strictly universal statement of the simplest kind has the form 'All cases of P are cases of Q'; a statistical probability statement of the simplest kind has the form 'The probability for a case of P to be a case of Q is r.' While the former implies an assertion about any particular instance of P—namely, that it is also an instance of Q—the latter implies no similar assertion concerning any particular instance of P or even concerning any finite set of such instances.[5] This circumstance gives rise to the second distinctive characteristic mentioned above: the statement E describing the occurrence under explanation or prediction or postdiction (for example, Johnny's catching the measles) is not logically deducible from the explanatory statements adduced (for example, (C_1) Johnny was exposed to the measles; (C_2) Johnny had not previously had the measles; (L) For persons who have not previously had the measles and are exposed to it, the probability is .92 that they will contract the disease); rather, on the assumption that the explanatory statements adduced are true, it is very likely, though not certain, that E is true as well. This kind of argument, therefore, is inductive rather than strictly deductive in character: it calls for the acceptance of E on the basis of other statements which constitute only partial, if strongly supporting, grounds for it. An argument of this kind—no matter whether it is used for explanation, prediction, or postdiction, or for yet another purpose—will be called an *inductive systematization*. In particular, we will assume of an inductive systematization that the conclusion is not logically implied by the premises.[6] Again, the procedure of establishing an argument of the kind just described will also be called inductive systematization.

By way of further illustration, let us note here two explanatory arguments which are of the inductive kind just characterized. They are adduced by von Mises in a statement to the effect that the everyday notion of causal explanation will eventually adjust itself to changes in the logical form of scientific theories (especially to the use of statistical probability statements as explanatory principles): "We think," von Mises says, that "people will gradually come to be satisfied by causal statements of this kind: It is *because* the die was loaded that the 'six' shows more frequently (but we do not know what the next number will be); or: *Because* the vacuum was heightened and the voltage increased, the radiation became more intense (but we do not know the precise number of scintillations that will occur in the next minute)."[7] Clearly, both of these statements can be construed as inductive explanations of certain physical phenomena.

All the cases of scientific systematization we have considered share this characteristic: they make use of general laws or general principles either of strictly universal or of statistical form. These general laws have the function of establishing systematic connections among empirical facts in such a way that with their help some empirical occurrences may be inferred, by way of explanation, prediction, or postdiction, from other such occurrences. When, in an explanation, we say that the event described by E occurred "because" of the circumstances detailed in C_1, C_2 . . . C_k, that phrase has significance if it can be construed as referring to general laws which render C_1, C_2 . . . C_k relevant to E in the sense that, granted the truth of the former, they make the truth of the latter either certain (as in a deductive systematization) or inductively probable (as in an inductive systematization). It is for this reason that the establishment of general laws is of crucial importance in the empirical sciences.

2. Observables and Theoretical Entities

Scientific systematization is ultimately aimed at establishing explanatory and predictive order among the bewilderingly complex "data" of our experience, the phenomena that can be directly "observed" by us. It is a remarkable fact, therefore, that the greatest advances in scientific systematization have not been accomplished by means of laws referring explicitly to *observables*, i.e., to things and events which are ascertainable by direct observation, but rather by means of laws that speak of various *hypothetical, or theoretical, entities*, i.e., presumptive objects, events, and attributes which cannot be perceived or otherwise directly observed by us.

For a fuller discussion of this point, it will be helpful to refer to the familiar distinction between two levels of scientific systematization: the level of *empirical generalization*, and the level of *theory formation*.[8] The early stages in the development of a scientific discipline usually belong to the former level, which is characterized by the search for laws (of universal or statistical form) which establish connections among the directly observable aspects of the subject matter under study. The more advanced stages belong to the second level, where research is aimed at comprehensive laws, in terms of hypothetical entities, which will account for the uniformities established on the first level. On the first level, we find everyday physical generalizations such as 'Where there is light there is heat,' 'Iron rusts in damp air,' 'Wood floats on water, iron sinks in it'; but we might assign to it also such more precise quantitative laws as Galileo's, Kepler's, Hooke's, and Snell's laws, as well as botanical and zoological generalizations about the concomitance of certain observable anatomical, physical, functional, and other characteristics in the members of a given species; generalizations in psychology that assert correlations among diverse observable aspects of learning, of perception, and so forth; and various descriptive generalizations in economics, sociology, and anthropology. All these generalizations, whether of strictly universal or of statistical form, purport to express regular connections among directly observable phenomena, and they lend themselves, therefore, to explanatory, predictive, and postdictive use.

On the second level, we encounter general statements that refer to electric, magnetic, and gravitational fields, to molecules, atoms, and a variety of subatomic particles; or to ego, id, superego, libido, sublimation, fixation, and transference; or to various not directly observable entities invoked in recent learning theories.

In accordance with the distinction here made, we will assume that the (extra-logical) vocabulary of empirical science, or of any of its branches, is divided into two classes: *observational terms* and *theoretical terms.* In regard to an observational term it is possible, under suitable circumstances, to decide by means of direct observation whether the term does or does not apply to a given situation.

Observation may here be construed so broadly as to include not only perception, but also sensation and introspection; or it may be limited to the perception of what in principle is publicly ascertainable, i.e., perceivable also by others. The subsequent discussion will be independent of how narrowly or how liberally the notion of observation is construed; it may be worth noting, however, that empirical science aims for a system of publicly testable statements, and that, accordingly, the observational data whose correct prediction is the hallmark of a successful theory are at least thought of as couched in terms on whose applicability in a given situation different individuals can decide, with high agreement, by means of direct observation. Statements which purport to describe readings of measuring instruments, changes in color or odor accompanying a chemical reaction, utterances made, or other kinds of overt behavior shown by a given subject under specified observable conditions—these all illustrate the use of *intersubjectively applicable* observational terms.[9]

Theoretical terms, on the other hand, usually purport to refer to not directly observable entities and their characteristics; they function, in a manner soon to be examined more closely, in scientific theories intended to explain empirical generalizations.

The preceding characterization of the two vocabularies is obviously vague; it offers no precise criterion by means of which any scientific term may be unequivocally classified as an observational term or as a theoretical one. But no such precise criterion is needed here; the questions to be examined in this essay are independent of precisely where the dividing line between the terms of the observational and the theoretical vocabularies is drawn.

3. Why Theoretical Terms?

The use of theoretical terms in science gives rise to a perplexing problem: Why should science resort to the assumption of hypothetical entities when it is interested in establishing predictive and explanatory connections among observables? Would it not be sufficient for the purpose, and much less extravagant at that, to search for a system of general laws mentioning only observables, and thus expressed in terms of the observational vocabulary alone?

Many general statements in terms of observables have indeed been formulated; they constitute the empirical generalizations mentioned in the preceding section. But, vexingly, many, if not all, of them suffer from definite shortcomings: they usually have a rather limited range of application; and even within that range, they have exceptions, so that actually they are not true general statements. Take, for example, one of our earlier illustrations of an empirical generalization:

(3.1) Wood floats on water; iron sinks in it.

This statement has a narrow range of application in the sense that it refers only to wooden and iron objects and concerns their floating behavior only in regard to water.[10] And, what is even more important, it has exceptions: certain kinds of wood will sink in water, and a hollow iron sphere of suitable dimensions will float on it.

As the history of science shows, flaws of this kind can often be remedied by attributing to the phenomena under study certain further constituents or characteristics which, though not open to direct observation, are connected in specified ways with the observable aspects of the subject matter under investigation, and which make it possible to establish systematic connections among the latter. By way of illustration—though it is admittedly an oversimplified one—consider the sentence (3.1). A much more satisfactory generalization is obtained by means of the concept of the specific gravity of a body x, which is definable as the quotient of its weight and its volume:

(3.2) Def.: $s(x) = w(x)/v(x)$

Let us assume that w and v have been characterized operationally, i.e., in terms of the directly observable outcomes of specified measuring procedures, and that therefore they are counted among the observables. Then s, as determined by (3.1), might be viewed as a characteristic that is less directly observable; and, just for the sake of obtaining a simple illustration, we will classify s as a hypothetical entity. For s, we may now state the following generalization, which is a corollary of the principle of Archimedes:

(3.3) A solid body floats on a liquid if its specific gravity is less than that of the liquid.

This statement avoids, first of all, the exceptions we noted above as refuting (3.1): it predicts correctly the behavior of a piece of heavy wood and of a hollow iron sphere. Moreover, it has a much wider scope: it refers to any kind of solid object and concerns its floating behavior in regard to any liquid. Even the new generalization has certain limitations, of course, and thus invites further improvement. But instead of pursuing this process, let us now examine more closely the way in which a systematic connection among observables is achieved, in our illustration, by the law (3.3), which involves a detour through the domain of unobservables.

Suppose that we wish to predict whether a certain solid object b will float on a given body l of liquid. We will then first have to ascertain, by appropriate operational procedure, the weight and the volume of b and l. Let the results of these measurements be expressed by the following four statements O_1, O_2, O_3, O_4:

(3.4) $(O_1)\ w(b) = w_1;\ (O_2)\ v(b) = v_1$
$(O_3)\ w(l) = w_2;\ (O_4)\ v(l) = v_2$

where w_1, w_2, v_1, v_2, are certain positive real numbers. By means of the definition (3.2), we can infer, from (3.4), the specific gravities of b and l:

(3.5) $s(b) = w_1/v_1;\ s(l) = w_2/v_2$

Suppose now that the first of these values is less than the second; then (3.4), via (3.5) implies that

(3.6) $s(b) < s(l)$

By means of the law (3.3), we can now infer that

(3.7) b floats on l

This sentence will also be called O_5. The sentences O_1, O_2, O_3, O_4, O_5 then share the characteristic that they are expressed entirely in terms of the observational vocabulary; for on our assumption, 'w' and 'v' are observational terms, and so are 'b' and 'l', which name certain observable bodies; finally, 'floats on' is an observational term because under suitable circumstances, direct observation will show whether a given observable object floats on a given observable liquid. On the other hand, the sentences (3.2), (3.3), (3.5), and (3.6) lack that characteristic, for they all contain the term 's', which, in our illustration, belongs to the theoretical vocabulary.

The systematic transition from the "observational data" listed in (3.4) to the prediction (3.7) of an observable phenomenon is schematized in the accompanying diagram. Here, an arrow represents a deductive in-

(3.8)

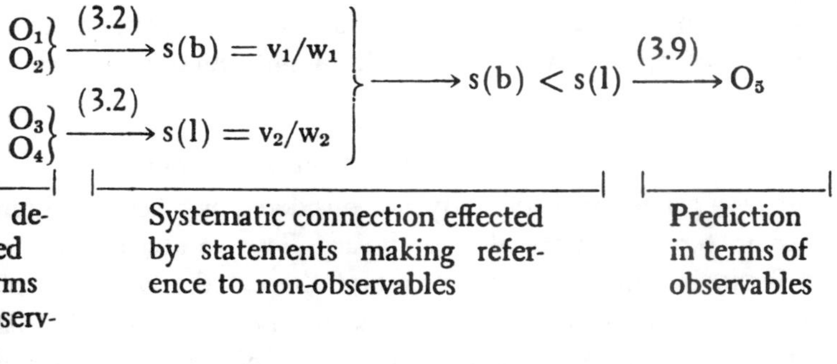

ference; mention, above an arrow, of a further sentence indicates that the deduction is effected by means of that sentence, i.e., that the conclusion stated at the right end follows logically from the premises listed at the left, taken in conjunction with the sentence mentioned above the arrow. Note that the argument just considered illustrates the schema (1.1), with O_1, O_2, O_3, O_4 constituting the statements of particular facts, the sentences (3.2) and (3.3) taking the place of the general laws, and O_5 that of E.[11]

Thus, the assumption of non-observable entities serves the purposes of systematization: it provides connections among observables in the form of laws containing theoretical terms, and this detour via the domain of hypothetical entities offers certain advantages, some of which were indicated above.

In the case of our illustration, however, brief reflection will show that the advantages obtained by the "theoretical detour" could just as well have been obtained without ever resorting to the use of a theoretical term. Indeed, by virtue of the definition (3.2), the law (3.3) can be restated as follows:

(3.3′) A solid body floats on a liquid if the quotient of its weight and its volume is less than the corresponding quotient for the liquid.

This alternative version clearly shares the advantages we found (3.3) to have over the crude generalization (3.1); and, of course, it permits the deductive transition from O_1, O_2, O_3, O_4 to O_5 just as well as does (3.3) in conjunction with (3.2).

The question arises therefore whether the systematization achieved by general principles containing theoretical terms can always be duplicated by means of general statements couched exclusively in observational terms. To prepare for an examination of this important problem, we must first consider more closely the form and function of a scientific theory.

4. Structure and Interpretation of a Theory

Formally, a scientific theory may be considered as a set of sentences expressed in terms of a specific vocabulary. The vocabulary, V_T, of a theory T will be understood to consist of the extralogical terms of T, i.e., those which do not belong to the vocabulary of pure logic. Usually, some of the terms of V are defined by means of others; but, on pain of a circle or an infinite regress, not all the terms of V can be so defined. Hence, V may be assumed to be divided into two subsets: *primitive terms*—those for which no definition is specified—and *defined terms*. Analogously, many of the sentences of a theory are derivable from others by means of the principles of deductive logic (and of the definitions of the defined terms); but, on pain of a vicious circle or an infinite regress in the deduction, not all of the theoretical sentences can be thus established. Hence, the set of sentences asserted by T falls into two subsets: *primitive sentences*, or *postulates* (also called *axioms*), and *derivative sentences*, or *theorems*. Henceforth, we will assume that theories are given in the form of axiomatized systems as here described; i.e., by listing, first, the primitive and the derivative terms and the definitions for the latter, second, the postulates. In addition, the theory will always be thought of as formulated within a linguistic framework of a clearly specified logical structure, which determines, in particular, the rules of deductive inference.

The classical paradigms of deductive systems of this kind are the axiomatizations of various mathematical theories, such as Euclidean, and various forms of non-Euclidean geometry, and the theory of groups and other branches of abstract algebra; [12] but by now, a number of theories in empirical science have likewise been put into axiomatic form, or approximations thereof; among them, parts of classical and relativistic mechanics,[13] certain segments of biological theory [14] and some theoretical systems in psychology, especially in the field of learning; [15] in economic theory, the concept of utility, among others, has received axiomatic treatment.[16]

If the primitive terms and the postulates of an axiomatized system have been specified, then the proof of theorems, i.e., the derivation of further sentences from the primitive ones—can be carried out by means of the purely formal canons of deductive logic, and thus, without any reference to the meanings of the terms and sentences at hand; indeed, for the deductive development of an axiomatized system, no meanings need be assigned at all to its expressions, primitive or derived.

However, a deductive system can function as a theory in empirical science only if it has been given an *interpretation* by reference to empirical phenomena. We may think of such interpretation as being effected by the specification of a set of *interpretative sentences*, which connect certain terms of the theoretical vocabulary with observational terms.[17] The character of these sentences will be examined in considerable detail in subsequent sections; at present may be mentioned, merely as an example, that interpretative sentences might take the form of so-called operational definitions, i.e., of statements specifying the meanings of theoretical terms with the help of observational ones; of special importance among these are rules for the measurement of theoretical quantities by reference to observable responses of measuring instruments or other indicators.

The manner in which a theory establishes explanatory and predictive connections among statements couched in observational terms can now be illustrated in outline by the following example. Suppose that the Newtonian theory of mechanics is used to study the motions, under the exclusive influence of their mutual gravitational attraction, of two bodies, such as the components of a double-star system, or the moon and a rocket coasting freely 100 miles above the moon's surface. On the basis of appropriate observational data, each of the two bodies may be assigned a certain mass, and, at a given instant t_0, a certain position and velocity in some specified frame of reference. Thus, a first step is taken which leads, via interpretative sentences in the form of rules of measurement, from certain statements $O_1, O_2 \ldots O_k$ which describe observable instrument readings, to certain theoretical statements, say $H_1, H_2 \ldots H_6$ which assign to each of the two bodies a specific numerical value of the theoretical quantities mass, position, and velocity. From these statements, the law of gravitation, which is couched entirely in theoretical terms, leads to a further theoretical statement, H_7, which specifies the force of the gravitational attraction the two bodies exert upon each other at t_0; and H_7 in conjunction with the preceding theoretical statements and the laws of Newtonian mechanics implies, via a deductive argument involving the principles of the calculus, certain statements H_8, H_9, H_{10}, H_{11}, which give the positions and velocities of the two objects at a specified later time, say t_1. Finally, use in reverse of the interpretative sentences leads, from the last four theoretical statements, to a set of sentences $O'_1, O'_2 \ldots O'_m$, which describe observable phenomena, namely, instrument readings that are indicative of the predicted positions and velocities.

By means of a schema analogous to (3.8), the procedure may be represented as follows:

$$(4.1)\quad \{O_1, O_2 \ldots O_k\} \xrightarrow{R} \{H_1, H_2 \ldots H_6\} \xrightarrow{G} \{H_1, H_2 \ldots H_6, H_7\} \xrightarrow{LM} \{H_8, H_9, H_{10}, H_{11}\} \xrightarrow{R} \{O'_1, O'_2 \ldots O'_m\}$$

Here, R is the set of the rules of measurement for mass, position, and velocity; these rules constitute the interpretative sentences; G is Newton's law of gravitation, and LM are the Newtonian laws of motion.

In reference to psychology, similar schematic analyses of the function of theories or of hypotheses involving "intervening variables" have repeatedly been presented in the methodological literature.[18] Here, the observational data with which the procedure starts usually concern certain observable aspects of an initial state of a given subject, plus certain observable stimuli acting upon the latter; and the final observational statements describe a response made by the subject. The theoretical statements mediating the transition from the former to the latter refer to various hypothetical entities, such as drives, reserves, inhibitions, or whatever other not directly observable characteristics, qualities, or psychological states are postulated by the theory at hand.

5. The Theoretician's Dilemma

The preceding account of the function of theories raises anew the problem encountered in section 3, namely, whether the theoretical detour, through a domain of not directly observable things, events, or characteristics cannot be entirely avoided. Assume, for example, that—as will often be the case—the interpretative sentences as well as the laws asserted by the theory have the form of equations which connect certain expressions in terms of theoretical quantities either with other such expressions, or with expressions in terms of observable quantities. Then the problem can be stated in Hull's succinct formulation: "If you have a secure equational linkage extending from the antecedent observable conditions through to the consequent observable conditions, why, even though to do so might not be positively pernicious, use several equations where one would do?" [19] Skinner makes the same point in more general form when he criticizes the construction, in psychological theories, of causal chains in which a first link consisting of an observable and controllable event is connected with a final ("third") one of the same kind by an intermediate link which usually is not open to observation and control. Skinner argues: "Unless there is a weak spot in our causal chain so that the second link is not lawfully determined by the first, or the third by the second, then the first and third links must be lawfully related. If we must always go back beyond the second link for prediction and control, we may avoid many tiresome and exhausting digressions by examining the third link as a function of the first." [20]

The conclusion suggested by these arguments might be called *the paradox of theorizing*. It asserts that if the terms and the general principles of a scientific theory serve their purpose, i.e., if they establish definite connections among observable phenomena, then they can be dispensed with since any chain of laws and interpretative statements establishing such a connection should then be replaceable by a law which directly links observational antecedents to observational consequents.

By adding to this crucial thesis two further statements which are obviously true, we obtain the premises for an argument in the classical form of a dilemma:

(5.1.) If the terms and principles of a theory serve their purpose they are unnecessary, as just pointed out, and if they don't serve their purpose they are surely unnecessary. But given any theory, its terms and principles either serve their purpose or they don't. Hence, the terms and principles of any theory are unnecessary.

This argument, whose conclusion accords well with the views of extreme methodological behaviorists in psychology, will be called the *theoretician's dilemma.*

However, before yielding to glee or to gloom over the outcome of this argument, it will be well to remember that the considerations adduced so far in support of the crucial first premise were formulated rather sketchily. In order to form a more careful judgment on the issue, it will therefore be necessary to inquire whether the sketch can be filled in so as to yield a cogent argument. To this task we now turn.

6. Operational Definitions and Reduction Sentences

It will be well to begin by considering more closely the character of interpretative sentences. In the simplest case, such a sentence could be an *explicit definition* of a theoretical expression in terms of observational ones, as illustrated by (3.2). In this case, the theoretical term is unnecessary in the strong sense that it can always be avoided in favor of an observational expression, its definiens. If all the primitives of a theory T are thus defined, then clearly T can be stated entirely in observational terms, and all its general principles will indeed be laws that directly connect observables with observables.

This would be true, in particular, of any theory that meets the standards of operationism in the narrow sense that each of its terms is introduced by an explicit definition which states an observable response whose occurrence is necessary and sufficient, under specified observable test conditions, for the applicability of the term in question. Suppose, for example, that the theoretical term is a one-place predicate, or property term, 'Q'. Then an operational definition of the kind just mentioned would take the form

(6.1) Def. $Qx \equiv (Cx \supset Ex)$

i.e., an object x has (by definition) the property Q if and only if it is such that if it is under test conditions of kind C, then it exhibits an effect, or response of kind E. Tolman's definition of expectancy of food provides an illustration: "When we assert that a rat expects food at L, what we assert is that *if* (1) he is deprived of food, (2) he has been trained on path P, (3) he is now put on path P, (4) path P is now blocked, and (5) there are other paths which lead away from path P, one of which points directly to location L, *then* he will run down the path which points directly to location L." [21] We can obtain this formulation by replacing, in (6.1), 'Qx' by 'rat x expects food at location L,' 'Cx' by the conjunction of the conditions (1), (2), (3), (4), (5) for rat x, and 'Ex' by 'x runs down the path which points directly to location L.'

However, as has been shown by Carnap in a now classical argument,[22] this manner of defining scientific terms, no matter how natural it may seem, encounters a serious difficulty. For on the standard extensional interpretation, a conditional sentence, such as the definiens in (6.1), is false only if its antecedent is true and its consequent false. Hence, for any object which does not satisfy the test conditions C, and for which therefore the antecedent of the definiens is false, the definiens as a whole is true; consequently, such an object will be assigned the property Q. In terms of our illustration: of any rat not exposed to the conditions (1)–(5) just stated, we would have to say that he expected food at L—no matter what kind of behavior the rat might exhibit.

One way out of this difficulty is suggested by the following consideration. In saying that a given rat expects food at L, we intend to attribute to the animal a state or a disposition which, under circumstances (1)–(5), will cause the rat to run down the path pointing directly to L; hence, in a proper operational definition, E must be tied to C nomologically, i.e., by virtue of general laws of the kind expressing causal connections. The extensional 'if . . . then . . .'—which requires neither logical nor nomological necessity of connection—would therefore have to be replaced in (6.1) by a stricter, nomological counterpart that might be worded perhaps as 'if . . . then, with causal necessity, . . .' However, the ideas of law and of causal or nomological necessity as here invoked are not clear enough at present to make this approach seem promising.[23]

Carnap [24] has proposed an alternative way of meeting the difficulty encountered by definitions of the form (6.1); it consists in providing a partial rather than a complete specification of meaning for 'Q'. This is done by means of so-called reduction sentences; in the simplest case, (6.1) would be replaced by the following *bilateral reduction sentence:*

(6.2) $Cx \supset (Qx \equiv Ex)$

i.e., if an object is under test conditions of kind C, then it has the property Q if and only if it exhibits a response of kind E. Here, the use of extensional connectives no longer has the undesirable aspects it exhibited in (6.1); if an object is not under test conditions C, then the entire formula (6.2) is true of it, but this implies nothing as to whether the object does, or does not, have the property Q. On the other hand, while (6.1) offers a full explicit definition of 'Q', (6.2) specifies the meaning of 'Q' only partly, namely, for just those objects that meet condition C; for those which don't, the meaning of 'Q' is left unspecified. In our illustration, for example, (6.3) would specify the meaning of 'x expects food at L' only for rats that meet conditions (1)–(5); for them, running down the path which pointed to L would be a necessary and sufficient condition of food expectancy. In reference to rats that don't meet the test conditions (1)–(5), the meaning of 'x expects food at L' would be left open; it could be further specified subsequently by means of additional reduction sentences.

In fact, it is this interpretation which is required for Tolman's concept of food expectancy. For while the passage quoted above seems to have exactly the form (6.1), this construal is ruled out by the following sentence which immediately follows the one quoted earlier: "When we assert that he does *not* expect food at location L, what we assert is that, under the same conditions, he will *not* run down the path which points directly to location L." The total interpretation thus given to 'rat x expects food at L' is most satisfactorily formulated in terms of a sentence of the form (6.2), in the manner outlined in the preceding paragraph.[25]

As this example vividly illustrates, reduction sentences offer an excellent way of formulating precisely the intent of operational definitions. By construing the latter as merely partial specifications of meaning, this approach treats theoretical concepts as "open"; and the provision for a set of different, and mutually supplementary, reduction sentences for a given term reflects the availability, for most theoretical terms, of different operational criteria of application, pertaining to different contexts.[26]

It should be noted, however, that while an analysis in terms of reduction sentences construes theoretical terms as not fully defined by reference to observables, it does not prove that a full explicit definition in observational terms *cannot* be achieved for theoretical expressions.

And indeed, it seems questionable whether a *proof* to this effect could even be significantly asked for. The next section deals with this issue in some detail.

7. On the Definability of Theoretical Terms by Means of an Observational Vocabulary

The first, quite general, point to be made here is this: the definition of any term, say 'v', by means of a set V of other terms, say 'v_1', 'v_2' . . . 'v_n', will have to state, in its definiens, a necessary and sufficient condition for the applicability of 'v' expressed in terms of some or all of the members of V. And in order to be able to judge whether this can be done in a given case, we will have to know how the terms under consideration are to be understood. For example, the vocabulary consisting of the terms 'male' and 'offspring of' permits the formulation of a necessary and sufficient condition of application for the term 'son of' in its biological, but not in its legal sense. How the given terms are to be understood can be indicated by specifying a set U of sentences which are to be considered as true, and which connect the given terms with each other and perhaps with other terms. Thus, U will be a set of sentences containing 'v', 'v_1' . . . 'v_n' and possibly also other extra-logical constants. For example, in the case of the biological use of the terms 'son', 'male', and 'offspring', in reference to humans, the following set of sentences—let us call it U_1—might be given: 'Every son is male,' 'No daughter is male,' 'x is an offspring of y if and only if x is a son or a daughter of y.'

Generally, the sentences of U specify just what assumptions are to be made, in the search for a definition, concerning the concepts under consideration; and the problem of definability now turns into the question whether it is possible to formulate, in terms of v_1, v_2 . . . v_n, a condition which, *in virtue of the assumptions included in U,* will be both necessary and sufficient for v. Thus, using an idea set forth and developed technically by Tarski,[27] we see that the concept of definability of 'v' by means of 'v_1', 'v_2' . . . 'v_n' acquires a precise meaning only if it is explicitly relativized by reference to a set U of specifying assumptions. That precise meaning may now be stated as follows:

(7.1) 'v' is definable by means of the vocabulary V = {'v_1', 'v_2' . . . 'v_n'} relative to a finite set U of statements containing, at least, 'v' and all the elements of V if from U there is deducible at least one sentence stating a necessary and sufficient condition for v in terms of no other extralogical constants than the members of V.

If all the terms under study are one-place predicates of the first order, for example, then a sentence of the required kind could most simply be stated in the form

(7.2) $v(x) \equiv D(x, v_1, v_2 \ldots v_n)$,

where the expression on the right-hand side stands for a sentential function whose only free variable is 'x', and which contains no extralogical constants other than those included in V.

Similarly, in the case of our illustration, the set U_1 specified above implies the statement:

$$x \text{ is a son of } y \equiv (x \text{ is male and } x \text{ is an offspring of } y),$$

so that, relatively to U_1, 'son' is definable as 'male offspring'.

We should add here an amplifying remark. A definition, when it is not simply a convention introducing an abbreviatory notation (such as the convention to let 'x^5' be short for '$x \cdot x \cdot x \cdot x \cdot x$') is usually considered as stating the *synonymy* of two expressions, or, as it is often put, the *identity of their meanings*. Now the question of the definability of a given term 'v' by means of a set V of other terms surely is not simply one of notational fiat; and indeed it will normally be construed as concerning the possibility of expressing the meaning which the term 'v' possesses with the help of the meanings of the members of V. If this conception is adopted, then naturally the information needed to answer the question of definability will concern the *meanings* of 'v' and of the members of V; accordingly, the statements in U, which provide this information, will then be required not simply to be true, but to be analytic, i.e., true by virtue of the intended meanings of the constituent terms. In this case, the statements in U would have the character of meaning postulates in the sense of Kemeny and Carnap.[28]

But in a study of the definability of theoretical expressions by means of observational terms, it is neither necessary nor even advisable to construe definition in this intensional manner. For, first of all, the idea of meaning, and related notions such as those of analyticity and synonymy, are by no means as clear as they have long been considered to be,[29] and it will be better, therefore, to avoid them when this is possible.

Secondly, even if those concepts are accepted as clearly intelligible, the definability of a theoretical term still cannot be construed exclusively as the existence of a synonymous expression containing only observational terms: it would be quite sufficient if a coextensive (rather than a strictly cointensive, or synonymous) expression in terms of observables were forthcoming. For such an expression would state an empirically necessary and sufficient observational condition of applicability for the theoretical term; and this is all that is required for our purposes. In fact, the sentence stating the condition—which might have the form (7.2), for example—can then be given the status of a truth-by-definition, by a suitable reformalization of the theory at hand.

It is of interest to note here that a necessary and sufficient observational condition for a theoretical term, say 'Q', might be inductively discovered even if only a partial specification of the meaning of 'Q' in terms of observables were available. Suppose, for example, that a set of alternative conditions of application for 'Q' has been specified by means of bilateral reduction sentences:

(7.3) $C_1x \supset (Qx \equiv E_1x)$
$C_2x \supset (Qx \equiv E_2x)$
. . . .
$C_nx \supset (Qx \equiv E_nx)$

where all predicates except 'Q' are observational. Suppose further that suitable investigations lead to the following empirical generalizations:

(7.4) $C_1x \supset (Ox \equiv E_1x)$
$C_2x \supset (Ox \equiv E_2x)$
. . . .
$C_nx \supset (Ox \equiv E_nx)$

where 'Ox' stands for a sentential function in 'x' which contains no non-observational extralogical terms. These findings, in combination with (7.3), would inductively support the hypothesis

(7.5) $Qx \equiv Ox$

which presents a necessary and sufficient observational condition for Q. However, (7.5) even if true (its acceptance involves the usual "inductive risk") clearly does not express a synonymy; if it did, no empirical investigations would be needed in the first place to establish it. Rather, it states that, as a matter of empirical fact, 'Q' is coextensive with 'O', or, that O is an empirically necessary and sufficient condition for Q.[30]

And if we wish, we may then imagine the theory-plus-interpretation at hand to be thrown into the form of a deductive system in which (7.5) becomes a definitional truth, and (7.3) assumes the character of a set of empirical statements equivalent to those listed in (7.4).

It might be mentioned here in passing that a similarly broad extensional interpretation of definability is called for also in the context of the problem whether a given scientific discipline, such as psychology, can be "reduced" to another, such as biology or even physics and chemistry. For one component of this problem is the question whether the terms of the first discipline can be defined by means of those of the latter; and what is wanted for this purpose is again a set of empirical hypotheses providing, for each psychological term, a necessary and sufficient condition of application expressed in the vocabulary of biology, or of physics and chemistry.

When we say, for example, that the concepts of the various chemical elements are definable in physical terms by a characterization of the specific ways in which their molecules are composed of elementary physical particles, we are clearly referring to results of experimental research rather than of a mere analysis of what is *meant* by the terms naming the various elements. If the latter were the case, it would indeed be quite incomprehensible why the problems pertaining to the definability of scientific terms should present any difficulty, and why they should be the objects of much conjecture and controversy.

The preceding considerations have important implications for our question whether all theoretical terms in empirical science can be defined in terms of observables. First of all, they show that the question as stated is elliptical: to complete it, we have to specify some set U of statements as referred to in (7.1). What set could reasonably be chosen for this purpose? One natural choice would be the set of all statements, in theoretical or observational terms, that are accepted as presumably true by contemporary science. Now, this pragmatic-historical characterization is by no means precise and unambiguous; there is a wide border area containing statements for which it cannot be clearly determined whether they are accepted by contemporary science. But no matter how the claims of these border-area statements are adjudicated, and no matter where—within reason—the borderline between observational and theoretical terms is drawn, it is at least an open question whether the set of presently accepted scientific statements implies for every theoretical term a necessary and sufficient condition of applicability in terms of observables. Certainly those who have asserted such definability have not supported their claim by actually deducing such conditions, or by presenting cogent general reasons for the possibility of doing so.

There is another way in which the claim of definability may be construed, namely as the assertion that as our scientific knowledge becomes more comprehensive, it will eventually be possible to deduce from it necessary and sufficient conditions of the required kind. (This is the sense in which definability is usually understood by those who claim the eventual definability of the concepts of psychology in terms of those of biology or of physics and chemistry; for that all the requisite definition statements—even in an extensional, empirical sense—cannot be deduced from current psychological, biological, physical, and chemical principles seems clear.[31]) But to assert definability of a theoretical term in this sense is to make a twofold claim: first, that the term in question will not be abandoned in the further development of scientific theorizing; and second, that general laws will be discovered which establish certain necessary and sufficient conditions, expressible in observational terms, for the applicability of the theoretical term at hand. Clearly, the truth of these claims cannot be established by philosophic arguments, but at best by the results of further scientific research.

Despite the precariousness of the problem, various claims and counterclaims have been advanced by philosophers of science and by methodologically interested scientists concerning the possibility of defining theoretical terms by reference to observables.

Some among the philosophers have simply urged that nothing short of explicit definition in terms of a vocabulary that is clearly understood can provide an acceptable method of introducing new terms into the language of science; and the argument supporting this view is to the effect that otherwise the new terms are not intelligible.[32] To this question we will return later. The protagonists of this view do not make an assertion, then, about the actual definability of the theoretical terms used in contemporary empirical science; rather, they stress the importance of clarifying the ideas of science by restating them, as far as possible, in a language with a clear and simple logical structure, and in such a way as to introduce all theoretical terms by means of suitable definitions.

Other writers, however, have argued, in effect, that scientific theories and the way in which they function have certain pervasive logical or methodological characteristics which presumably are not affected by changes in scientific knowledge, and which provide a basis on which the question as to the definability of theoretical terms can be settled without the need either to examine all the statements accepted by contemporary science or to wait for the results of further research.

An illuminating example of a careful use of this type of procedure is provided by Carnap's argument, referred to in the beginning of section 6 above, which shows that definitions of the form (6.1) cannot serve to introduce scientific concepts of the kind they were meant to specify. The argument is limited, however, in the sense that it does not show (and does not claim to show) that an explicit definition of theoretical terms by means of observational ones is generally impossible.

Recently,[33] Carnap has extended his examination of the problem in the following direction. Suppose that a given object, b, exhibits this kind of lawful behavior: whenever b is under conditions of a certain observable kind C, then it shows a response of a specified observable kind E. We then say that b has the disposition to react to C by E; let us call this dispositional property Q for short. Clearly, our earlier discussion in section 6 concerns the problem of precisely defining 'Q' in terms of 'C' and 'E'; we noted there, following Carnap, that we will either have to resign ourselves to a partial specification of meaning for 'Q' by means of the bilateral reduction sentence (6.2); or, if we insist on an explicit complete definition, we will have to use causal modalities in the definiens.

But no matter which of these alternative courses is chosen, the resulting disposition term 'Q' has this characteristic: if a given object b is under condition C and fails to show response E, or briefly, if Cb but ~Eb, then this establishes conclusively that b lacks the property Q, or briefly that ~Qb. This characteristic, Carnap argues, distinguishes "pure disposition terms," such as 'Q', from the theoretical terms used in science; for though the latter are connected with the observational vocabulary by certain interpretative sentences—Carnap calls them C-rules—those rules will not, in general, permit a set of observational data (such as 'Cb' and '~Eb' above) to constitute conclusive evidence for or against the applicability of the theoretical term in a given situation. There are two reasons for this assertion. First, the interpretative sentences for a given theoretical term provide an observational interpretation only within a certain limited range; thus, for example, in the case of the theoretical term 'mass', no C-rule is directly applicable to a sentence S_m ascribing a certain value of mass to a given body, if the value is either so small that the body is not directly observable or so large that the observer cannot "manipulate the body." [34]

And second, a direct observational interpretation for a theoretical term always involves the tacit understanding that the occurrence or absence of the requisite observable response in the specified test situation is to serve as a criterion only if there are no disturbing factors, or, provided that "the environment is in a normal state." [35] Thus, for example, a rule of correspondence might specify the deflection of a magnetic needle as an observable symptom of an electric current in a nearby wire, but with the tacit understanding that the response of the needle is to count only if there are no disturbing factors, such as, say, a sudden magnetic storm.

Generally, then, Carnap holds that "if a scientist has decided to use a certain term 'M' in such a way, that for certain sentences about M, any possible observational results can never be absolutely conclusive evidence but at best evidence yielding a high probability," then the appropriate place for 'M' is in the theoretical vocabulary.[36]

Now we should note, first of all, that if Carnap's arguments are sound, they establish that the theoretical terms of science cannot be construed as pure disposition terms, and thus even if, by the use of causal modalities, explicit definitions of the latter should be achieved, this method would be unavailing for theoretical terms. But the arguments do not show—and are not claimed to show—that theoretical terms can in no way be explicitly defined in terms of observables. In fact, if Carnap's statement quoted in the preceding paragraph is accepted, then many terms that can be explicitly defined by means of the observational vocabulary must be qualified as theoretical. For example, let 'R' be a two-place observational predicate, and let a one-place predicate 'M_1' be defined as follows:

(7.6) Def. $M_1x \equiv (\exists y)\, Rxy$

i.e., an object x has the property M_1 just in case it stands in relation R to at least one object y. If, for example, 'Rxy' stands for 'x is less heavy than y', then M_1 is the property of being exceeded in weight by at least one object, or, of not being the heaviest of all objects.

Let us assume, as customary, that the domain of objects under study is infinite or at least has not been assigned any definite maximum number of elements. Consider now the possibility of conclusive observational evidence for or against the sentence 'M_1a', which attributes M_1 to a certain object a. Obviously, a single observational finding, to the effect that a bears R to a certain object b, or that Rab, would suffice to verify 'M_1a' completely. But no finite set of observational data—'Raa', 'Rab', 'Rac', and so forth—would suffice for a conclusive refutation of 'M_1a'. According to Carnap's criterion, therefore, 'M_1', though defined in terms of the observational predicate 'R', might have to be classified as a theoretical term.

But possibly, in the passage quoted above, Carnap meant to require of a theoretical term 'M' that for certain sentences about M no observational results can be conclusively verificatory or falsificatory evidence. Yet even terms meeting this requirement can be explicitly defined in terms of observables. Let 'S' be a three-place observational predicate; for example, 'Sxyz' might stand for 'x is farther away from y than from z.' And let 'M_2' be defined as follows:

(7.7) Def. $M_2x \equiv (\exists y)\,(z)\,(\sim(z = y) \supset Sxyz)$.

In our example, an object x has M_2 just in case there is an object y from which it is farther away than from any other object z. Consider now the sentence 'M_2a'. As is readily seen, no finite set of observational findings (all the relevant ones would have the form 'Sabc' or '$\sim$Sabc') can be conclusive evidence, either verificatory or falsificatory, concerning 'M_2a'. Hence, though explicitly defined in terms of the observational predicate 'S', the term 'M_2' is theoretical according to the criterion suggested by Carnap.

The preceding discussion illustrates an elementary but important point: when a term, say a one-place predicate 'Q', is defined in terms of observables, its definiens must state a necessary and sufficient condition for the applicability of 'Q', i.e., for the truth of sentences of the form 'Qb'. But even though that condition is then stated completely in observational terms, it still may not enable us to decide, on the basis of a finite number of observational findings, whether 'Q' applies to a given object b; for the truth condition for 'Qb' as provided by the definiens may not be equivalent to a finite set of sentences each of which expresses a potential observational finding.

To add one more example to those given before: suppose that the property term 'iron object' and the relation terms 'attracts' and 'in the vicinity of' are included in the observational vocabulary. Then the definition

(7.8) Def. x is a magnet ≡ x attracts every iron object in its vicinity

is in terms of observables; but the criterion it provides for an object b being a magnet cannot be expressed in terms of any finite number of observational findings; for to establish that b is a magnet, we would have to show that any piece of iron which, at any time whatever, is brought into the vicinity of b, will be attracted by b; and this is an assertion about an infinity of cases.

To express the idea more formally, let us assume that our observational vocabulary contains, in addition to individual names for observable objects, just first-order predicates of any degree, representing attributes (i.e., properties or relations) which are observable in the sense that a small number of direct observations will suffice, under suitable conditions, to ascertain whether a given object or group of objects exhibits the attribute in question.

Now let us adopt the following definitions: An *atomic sentence* is a sentence, such as 'Pa', 'Rcd', 'Sadg', which ascribes an observable attribute to a specified object or group of objects. A *basic sentence* is an atomic sentence or the negation of an atomic sentence. A *molecular sentence* is a sentence formed from a finite number of atomic sentences by means of truth-functional connectives. Basic sentences will be considered as included among the molecular sentences.

Basic sentences can be considered as the simplest statements describing potential results of direct observation: they assert that some specified set of (one or more) objects has, or lacks, such and such an observable attribute.

Now by virtue of a theorem of truth-functional logic, there exist for every molecular statement S certain finite classes of basic statements which imply S, and certain other such classes which imply the negation of S. Thus, the molecular sentence 'Pa v (∼Pa · Rab)' is implied by {'Pa'} and also by {'∼Pa', 'Rab'}, for example; whereas its negation is implied by the set {'∼Pa', '∼Rab'}. This shows that for each molecular sentence S, it is possible to specify a set of basic sentences whose truth would conclusively verify S, and also a set of basic sentences whose truth would verify the negation of S, and would thus conclusively refute S. Thus, a molecular sentence is capable both of conclusive observational verification and of conclusive observational falsification "in principle," i.e., in the sense that potential data can be described whose occurrence would verify the sentence, and others whose occurrence would falsify it; but not of course in the sense that the two kinds of data might occur jointly—indeed, they are incompatible with each other.

But molecular sentences are the only ones which have this characteristic. A sentence which contains quantifiers non-vacuously is not both verifiable and falsifiable in the sense just indicated.

Returning now to the definition of scientific terms, we may accordingly distinguish two kinds of explicit definitions, those with molecular and those with non-molecular definiens. In both cases, all the extralogical terms in the definiens belong to the observational vocabulary; but in the former, the definiens contains truth-functional connectives as the only logical terms; in the latter, it also contains quantifiers. The definitions (7.6), (7.7), and (7.8) above are of the non-molecular type; the molecular type is illustrated by

(7.9) $\text{Son } xy \equiv (\text{Male } x \cdot \text{Offspring } xy)$.

We may say now that all and only definitions with molecular definiens provide finite *observational criteria* of application for the terms they define: in the case of a non-molecular definiens, the ascription of the defined attribute to a given case cannot be conclusively based on a finite set of observational findings.

Definition by means of a molecular observational definiens does not suffice to introduce all non-observational terms of science. That this is so might seem to be clear from the preceding discussion, for surely terms of the kind defined in (7.6) and (7.7) will be permissible and indeed needed in science. This reason is not conclusive, however. For the circumstance that a term, say 'M', is originally introduced by a definition with non-molecular definiens does not preclude the possibility that 'M' is in fact coextensive with a certain observational predicate, or some molecular compound of such predicates, say 'O_M'. And if this is the case, it might be discovered by further research, and 'M' could then be (re-)defined by 'O_M'.

There is another argument, however, which shows that, granting certain plausible general assumptions concerning the observational vocabulary, not all non-observational scientific terms can be defined by means of molecular observational expressions. We will now briefly consider this argument.[37] The observational vocabulary will be assumed to be finite. It may contain individual names designating certain observable objects; first-order predicate terms with any finite number of places, representing properties and relations of observable objects; and also functors, i.e., terms expressing quantitative aspects—such as weight in grams, volume in cubic centimeters, age in days—of the observable objects. However, in accordance with our assumption of observability, we will suppose that each of the functors can take on only a finite number of different values; this corresponds to the fact that only a finite number of different weights, for example, can be ascertained and distinguished by direct observation.

In contrast to these functors in the observational vocabulary, the theoretical vocabulary of physics, for example, contains a large number of functors whose permissible values range over all real numbers or over all real numbers within a certain interval.

Thus, for example, the distance between two points may theoretically have any non-negative value whatever. Now a definition of the required kind for a theoretical functor would have to specify, for each of its permissible values, a finite observational criterion of application. For example, in the case of the theoretical functor 'length', a necessary and sufficient condition, in the form of a finite observational criterion, would have to be forthcoming for each of the infinitely many statements of the form 'The distance, in centimeters, between points x and y is r' or briefly, '$l(x,y) = r$', where r is some real number.

Hence, we would have to specify, for each value of 'r', a corresponding finitely ascertainable configuration of observables. But this is impossible because in view of the limits of discrimination in direct observation, only a finite, though very large, number of finitely observable configurations can altogether be ascertained and distinguished.

However, if we drop the requirement of a *molecular* equivalent, in observational terms, for each permissible value of a theoretical functor, then an infinity of different values may become available.[38] Consider, for example, the functor 'the number of cells contained in organism y'. If 'x is a cell', 'y is an organism', and 'x is contained in y' are admitted as observational expressions, then it is possible to give a separate criterion of applicability, in terms of observables, for each of the infinitely many values 1, 2, 3 . . . which that functor may theoretically assume.[39] This can be done by means of the Frege-Russell analysis of cardinal numbers. For n = 1, for example, the necessary and sufficient condition is the following:

(7.10) $(\exists\mu)\,(v)\,(y \text{ is an organism} \cdot ((v \text{ is a cell} \cdot v \text{ is contained in } y) \supset (v = \mu)))$.

All extralogical constants occurring in this definiens are, on our assumption, observation terms; but the expression contains two quantifiers and thus is clearly non-molecular. Generally, the definiens for any value n will require n + 1 quantifiers.

Thus, the reach of explicit definition in terms of observables, even in the first-order functional calculus, is greatly extended if quantification is permitted in the definiens. And if stronger logical means are countenanced, vast further extensions may be obtained. For example, the functor 'the number of cells contained in y' can be explicitly defined by the single expression

(7.11) $\hat{\alpha}\,(\alpha \text{ sim } \hat{x}\,(x \text{ is a cell} \cdot x \text{ is contained in } y))$.

Here, the circumflex accent is the symbol of class abstraction, and 'sim' the symbol for similarity of classes (in the sense of one-to-one matchability of their elements); both of these symbols belong to the vocabulary of pure logic (more specifically, of the logic of classes).

So far, we have examined only functors whose values are integers. Can functors with rational and even irrational values be similarly defined in terms of observables? Consider, for example, the theoretical functor 'length in centimeters'. Is it possible to express, in observational terms, a necessary and sufficient condition for

(7.12) $l(x,y) = r$

for every non-negative value of r? We might try to develop a suitable definition which would correspond to the fundamental method of measuring length by means of rigid rods. And indeed, if our observational vocabulary contains a name for the standard meter bar, and furthermore the (purely qualitative) terms required to describe the fundamental measuring procedure, it is possible to state, for any specified rational or irrational value of r, a necessary and sufficient condition for (7.12). However, the definiens will in most cases be far from molecular in form: it will be teeming with symbols of quantification over individuals and over classes and relations of various types.[40] We will indicate, in brief outline, the way in which such definitions may be obtained. Expressions assumed to belong to the observational vocabulary will be italicized.

First, *the segment determined by points* x,y will be said to have a length of 100 centimeters if it is *congruent with* (i.e., can be made to coincide with) *the segment marked off on the standard meter bar*. This definiens is even of molecular form. But consider next the observational criterion for $l(x,y) = .25$. It may be stated in the following form, which begins with a quantificational phrase: there are four *segments*, each *marked off on a rigid body*, such that (i) all four are *congruent with* each other; (ii) their *sum* (i.e., the segment obtained by placing them end to end along a straight line) is *congruent* with *the segment marked off on the standard meter bar*; (iii) any of the four *segments* is *congruent with* the *segment determined by points* x,y. Analogously, an explicit observational definiens can be formulated for any other value of n that is a rational multiple of 100, and hence, for any rational value of n.

Next, the consideration that an irrational number can be construed as the limit of a sequence of rational numbers yields the following necessary and sufficient condition for $l(x,y) = r$, where r is irrational: the *segment determined by points* x,y contains an infinite sequence of *points* $x_1, x_2, x_3 \ldots$ such that (i) x_1 is *between* x and y, x_2 *between* x_1 and y, and so forth; (ii) given any *segment* S of rational length, there is a *point*, say x_n, in the sequence such that the *segments determined by* x_n and y, $x_n + 1$ and y, and so forth are all *shorter than* S, (iii) the lengths of the *segments determined by* x and x_1, x and x_2, and so forth, form a sequence of rational numbers with the limit r.

Finally, the idea underlying the preceding definition can be used to formulate an explicit definiens for the expression '$l(x,y)$' in such a way that its range of values is the set of all non-negative numbers.

Definitions of the kind here outlined are attainable only at the cost of using a strong logical apparatus, namely, a logic of sets adequate for the development of the theory of real numbers.[41] This price will be considered too high by nominalists, who hold that many of the logical concepts and principles here required, beginning with the general concept of set, are intrinsically obscure and should not, therefore, be used in a purported explication of the meanings of scientific terms. This is not the place to discuss the nominalistic strictures, however; and besides, it would no doubt be generally considered a worthwhile advance in clarification if for a set of theoretical scientific expressions explicit definitions in terms of observables can be constructed at all.

Another objection that might be raised against the definitional procedure here outlined is that it takes a schematic and oversimplified view of the fundamental measurement of length, and that it is rather liberal in construing as observational certain terms needed in the definiens, such as 'rigid body' and 'point'. This is quite true. By including the term 'point' in the observational vocabulary, for example, we construed points as directly observable physical objects; but our observational criterion for two points x,y determining a segment of irrational length required that there should be an infinite sequence of other points between x and y. This condition is never satisfied by the observable "points" in the form of small physical objects, or marks on rigid bodies, which are used in the fundamental measurement of length. As a consequence, the actual performance of fundamental measurement as represented in the above definition will never yield an irrational value for the length of a segment. But this does not show that no meaning has been assigned to irrational lengths; on the contrary, our outline of the definition shows that a meaning can indeed be formulated in observational terms for the assignment of any specified irrational value to the length of a physical line segment, as well as for the functor 'length in cm' in general.

However, the concept of length thus defined is not adequate for a physical theory which incorporates geometry, say in its Euclidean form. For the latter requires that certain segments which are well accessible to direct measurement—such as the diagonal of a square whose sides have a length of 100 centimeters—have an irrational length value; and statements to this effect will always turn out to be false if the criterion just discussed is made strictly definitory of length; for that procedure, as we noted, will always yield a rational value for the length of a given segment.

What the preceding argument about quantitative terms (represented by functors) shows, then, is this: the fact that the set of permissible values of a theoretical functor term is denumerably infinite or even has the cardinality of the continuum does not exclude the possibility of an explicit definition for it by means of a finite vocabulary containing only qualitative terms which are, by reasonably liberal standards, observational in character. The argument does not show, however, that such a definition is available for every functor term required by science (even our illustrative definition of 'length' turned out not to meet the needs of theoretical physics); and indeed, as was pointed out early in this section, a general proof to this effect cannot be expected.

A number of writers have taken the position that even if in principle theoretical terms could be avoided in favor of observational ones, it would be practically impossible or—what is more serious—methodologically disadvantageous or even stultifying to do so.

There is, for example, the answer given by Tolman and by Spence to the problem considered by Hull, which was mentioned in section 5 above: if intervening theoretical variables can establish a secure linkage between antecedent and consequent observable conditions, why should we not use just one functional connection that directly links antecedents and consequents? Spence adduces as one reason, also suggested by Tolman,[42] the following consideration: the mathematical function required to express the connection will be so complex that it is humanly impossible to conceive of it all at once; we can arrive at it only by breaking it down into a sequence of simpler functional connections, mediated by intervening variables. This argument, then, attributes to the introduction of unobservable theoretical entities an important practical role in the context of discovering interdependencies among observables, and presumably also in the context of actually performing the calculations required for the explanation or prediction of specific occurrences on the basis of those interdependencies.

An important methodological function is attributed to hypothetical entities in an interesting passage of Hull's essay on intervening variables in molar behavior theory.[43] The crucial point of Hull's argument is this: Suppose that in order to explain or predict the response of a subject in a given situation, we attribute to the subject, at the time t_1 of his response, a certain habit strength, which has the status of a hypothetical entity. That strength is, in Hull's theory, "merely a quantitative representation of the perseverative after-effects" of certain earlier observable events, such as observable stimuli received in temporally remote learning situations. Consequently, if reference to the hypothetical entity, habit strength, were avoided by linking the subject's observable response at t_1 directly to the observable stimuli received earlier then we would be invoking, as causal determinants for the response, certain observable events which at the time of the response, have long ceased to exist. And Hull rejects this notion, apparently inevitable when intervening hypothetical entities are eschewed, of causal action over a temporal distance: "it is hard to believe that an event such as stimulation in a remote learning situation can be causally active long after it has ceased to act on the receptors. I fully agree with Lewin that all the factors alleged to be causally influential in the determination of any other event must be in existence at the time of such causal action." [44] The hypothetical factor represented by the habit strength of the subject at the time t_1 of his response permits an explanation that accords with this principle.

Though the concluding part of the passage just quoted sounds quite metaphysical, the basic import of Hull's argument is methodological. It credits the assumption of explanatory hypothetical entities with an accomplishment that is well described by Feigl in another context: "the discontinuous and historical character (action at a spatial and/or temporal distance) of the phenomenalistically restricted account vanishes and is replaced by a spatio-temporally continuous (contiguous) and nomologically coherent formulation on the level of hypothetical construction." [45] Such spatio-temporally continuous theories appear to recommend themselves for at least two reasons: first, they possess a certain formal simplicity, which at present can hardly be characterized in precise terms, but which is reflected, for example, in the possibility of using the powerful and elegant mathematical machinery of the calculus for the deduction, from the postulates of the theory, of explanatory and predictive connections among particular occurrences. And second, as was mentioned in section 3, the past development of empirical science seems to show that explanatory and predictive principles asserting discontinuous connections among (spatio-temporally separated) observable events are likely to be found to have limited scope and various kinds of exceptions. The use of theories in terms of hypothetical entities frequently makes it possible to account for such exceptions by means of suitable assumptions concerning the hypothetical entities involved.

Another, more general, argument that must be considered here has been developed in a lucid and precise manner by Braithwaite,[46] who gives credit to Ramsey for the basic principle.[47] Braithwaite's main contention is that "theoretical terms can only be defined by means of observable properties on condition that the theory cannot be adapted properly to apply to new situations." [48] He elaborates this idea by reference to a simple, precisely formulated, miniature model of an interpreted theory. Without going into the details of that model—which would require too long a digression here—Braithwaite's claim can be adequately illustrated, it seems, by the following example: suppose that the term 'temperature' is interpreted, at a certain stage of scientific research, only by reference to the readings of a mercury thermometer. If this observational criterion is taken as just a partial interpretation (namely as a sufficient but not necessary condition), then the possibility is left open of adding further partial interpretations, by reference to other thermometrical substances which are usable above the boiling point or below the freezing point of mercury; and this permits a vast increase in the range of application of such laws as those connecting the temperature of a metal rod with its length or with its electric resistance, or the temperature of a gas with its pressure or its volume. If, however, the original criterion is given the status of a complete definiens, then the theory is not capable of such expansion; rather, the original definition has to be abandoned in favor of another one, which is incompatible with the first.[49]

The concept of intelligence lends itself to a similar argument: if test criteria which presuppose, on the part of the subject, the ability to read or at least to use language extensively are accorded the status of full definitions, then difficulties of the sort just indicated arise when the concept and the corresponding theory are to be extended to very young children or to animals.

However, the argument here outlined can hardly be said to establish what is claimed, namely that "A theory which it is hoped may be expanded in the future to explain more generalizations than it was originally designed to explain must allow more freedom to its theoretical terms than would be given them were they to be logical constructions out of observable entities" [50] (and thus defined in terms of the latter). For clearly, the procedure of expanding a theory at the cost of changing the definitions of some theoretical terms is not logically faulty; nor can it even be said to be difficult or inconvenient for the scientist, for the problem at hand is rather one for the methodologist or the logician, who seeks to give a clear "explication" or "logical reconstruction" of the changes occurring in an expansion of a given theory. And in the type of case discussed by Braithwaite, for example, this can be done in alternative ways—either in terms of additions to the original partial interpretation, or in terms of a total change of definition for some theoretical expressions. And if it is held that this latter method constitutes, not an expansion of the original theory, but a transition to a new one, this would raise more a terminological question than a methodological objection.

But though the above argument against definition does not have the intended systematic weight, it throws into relief an important heuristic aspect of scientific theorizing: when a scientist introduces theoretical entities such as electric currents, magnetic fields, chemical valences, or subconscious mechanisms, he intends them to serve as explanatory factors which have an existence independent of the observable symptoms by which they manifest themselves; or, to put it in more sober terms: whatever observational criteria of application the scientist may provide are intended by him to describe just symptoms or indications of the presence of the entity in question, but not to give an exhaustive characterization of it. The scientist does indeed wish to leave open the possibility of adding to his theory further statements involving his theoretical terms—statements which may yield new interpretative connections between theoretical and observational terms; and yet he will regard these as additional assumptions about the same hypothetical entities to which the theoretical terms referred before the expansion. This way of looking at theoretical terms appears to have definite heuristic value. It stimulates the invention and use of powerfully explanatory concepts for which only some links with experience can be indicated at the time, but which are fruitful in suggesting further lines of research that may lead to additional connections with the data of direct observation.[51]

The survey made in the present section has yielded no conclusive argument for or against the possibility of explicitly defining all theoretical terms of empirical science by means of a purely observational vocabulary; and in fact we have found strong reasons to doubt that any argument can settle the question once and for all.

As for the theoretical terms currently in use, we are quite unable at present to formulate observational definientia for all of them, and thus to make them, in principle, unnecessary. In effect, therefore, most theoretical terms are presently used in science on the basis of only a partial experiential interpretation; and this use, as we noted, appears to offer distinct heuristic advantages.

In view of the importance that thus attaches to the idea of partial interpretation, we will now consider somewhat more closely what kind of formal account might be given of it, and we will then turn to the question whether, or in what sense, the verdict of dispensability as proclaimed by the "theoretician's dilemma," applies also to theoretical terms which have been only partially interpreted, and which, therefore, cannot be dispensed with simply by virtue of definition.

8. Interpretative Systems

Carnap's theory of reduction sentences is the first systematic study of the logic of partial definition. The introduction of a term by means of a chain of reduction sentences differs in two significant respects from the use of a chain of definitions. First, it specifies the meaning of the term only partially and thus does not provide a way of eliminating the term from all contexts in which it may occur. Second, as a rule, it does not amount just to a notational convention, but involves empirical assertions. If, for example, the term 'Q' is introduced by the two reduction sentences

(8.1) $C_1x \supset (Qx \equiv E_1x)$
(8.2) $C_2x \supset (Qx \equiv E_2x)$

then the following empirical law is asserted by implication:

(8.3) $(x)((C_1x \cdot E_1x) \supset (C_2x \supset E_2x))$

i.e., roughly speaking: any object that shows a positive response under the first test condition will, when put into the second test condition, show a positive response as well. Thus, a chain of reduction sentences for a given term normally combines two functions of language that are often considered as sharply distinct: the stipulative assignment of meaning, and the assertion or description of empirical fact.

Reduction sentences, as we saw earlier, are very well suited for the formulation of operational criteria of application as partial definitions. But they are subject to rather severe limitations as to logical form and thus do not seem sufficient to provide a satisfactory general schema for the partial interpretation of theoretical terms.[52] A broader view of interpretation is suggested by Campbell's conception of a physical theory as consisting of a "hypothesis," represented by a set of sentences in theoretical terms, and a "dictionary," which relates the latter to concepts of experimental physics (which must be interconnected by empirical laws).[53] In contrast to the standard conception of a dictionary, Campbell's dictionary is assumed to contain, not definitions for the theoretical terms, but statements to the effect that a theoretical sentence of a certain kind is true if and only if a corresponding empirical sentence of a specified kind is true.[54] Thus, rather than definitions, the dictionary provides rules of translation; and partial rules at that, for no claim is made that a translation must be specified for each theoretical statement or for each empirical statement.

This latter feature accords well, for example, with the consideration that a particular observable macro-state of a given physical system may correspond to a large number of theoretically distinguishable micro-states; so that, for a theoretical sentence describing just one of those micro-states, the sentence describing the corresponding macro-state does not express a necessary and sufficient condition, and hence provides no translation.[55]

The statements in Campbell's dictionary evidently do not have the character of reduction sentences; they might be formulated, however, as biconditionals in which a sentence in theoretical terms is connected, by an "if and only if" clause, with a sentence in observational terms.

In other contexts, neither reduction sentences nor such biconditionals seem to be adequate. For as a rule, the presence of a hypothetical entity H, such as a certain kind of electric field, will have observable symptoms only if certain observational conditions, O_1, are satisfied, such as the presence of suitable detecting devices, which will then have to show observable responses, O_2. A sentence stating this kind of criterion would have the character of a generalized reduction sentence; it might be put into the form

(8.4) $O_1 \supset (H \supset O_2)$

where 'O_1' and 'O_2' are sentences—possibly quite complex ones—in terms of observables, and 'H' is a sentence which is expressed in theoretical terms.

But there is no good reason to limit interpretative statements to just the three types here considered. In order to obtain a general concept of partial interpretation, we will now admit as interpretative statements any sentences, of whatever logical form, which contain theoretical and observational terms. On the assumption that the theoretical and observational statements of empirical science are formulated within a specified logical framework, this idea can be stated more precisely and explicitly as follows:

(8.5) Let T be a theory *characterized by* a set of postulates in terms of a finite *theoretical vocabulary* V_T, and let V_B be a second set of extra-logical terms, to be called the *basic vocabulary*, which shares no term with V_T. By an *interpretative system* for T with the basis V_B we will then understand a set J of sentences which (i) is finite, (ii) is logically compatible with T, (iii) contains no extra-logical term that is not contained in V_T or V_B, (iv) contains every element of V_T and V_B essentially, i.e., it is not logically equivalent to some set of sentences in which some term of V_T or V_B does not occur at all.[56]

In applying the concept here defined to the analysis of scientific theories, we will have to assume, of course, that V_B consists of terms which are antecedently understood. They might be observational terms, in the somewhat vague sense explained earlier; but we need not insist on this. One might well take the view, for example, that certain disposition terms such as 'malleable', 'elastic', 'hungry', and 'tired' are not strictly observation terms, and are not known to be explicitly definable by means of observation terms; and yet, such terms might be taken to be well understood in the sense that they are used with a high degree of agreement by different competent observers. In this case, it would be quite reasonable to use these terms in interpreting a given theory, i.e., to admit them into V_B.

Campbell's conception of the function of his "dictionary" illustrates this possibility very well and shows that it comes closer to actual scientific procedure. Campbell specifies that the interpretation provided by the dictionary must be in terms of what he calls "concepts," such as the terms 'temperature', 'electrical resistance', 'silver', and 'iron' as used in experimental physics and chemistry. These are hardly observational in the narrow sense, for they are specifically conceived as representing clusters of empirical laws: "Thus, if we say anything about electrical resistance we assume that Ohm's Law is true; bodies for which Ohm's Law is not true, gases for example, have no electrical resistance." [57] And when we qualify a given body as silver, we attribute to it certain kinds of lawful behavior concerning solubility in various liquids, melting point, and so forth. But even though one might not wish to qualify these terms as observational, one may still consider them as well understood, and as used with high intersubjective agreement, by scientific experimenters; and thus, they might be admitted into V_B.

Interpretative systems as just defined include as special cases all the types of interpretation we considered earlier, namely, interpretation by explicit definitions for all theoretical terms, by chains of reduction sentences, by biconditional translation statements in the sense of Campbell's dictionary, and by generalized reduction sentences of the form (8.4); but of course they also allow for interpretative statements in a large variety of other forms.

Interpretative systems have the same two characteristics which distinguish chains of reduction sentences from chains of definitions: First, an interpretative system normally effects only a partial interpretation of the terms in V_T; i.e., it does not lay down (by explicit statement or by logical implication), for every term in V_T, a necessary and sufficient condition of application in terms of V_B. Second, like a chain of reduction sentences for a given theoretical term, an interpretative system will normally not be purely stipulative in character, but will imply certain statements in terms of V_B alone which are not logical truths, and which, on the conception of V_B as consisting of antecedently understood empirical terms, may be viewed as expressing empirical assertions. Thus, here again, we find a combination of the stipulative and the descriptive use of language.

But, to turn to a third point of comparison, an interpretative system need not provide an interpretation—complete or incomplete—for each term in V_T individually. In this respect it differs from a set of definitions, which specifies for each term a necessary and sufficient condition, and from a set of reduction sentences, which provides for each term a necessary and a—usually different—sufficient condition. It is quite possible that an interpretative system provides, for some or even all of the terms in V_T, no necessary or no sufficient condition in terms of V_B, or indeed neither of the two; instead, it might specify, by explicit statement or by logical implication, sufficient or necessary conditions in terms of V_B only for certain expressions containing several terms of V_T—for example, in the manner of Campbell's dictionary.

As a rule, therefore, when a theory T is interpreted by an interpretative system J, the theoretical terms are not dispensable in the narrow sense of eliminability from all contexts, in favor of defining expressions in terms of V_B. Nor are they generally dispensable in the sense that J provides, for every sentence H that can be formed by means of V_T, a "translation" into terms of V_B, i.e., a sentence O in terms of V_B such the the biconditional $H \equiv O$ [58] is logically deducible from J.

Are theoretical terms, then, altogether indispensable on this broad conception of interpretation so that the "paradox of theorizing" formulated in section 5 no longer applies to them? We consider this question in the next section.

9. Functional Replaceability of Theoretical Terms

The systematizing function of a theory T, as interpreted by an interpretative system J will consist in permitting inferences from given "data" in terms of V_B to certain other (e.g., predictive) statements in terms of V_B. If O_1 is the statement expressing the data, O_2 the inferred statement, then the connection may be symbolized thus:

(9.1) $(O_1 \cdot T \cdot J) \longrightarrow O_2$

Here, as in similar contexts below, 'T' stands for the set of postulates of the theory at hand; the arrow represents deductive implication.

Now, (9.1) holds if and only if $T \cdot J$ implies the sentence $O_1 \supset O_2$; so that (9.1) is tantamount to

(9.2) $(T \cdot J) \longrightarrow (O_1 \supset O_2)$

Whatever systematization is achieved among the V_B-sentences is clearly accomplished by T in conjunction with J. It will be convenient therefore to consider the postulates of T together with the sentences of J as the postulates of a deductive system T', which will be called an *interpreted theory*. Its vocabulary $V_{T'}$ will be the sum of V_T and V_B.

What was noted in connection with (9.1) and (9.2) may now be restated thus: If an interpreted theory T' establishes a deductive transition from O_1 to O_2, i.e., if

(9.3) $(O_1 \cdot T') \longrightarrow O_2$

then

(9.4) $T' \longrightarrow (O_1 \supset O_2)$

and conversely, where T' is the set of postulates of the interpreted theory.

Now it can readily be shown that an interpreted theory T' establishes exactly the same deductive connections among V_B-sentences as does the set of all those theorems of T' which are expressible in terms of

V_B alone; we will call this the set of *V_B-theorems,* or *V_B-consequences,* of T′, and will designate it by '$O_{T'}$'. This means, that for all purposes of deductive systematization, T′ is, as we will say, functionally equivalent to the set $O_{T'}$, which contains not a single theoretical term.

The proof is as follows: The deductive transition, represented in (9.3), from O_1 to O_2 can be achieved just as well by using, instead of T′, simply the sentence $O_1 \supset O_2$, which by virtue of (9.4) belongs to $O_{T'}$; for we have, by *modus ponens,*

(9.5) $(O_1 \cdot (O_1 \supset O_2)) \longrightarrow O_2$

And since $O_{T'}$ surely contains all the V_B-sentences of the form $O_1 \supset O_2$ that are implied by T′, the set $O_{T'}$ suffices to effect all the deductive systematizations achievable by means of T′. On the other hand, $O_{T'}$ is no stronger in this respect than T′; for $O_{T'}$ permits the deductive transition from O_1 to O_2 only if it implies $O_1 \supset O_2$; but in this case, T′ also implies $O_1 \supset O_2$, which means, in view of the equivalence of (9.4) with (9.3), that T′ will permit the deductive transition from O_1 to O_2.

Thus, *the deductive systematization that an interpreted theory T′ achieves among sentences expressed in terms of a basic vocabulary V_B is exactly the same as that accomplished by the set $O_{T'}$ of those statements (theorems) of T′ which can be expressed in terms of V_B alone.* In this sense, the theoretical terms used in T can be dispensed with.

But $O_{T'}$ is normally an unwieldy infinite set of statements, and the question arises therefore whether there is some generally applicable method of making it more manageable and perspicuous by putting it into the form of an axiomatized theoretical system T'_B, which would be formulated in terms of V_B alone. A theorem in formal logic which was recently proved by Craig shows that this is indeed the case, provided only that T′ satisfies certain extremely liberal and unconfining conditions.[59]

Thus Craig's theorem has a definite bearing upon the problems raised by the "paradox of theorizing," which was stated in section 5 in somewhat vague terms. The theorem at hand points out one way in which the "paradox" can be given a clear and precise interpretation and a rigorous proof: It shows that for any theory T′ using both theoretical terms and non-theoretical, previously understood ones, there exists, under certain very widely satisfied conditions, an axiomatized theoretical system T'_B which uses only the non-theoretical terms of T′ and yet is functionally equivalent with T′ in the sense of effecting, among the sentences expressible in the non-theoretical vocabulary, exactly the same deductive connections as T′.

Should empirical science then avail itself of this method and replace all its theories involving assumptions about hypothetical entities by functionally equivalent theoretical systems couched exclusively in terms which have direct observational reference or which are, at any rate, clearly understood? There are various reasons which make this inadvisable in consideration of the objectives of scientific theorizing.

To begin with, let us consider the general character of Craig's method. Disregarding many subtle points of detail, the procedure may be described as follows: By means of a constructive procedure, Craig arranges all the V_B-theorems of T′ in a sequence. This sequence is highly redundant, for it contains, for any sentence occurring in it, also all its logical equivalents (as far as they are expressible in V_B). Craig prescribes a procedure for eliminating many, though not even all, of these duplications. The remaining sequence therefore still contains each V_B-theorem of T′ in at least one of its various equivalent formulations. Finally, all the sentences in this remaining sequence are made postulates of T'_B. Thus, the set of V_B-theorems of T′ is "axiomatized" in T'_B only in a rather Pickwickian sense, namely by making every sentence of the set, in some of its many equivalent formulations, a postulate of T'_B; whereas normally, the axiomatization of a set of sentences is intended to select as postulates just a small subset from which the rest can then be logically derived as theorems. In this manner, the axiomatization expresses the content of the whole set "in a form which is psychologically or mathematically more perspicuous." [60] And since Craig's method in effect includes all sentences that are to be axiomatized among the postulates of T'_B, the latter, as Craig himself puts it, "fail to simplify or to provide genuine insight." [61]

The loss in simplicity which results from discarding the theoretical terms of T′ is reflected in the circumstance that the set of postulates which Craig's method yields for T'_B is always infinite. Even in cases where actually there exists some finite subset of $O_{T'}$ of V_B-theorems of T′ from which all the rest can be deduced, Craig's procedure will not yield such a subset: that is the price of the universal applicability of Craig's method.

Now there are cases where an infinity of postulates may not be excessively unwieldy; notably when the axioms are specified by means of axiom-schemata,[62] i.e., by stipulations to the effect that any sentence that has one of a finite number of specified forms (such as '$x = x$', for example) is to count as an axiom. But the manner in which the axioms, or postulates, of T'_B are specified by Craig's method is vastly more intricate, and the resulting system would be practically unmanageable—to say nothing of the loss in heuristic fertility and suggestiveness which results from the elimination of the theoretical concepts and hypotheses. For empirical science, therefore, this method of dispensing with theoretical expressions would be quite unsatisfactory.

So far, we have examined the eliminability of theoretical concepts and assumptions only in the context of deductive systematization: we considered an interpreted theory T′ exclusively as a vehicle of establishing deductive transitions among observational sentences. However, such theories may also afford means of inductive systematization (in the sense outlined in section 1); and an analysis of this function will yield a further argument against the elimination of theoretical expressions by means of Craig's method.

By way of illustration, we will refer to the following example, which is deliberately oversimplified in order the more clearly to exhibit the essentials: we will assume that V_T contains the term 'white phosphorus', or 'P' for short, and that the interpretative system incorporated into T′ states no sufficient observational conditions of application for it, but several necessary ones. These will be assumed to be independent of each other in the sense that, though in the case of white phosphorus they occur jointly, any one of them will in certain other cases occur in the absence of one or more of the others. Those necessary conditions might be the following: white phosphorus has a garlic-like odor; it is soluble in turpentine, in vegetable oils, and in ether; it produces skin burns. In symbolic notation:

(9.6) $(x)(Px \supset Gx)$
(9.7) $(x)(Px \supset Tx)$
(9.8) $(x)(Px \supset Vx)$
(9.9) $(x)(Px \supset Ex)$
(9.10) $(x)(Px \supset Sx)$

All predicates other than 'P' that occur in these sentences will belong, then, to V_B.

Now let V_T contain just one term in addition to 'P', namely 'has an ignition temperature of 30° C', or 'I' for short; and let there be exactly one interpretative sentence for 'I', to the effect that if an object has the property I then it will burst into flame if surrounded by air in which a thermometer shows a reading above 30° C. This property will be considered as observable and will be represented by the predicate 'F' in V_B. The interpretative sentence for 'I', then, is

(9.11) $(x)(Ix \supset Fx)$

Finally, we will assume that the theoretical part of T′ contains one single postulate, namely,

(9.12) $(x)(Px \supset Ix)$

i.e., white phosphorus has an ignition temperature of 30° C. Let the seven sentences (9.6)–(9.12) represent the total content of T′.

Then, as is readily seen, T′ has no consequences in terms of V_B except for purely logical truths; consequently, T′ will permit a deductive transition from one V_B-sentence to another only if the latter is logically implied by the former, so that T′ is not required to establish the connection. In other words: T′ effects no deductive systematization among V_B-sentences at all. Nevertheless, T′ may play an essential role in establishing, among the V_B-sentences, certain explanatory or predictive connections of an inductive kind. Suppose, for example, that a certain object, b, has been found to have all the characteristics G, T, V, E, S. In view of the sentences (9.6)–(9.10), according to which these characteristics are symptomatic of P, it might then well be inferred that b is white phosphorus. This inference would be inductive rather than deductive, and part of its strength would derive from the mutual independence which we assumed to exist among those five observable symptoms of white phosphorus. The sentence 'Pb' which has thus been arrived at inductively leads, via (9.12), to the prediction 'Ib', which in turn, in virtue of (9.11), yields the forecast 'Fb'. Thus, T′ permits the transition from the observational data 'Gb', 'Tb', 'Vb', 'Eb', 'Sb' to the observational prediction 'Fb'. But the transition requires an inductive step, consisting of the acceptance of 'Pb' on the strength of the five data sentences, which support, but which, of course, do not conclusively establish, 'Pb'.

On the other hand, the system T'_B obtained by Craig's method does not lend itself to this inductive use; in fact, all its sentences are logical truths and thus T'_B makes no empirical assertion at all, for, as was noted above, all the V_B-theorems of T′ are logically true statements.[63]

Thus, if the systematizing use of an interpreted theory T′ is conceived as involving inductive as well as deductive procedures, then the corresponding system T'_B cannot, in general, replace T′.

Another, inductively simpler, method of obtaining a functional equivalent, in observational terms, of a given interpreted theory T′ is provided by an idea of Ramsey's. In effect, the method amounts to treating all theoretical terms as existentially quantified variables, so that all the extra-logical constants that occur in Ramsey's manner of formulating a theory belong to the observational vocabulary.[64] Thus, for example, the interpreted theory determined by our formulas (9.6)–(9.12) would be expressed by the following sentence, which we will call the *Ramsey-sentence associated with the given theory*.

(9.13) $(\exists\phi)(\exists\psi)(x)((\phi x \supset \psi x) \cdot (\phi x \supset (Gx \cdot Tx \cdot Vx \cdot Ex \cdot Sx)) \cdot (\psi x \supset Fx))$

This sentence is equivalent to the expression obtained by conjoining the sentences (9.6)–(9.12), replacing 'P' and 'I' throughout by the variables 'ϕ' and 'ψ' respectively, and prefixing existential quantifiers with regard to the latter. Thus, (9.13) asserts that there are two properties, ϕ and ψ, otherwise unspecified, such that any object with the property ϕ also has the property ψ, and any object with the property ϕ also has the (observable, fully specified) properties G, T, V, E, S, and any object with the property ψ also has the (observable, fully specified) property F.

An interpreted theory T′ as we have been conceiving it is not, of course, logically equivalent with its associated Ramsey-sentence any more than it is logically equivalent with the associated Craig-system T'_B; in fact, each of the two is implied by, but does not in turn imply, T′. But though the Ramsey-sentence contains, apart from variables and logical constants, only terms from V_B, it can be shown to imply exactly the same V_B-sentences as does T′; hence, it establishes exactly the same deductive transitions among V_B-sentences as does T′. In this respect then, the Ramsey-sentence associated with T′ is on a par with the Craig-system T'_B obtainable from T′. But its logical apparatus is more extravagant than that required by T′ or by T'_B. In our illustration, (9.6)–(9.12), for example, T′ and T'_B contain variables and quantifiers only with respect to individuals (physical objects), whereas the Ramsey-sentence (9.13) contains variables and quantifiers also for properties of individuals; thus, while T′ and T'_B require only a first-order functional calculus, the Ramsey-sentence calls for a second-order functional calculus.

But this means that the Ramsey-sentence associated with an interpreted theory T′ avoids reference to hypothetical entities only in letter—replacing Latin constants by Greek variables—rather than in spirit. For it still asserts the existence of certain entities of the kind postulated by T′, without guaranteeing any more than does T′ that those entities are observable or at least fully characterizable in terms of observables. Hence, Ramsey-sentences provide no satisfactory way of avoiding theoretical concepts.

And indeed, Ramsey himself made no such claim. Rather, his construal of theoretical terms as existentially quantified variables appears to have been motivated by considerations of the following kind: If theoretical terms are treated as constants which are not fully defined in terms of antecedently understood observational terms, then the sentences that can formally be constructed out of them do not have the character of assertions with fully specified meanings, which can be significantly held to be either true or false; rather, their status is comparable to that of sentential functions, with the theoretical terms playing the role of variables. But of a theory, we want to be able to predicate truth or falsity, and the construal of theoretical terms as existentially quantified variables yields a formulation which meets this requirement and at the same time retains all the intended empirical implications of the theory.

This consideration raises a further problem concerning the status of partially interpreted theories; that problem will be discussed in the next section.

10. On Meaning and Truth of Scientific Theories

The problem suggested by Ramsey's approach is this: if, in the manner of section 8, we construe the theoretical terms of a theory as extralogical constants for which the system J provides only a partial interpretation in terms of the antecedently understood vocabulary V_B, can the sentences formed by means of the theoretical vocabulary nevertheless be considered as statements which make definite assertions, and which are either true or false?

The question might seem to come under the jurisdiction of semantics, and more specifically, of the semantical theory of truth. But this is not the case. What the semantical theory of truth provides (under certain conditions) is a general definition of truth for the sentences of a given language L. That definition is stated in a suitable metalanguage, M, of L and permits the formulation of a necessary and sufficient condition of truth for any sentence S of L. This condition is represented simply by a translation of S into M.[65] (To be suited for its purpose, M must therefore contain a translation of every sentence of L and must meet certain other conditions which are precisely specified in the semantical theory of truth.) But if the truth criteria thus stated in M are to be intelligible at all, then clearly all the translations of L-statements into M must be assumed to be significant to begin with. Instead of deciding the question of the significance of L-sentences, the semantical definition of truth presupposes that it has been settled antecedently.

For quite analogous reasons, semantics does not enable us to decide whether the theoretical terms in a given system T′ do, or do not, have semantical, or factual, or ontological reference—a characteristic which some writers have considered as distinguishing genuinely theoretical constructs from auxiliary or intervening theoretical terms.[66] One difficulty with the claims and counter-claims that have been made in this connection lies in the failure of the discussants to indicate clearly what they wish to assert by attributing ontological reference to a given term. Let us note here that from a purely semantical point of view, it is possible to attribute semantical reference to any term of a language L that is taken to be understood: the referent can be specified in the same manner as the truth condition of a given sentence in L, namely by translation into a suitable metalanguage. For example, using English as a metalanguage, we might say, in reference to Freud's terminology, that 'Verdraengung' designates repression, 'Sublimierung', sublimation, and so on. Plainly, this kind of information is unilluminating for those who wish to use existential reference as a distinctive characteristic of a certain kind of theoretical terms; nor does it help those who want to know whether, or in what sense, the entities designated by theoretical terms can be said actually to exist—a question to which we will return.

Semantics, then, does not answer the question raised at the beginning of this section; we have to look elsewhere for criteria of significance for theoretical terms and sentences.

Generally speaking, we might qualify a theoretical expression as intelligible or significant if it has been adequately explained in terms which we consider as antecedently understood. In our earlier discussion, such terms were represented by the vocabulary V_B (plus the terms of logic). But now the question arises, What constitutes an "adequate" explanation? No generally binding standards can be specified: the answer is ultimately determined by one's philosophical conscience. The logical and epistemological puritan will declare intelligible only what has been explicitly defined in terms of V_B; and he may impose further restrictions, in a nominalistic vein, for example, on the logical apparatus that may be used in formulating the definitions. Others will find terms introduced by reduction sentences quite intelligible, and still others will even countenance an interpretation as tenuous as that afforded by an interpretative system. One of the most important advantages of definition lies in the fact that it ensures the possibility of an equivalent restatement of any theoretical sentence in terms of V_B. Partial interpretation does not guarantee this; consequently it does not provide, for every sentence expressible in theoretical terms, a necessary and sufficient condition of truth that can be stated in terms which are antecedently understood. This, no doubt, is the basic difficulty that critics find with the method of partial interpretation.

In defense of partial interpretation, on the other hand, it can be said that to understand an expression is to know how to use it, and in a formal reconstruction the "how to" is expressed by means of rules. Partial interpretation as we have construed it provides such rules. These show, for example, what sentences in terms of V_B alone may be inferred from sentences containing theoretical terms; and thus they specify, in particular, a set of V_B-sentences that are implied, and hence indirectly asserted, by an interpreted theory T'. (If the set is empty, the theory does not fall within the domain of empirical science.) Conversely, the rules also show that sentences in theoretical terms may be inferred from V_B-sentences. Thus, there are close resemblances between our theoretical sentences and those sentences which are intelligible in the narrower sense of being expressible entirely in terms of V_B—a circumstance which militates in favor of admitting theoretical sentences into the class of significant statements.

It should be mentioned that if this policy is adopted, then we will have to recognize as significant (though not, of course, as interesting or worth investigating) certain interpreted systems which surely would not qualify as potential scientific theories. For example, let L be the conjunction of some finite number of empirical generalizations about learning behavior, formulated in terms of an observational vocabulary V_B, and let P be the conjunction of a finite number of arbitrary sentences formed out of a set V_T of arbitrarily chosen uninterpreted terms (for example, P might be the conjunction of the postulates of some axiomatization of elliptic geometry). Then, by making P the postulates of T and by choosing the sentence $P \supset L$ as the only member of our interpretative system J, we obtain an interpreted theory, which "explains" in a trivial way all the given empirical generalizations, since $T \cdot J$ plainly implies L. Yet, needless to say, T′ would not be considered a satisfactory learning theory.[67] The characteristic here illustrated does not vitiate our analysis of partial interpretation, since the latter does not claim that every partially interpreted theoretical system is a potentially interesting theory; and indeed, even the requirement of full definition of all theoretical terms by means of V_B still leaves room for similarly unrewarding "theories." Examples like our mock "learning theory" simply remind us that, in addition to having an empirical interpretation (which is necessary if there are to be any empirically testable consequences) a good scientific theory must satisfy various important further conditions; its V_B-consequences must be empirically well confirmed; it must effect a logically simple systematization of the pertinent V_B-sentences, it must suggest further empirical laws, and so forth.

If the sentences of a partially interpreted theory T′ are granted the status of significant statements, they can be said to be either true or false. And then the question, touched upon earlier in this section, as to the factual reference of theoretical terms, can be dealt with in a quite straightforward manner: To assert that the terms of a given theory have factual reference, that the entities they purport to refer to actually exist, is tantamount to asserting that what the theory tells us is true; and this in turn is tantamount to asserting the theory. When we say, for example, that the elementary particles of contemporary physical theory actually exist, we assert that there occur in the universe particles of the various kinds indicated by physical theory, governed by specified physical laws, and showing certain specific kinds of observable symptoms—again indicated by physical theory—of their presence in certain specified circumstances. But this is tantamount to asserting the truth of the (interpreted) physical theory of elementary particles. Similarly, asserting the existence of the drives, reserves, habit strengths, postulated by a given theory of learning amounts to affirming the truth of the system consisting of the statements of the theory and its empirical interpretation.[68]

Thus understood, the existence of hypothetical entities with specified characteristics and interrelations, as assumed by a given theory, can be examined inductively in the same sense in which the truth of the theory itself can be examined, namely, by an empirical investigation of its V_B-consequences.

According to the conception just outlined, we have to attribute factual reference to all the (extra-logical) terms of a theory if that theory is true; hence, this characteristic provides no basis for a semantical dichotomy in the theoretical vocabulary. Also, the factual reference, as here construed, of theoretical terms does not depend on whether those terms are avoidable in favor of expressions couched in terms of V_B alone. Even if all the theoretical terms of a theory T' are explicitly defined in terms of V_B, so that their use affords a convenient shorthand way of saying what could also be said by means of V_B alone, they will still have factual reference if what the theory says is true.

The preceding observations on truth and on factual reference in regard to partially interpreted theories rest on the assumption that the sentences of such theories are accorded the status of statements. For those who find this assumption unacceptable, there are at least two other ways of construing what we have called an interpreted theory. The first of these is Ramsey's method, which was described in the previous section. It has the very attractive feature of representing an interpreted theory in the form of a bona fide statement, which contains no extra-logical constants other than those contained in V_B, and still has exactly the same V_B-consequences as the interpreted theory stated in terms of incompletely interpreted theoretical constants. It is perhaps the most satisfactory way of conceiving the logical character of a scientific theory, and it will be objectionable mainly, or perhaps only, to those who, on philosophical grounds, are opposed to the ontological commitments[69] involved in countenancing all the requisite variables ranging over domains other than that of the individuals of the theory (such as, for example, the set of all quantitative characteristics of physical objects, or the set of all dyadic relations among them, or sets of such sets, and so forth).

Those finally, who, like the contemporary nominalists, reject such strong ontological commitments, may adopt a conception of scientific theories, not as significant statements, but as intricate devices for inferring, from intelligible initial statements, expressed in terms of an antecedently understood vocabulary V_B, certain other, again intelligible, statements in terms of that vocabulary.[70] The theoretical terms are then construed as meaningless auxiliary marks, which serve as convenient symbolic devices in the transition from one set of experiential statements to another. To be sure, the conception of laws and theories as extra-logical principles of inference does not reflect the way in which they are used by theoretical scientists. In publications dealing with problems of theoretical physics, or biology, or psychology, for example, sentences containing theoretical terms are normally treated on a par with those which serve to describe empirical data: together with the latter, they function as premises and as conclusions of deductive and of inductive arguments. And indeed, for the working scientist the actual formulation and use of theoretical principles as complex extra-logical rules of inference would be a hindrance rather than a help. However, the purpose of those who suggest this conception is not, of course, to facilitate the work of the scientist but rather to clarify the import of his formulations; and from the viewpoint of a philosophical analyst with nominalistic inclinations, for example, the proposed view of scientific sentences which by his standards are not intelligible statements, does represent an advance in clarification.

However, the question posed by the theoretician's dilemma can be raised also in regard to the two alternative conceptions of the status of a theory. Concerning Ramsey's formulation, we may ask whether it is not possible to dispense altogether with the existentially quantified variables which represent the theoretical terms, and thus to avoid the ontological commitment they require, without sacrificing any of the deductive connections that the Ramsey-sentence establishes among V_B-sentences. And in regard to theories conceived as inferential devices, we may ask whether they cannot be replaced by a functionally equivalent set of rules—i.e., one establishing exactly the same inferential transitions among V_B-sentences—which uses none of the "meaningless marks."

In both cases, Craig's theorem gives an affirmative answer by describing a general method for constructing the desired kind of equivalent (provided only that the original formulation meets the very weak requirements of that theorem). But again, in both cases, the result has the same shortcomings that were mentioned in section 8: First of all, the method would replace the Ramsey-sentence by an infinite set of postulates, or the body of inferential rules by an infinite set of rules, in terms of V_B, and would thus lead to a loss of economy. Second, the resulting system of postulates or of inferential rules would not lend itself to inductive prediction and explanation. And third, it would have the pragmatic defect, partly reflected already in the second point, of being less fruitful heuristically than the system using theoretical terms.

When scientists or methodologists claim that the theoretical terms of a given theory refer to entities which have an existence of their own, which are essential constituents or aspects of the world we live in, then, no matter what individual connotations they may connect with this assertion, the reasons they could adduce in support of it seem clearly to lie in the fact that those terms function in a well-confirmed theory which effects an economical systematization, both deductive and inductive, of a large class of particular facts and empirical generalizations, and which is heuristically fertile in suggesting further questions and new hypotheses. And as far as suitability for inductive systematization, along with economy and heuristic fertility, are considered essential characteristics of a scientific theory, theoretical terms cannot be replaced without serious loss by formulations in terms of observables only: the theoretician's dilemma, whose conclusion asserts the contrary, starts with a false premise.

NOTES

[1] This term was suggested by a passage in Reichenbach's volume (70), where the word 'postdictability' is used to refer to the possibility of determining "past data in terms of given observations" (p. 13). In a similar context, Ryle uses the term 'retrodict' (see, for example, (78, p. 124)), and Walsh speaks of the historians' "business to 'retrodict' the past: to establish, on the basis of present evidence, what the past must have been like" (93, p. 41). According to a remark in Acton's review of Walsh's book (*Mind*, vol. 62 (1953), pp. 564–565), the word 'retrodiction' was used in this sense already by J. M. Robertson in *Buckle and his Critics* (1895).

[2] More accurately: it can be deduced from the principles of Newtonian mechanics that Kepler's laws hold in approximation, namely, on the assumption that the forces exerted upon the planets by celestial objects other than the sun (especially other planets) are negligible.

[3] For a fuller presentation of this deductive schema of scientific explanation and prediction, see Popper (62, section 12, and 63, Chapter 25), and Hempel and Oppenheim (36).

[4] The distinction here made refers, then, exclusively to the form of the statements under consideration, and not to their truth status or to the extent to which they are

supported by empirical evidence. If it were established, for example, that actually only 80 per cent of all crows are black, this would not show that 'all crows are black,' or S_1 for short, was a statistical probability statement, but rather that it was a false statement of strictly universal form, and that 'The probability for a crow to be black is .8,' or S_2 for short, was a true statement of statistical form.

Furthermore, to be sure, neither S_1 nor S_2 can ever be established conclusively: they can only be more or less well supported by available evidence; each of them thus has a more or less high logical, or inductive, probability, relative to that evidence. But this again does not affect at all the fact that S_1 is of strictly universal and S_2 of statistical form.

[5] For a fuller discussion of this point, see, for example, Nagel (55, section 7), Reichenbach (71, sections 63–67), Cramer (18, Chapter 13).

[6] The explanatory and predictive use of statistical laws constitutes perhaps the most important type of inductive systematization; but the occurrence of such laws among the premises is not required by our general concept of inductive systematization. And indeed, as Carnap (11, pp. 574–575) has pointed out, it is sometimes possible to make predictions of an inductive character exclusively on the basis of information about a finite set of particular cases, without the mediation of any laws whatever. For example, information to the effect that a large sample of instances of P has been examined, that all of its elements have the characteristic Q, and that a certain case x, not included in the sample, is an instance of P, will lend high inductive support to the prediction that x, too, has the characteristic Q. Also, it is sometimes possible to base an inductive systematization on a set of premises which include one or more strictly universal statements, but no statistical laws. An example of such a systematization will be found in Section 9, in the prediction based on the formulas (9.6)–(9.12).

[7] Mises (54, p. 188). Whether it is advisable to refer to explanations of this kind as causal is debatable: since the classical conception of causality is intimately bound up with the idea of strictly universal laws connecting cause and effect, it might be better to reserve the term 'causal explanation' for some or all of those explanatory arguments of form (1.1) in which all the laws invoked are of strictly universal form.

[8] Northrop (59, Chapters III and IV), for example, presents this distinction very suggestively; he refers to the two levels as "the natural history stage of inquiry" and "the stage of deductively formulated theory." A very lucid and concise discussion of the idea at hand will be found in Feigl (20).

[9] In his essay on Skinner's analysis of learning (in Estes *et al.* (19)), Verplanck throws an illuminating sidelight on the importance, for the observational vocabulary (the terms of the data-language, as he calls it), of high uniformity of use among different experimenters. Verplanck argues that while much of Skinner's data-language is sound in this respect, it is "contaminated" by the inclusion, in its vocabulary, of two kinds of terms that are not suited for the description of objective scientific data. The first kind includes terms "that cannot be successfully used by many others"; the second kind includes certain terms that should properly be treated as higher order theoretical expressions.

The non-precise and pragmatic character of the requirement of intersubjective uniformity of use is nicely reflected in Verplanck's conjecture "that if one were to work with Skinner, and read his records with him, he would find himself able to make the same discriminations as does Skinner and hence eventually give some of them at least data-language status" (*loc. cit.*, p. 279n).

[10] It should be mentioned, however, that the idea of the range of application of a generalization is here used in an intuitive sense which it would be difficult to explicate. The range of application of (3.1), for example, might plausibly be held to be narrower than here indicated: it might be construed as consisting only of wooden-objects-placed-in-water, and iron-objects-placed-in-water. On the other hand, (3.1) may be equivalently restated thus: Any object whatever has the two properties of either not

being wood or floating on water, and of either not being iron or sinking in water. In this form, the generalization might be said to have the largest possible range of application, the class of all objects whatever.

[11] Since (3.2) was presented as a definition, it might be considered inappropriate to include it among the general laws effecting the predictive transition from O_1, O_2, O_3, O_4, to O_5. And indeed, it is quite possible to construe the concept of logical deduction as applied to (1.1) in such a way that it includes the use of any definition as an additional premise. In this case, (3.3) is the only law invoked in the prediction here considered. On the other hand, it is also possible to treat sentences such as (3.2), which are usually classified as purely definitional, on a par with other statements of universal form, which are qualified as general laws. This view is favored by the consideration, for example, that when a theory conflicts with pertinent empirical data, it is sometimes the "laws" and sometimes the "definitions" that are modified in order to accommodate the evidence. Our analysis of deductive systematization is neutral with respect to this issue.

[12] A lucid elementary discussion of the nature of axiomatized mathematical systems may be found in Cohen and Nagel (15, Chapter VI; also reprinted in Feigl and Brodbeck (24)). For an analysis in a similar vein, with special emphasis on geometry, see also Hempel (30). An excellent systematic account of the axiomatic method is given in Tarski (84, Chapters VI–X); this presentation, which makes use of some concepts of elementary symbolic logic, as developed in earlier chapters, includes several simple illustrations from mathematics. A careful logical study of deductive systems in empirical science, with special attention to the role of theoretical terms, is carried out in the first three chapters of Braithwaite (4), and a logically more advanced exposition of the axiomatic method, coupled with applications to biological theory, has been given by Woodger, especially in (98) and (99).

[13] See, for example, Hermes (37); Walker (92); McKinsey, Sugar, and Suppes (52); McKinsey and Suppes (53); Rubin and Suppes (75), and the further references given in these publications. An important pioneer work in the field is Reichenbach (68).

[14] See especially Woodger (98) and (99).

[15] See, for example, Hull *et al.* (39).

[16] For example, in von Neumann and Morgenstern (58, Chapter III and Appendix).

[17] Statements effecting an empirical interpretation of theoretical terms have been discussed in the methodological literature under a variety of names. For example, Reichenbach, who quite early emphasized the importance of the idea with special reference to the relation between pure and physical geometry, speaks of *coordinative definitions* (69, section 4; also 72, Chapter VIII); Campbell (7, Chapter VI; an excerpt from this chapter is reprinted in Feigl and Brodbeck (24)) and Ramsey (67, pp. 212–236) assume a *dictionary* connecting theoretical and empirical terms. (See also Section 8 below). Margenau (48, especially Chapter 4) speaks of *rules of correspondence*, and recently Carnap (14) has likewise used the general term '*correspondence rules*'. Northrop's *epistemic correlations* (59, especially Chapter VII) may be viewed as a special kind of interpretative statements. For a discussion of interpretation as a semantical procedure, see Carnap (10, sections 23, 24, 25), and Hutten (40, especially Chapter II). A fuller discussion of interpretative statements is included in sections 6, 7, 8 of the present essay.

[18] A lucid and concise presentation may be found, for example, in Bergmann and Spence (3).

[19] Hull (38, p. 284).

[20] Skinner (80, p. 35).

[21] Tolman, Ritchie, and Kalish (88, p. 15). See also the detailed critical analysis of Tolman's characterization of expectancy in MacCorquodale and Meehl (47, pp. 179–181).

[22] See Carnap (8, section 4).

[23] On this point, and on the general problem of explicating the concept of a law of nature, see Braithwaite (4, Chapter IX); Burks (6); Carnap (14, section 9); Goodman (28); Hempel and Oppenheim (36, Part III); Reichenbach (73).

[24] In his theory of reduction sentences, developed in Carnap (8). There is a question, however, whether certain conditions which Carnap imposes upon reduction sentences do not implicitly invoke causal modalities. On this point, see Hempel (34, section 3).

[25] And in fact, the total specification of meaning effected by the passages quoted is then summarized by the authors in their "definition" DF II, which has exactly the form (6.2) of a bilateral reduction sentence for 'rat x expects food at L.' (Tolman, Ritchie, and Kalish (88, p. 15).)

[26] For a fuller discussion, see Carnap (8, section 7) and Carnap (14, section 10).

[27] See Tarski (83, especially pp. 80–83).

[28] See Kemeny (41) and (42); Carnap (12).

[29] On this point, see especially Quine (65); Goodman (26); White (95) and (96, Part II). The significance of the notion of analyticity in special reference to theoretical statements is critically examined, for example, in Pap (60) and (61) and in Hempel (34). Arguments in defense of concepts such as analyticity and synonymy are advanced in the following articles, among others: Carnap (12, 13); Grice and Strawson (29); Martin (49); Mates (51); Wang (94).

[30] Since the reduction sentences (7.3) determine the meaning of 'Q' for those and only those cases which meet at least one of the conditions $C_1, C_2 \ldots C_n$, it might be argued that (7.4) in conjunction with (7.3) inductively supports, not (7.5), but only the following hypothesis, which limits the assertion of (7.5) to the cases just specified:

(7.5′) $(C_1x \vee C_2x \vee \ldots \vee C_nx) \supset (Qx \equiv Ox)$

and that what (7.5) asserts beyond this, namely,

(7.5″) $\sim(C_1x \vee C_2x \vee \ldots \vee C_nx) \supset (Qx \equiv Ox)$

constitutes a stipulation rather than an empirical hypothesis. But this does not alter the fact that (7.5), since it implies (7.5′), is empirical in character, and that its acceptance requires therefore inductive support.

[31] This point is discussed more fully in Hempel (32). On the problem of "reducing" the concepts of one discipline to those of another, the following publications have important bearing: Nagel (56) and (57); Woodger (100, pp. 271ff); Kemeny and Oppenheim (43).

[32] One writer who is impelled by his "philosophical conscience" to take this view is Goodman (see 27, Chapter I; 28, Chapter II, section 1). A similar position was taken by Russell when he insisted that physical objects should be conceived as "logical constructions" out of sense-data, and thus as definable in terms of the latter (see, for example, 77, Chapter VIII).

[33] See Carnap (14, especially sections 9, 10).

[34] Carnap (14, section 10).

[35] Carnap (14, section 10).

[36] Carnap (14, section 10). An idea which is similar in spirit, but not quite as clear in its content, has been put forward by Pap in (60) and in (61, sections 10–13 and 70), with the claim (not made by Carnap for his argument) that it establishes the "untenability" of the "thesis of explicit definability" (of theoretical terms by means of observational ones). Pap (60, p. 8). On the other hand, Bergmann (especially in 1 and 2) holds that many concepts of theoretical physics, including "even the particle notions of classical physics could, in principle, be introduced by explicit definitions. This, by the way, is also true of all the concepts of scientific psychology." Bergmann (2, section 1). In the same context Bergmann mentions that the method of partial

interpretation seems to be necessary in order to dissolve some of the puzzles concerning quantum theory. However, this strong assertion is supported chiefly by sketches of some sample definitions. Bergmann suggests, for example, that 'This place is in an electric field' can be defined by a sentence of the form 'If R_1 then R_2,' where R_1 stands for a sentence to the effect that there is an electroscope at the place in question, and R_2 stands "for the description of the behavior of the electroscope (in an electric field)." (1, pp. 98–99.) However, this kind of definition may be questioned on the basis of Carnap's arguments, which have just been considered. And in addition, even if unobjectionable, some examples cannot establish the general thesis at issue. Thus, the question remains unsettled.

[37] The basic idea was outlined earlier in Hempel (33, p. 30).

[38] I am grateful to Herbert Bohnert who, in a conversation some years ago, provided the stimulus for the development of the ideas here outlined concerning the definability of functors with infinitely many permissible values. Mr. Bohnert remarked on that occasion that explicit definition of such functors in terms of an observational vocabulary should be possible along lines indicated by the Frege-Russell theory of natural and of real numbers.

[39] If it should be objected that 'cell' and 'organism' are theoretical rather than observational terms, then they may be replaced, without affecting the crux of the argument, by terms whose observational character is less controversial, such as 'marble' and 'bag', for example.

[40] I was mistaken, therefore, when in an earlier essay I asserted that "no sentence expressible in observation terms and logical terms alone can state a sufficient condition (let alone a necessary and sufficient one) for a sentence of the form 'the length of interval i is r centimeters.'" Hempel (35, p. 68.) The consideration that led me to this assertion (*loc. cit.*, pp. 66–68) overlooked the point made above (between (7.7) and (7.8)), namely that a sentence expressed in observational and logical terms alone may represent a necessary and sufficient condition for the truth of a given theoretical sentence without providing *finite* observational truth criteria for that sentence: it may be non-molecular in form.

[41] The argument can readily be extended to functors taking complex numbers or vectors of any number of components as values. Our reasoning has relied essentially on the Frege-Russell method of defining the various kinds of numbers (integers, rational, irrational, complex numbers, etc.) in terms of the concepts of the logic of sets. For a detailed outline of the procedure, see Russell (76); fuller technical accounts may be found in many works on symbolic logic.

[42] See Tolman (87, as reprinted in Marx, 50, p. 89); and Spence (82, p. 65n).

[43] Hull (38).

[44] Hull (38, p. 285).

[45] Feigl (21, p. 40).

[46] Braithwaite (4, Chapter III).

[47] See the essay "Theories" (1929) in Ramsey (67).

[48] Braithwaite (4, p. 76).

[49] This point is also made in Carnap (8, section 7) in a discussion of the advantages of reduction sentences over definitions in the introduction of scientific terms. And Feigl eloquently argues in the same vein in his essay (23), in which the general principle is illustrated by various suggestive examples taken from physics and psychology.

[50] Braithwaite (4, p. 76).

[51] An excellent concise synopsis of various arguments in favor of invoking "hypothetical constructs" will be found in Feigl (21, pp. 38–41). Some aspects of the "semantic realism" concerning theoretical terms and sentences which Feigl presents in the same article are discussed in section 10 of the present essay.

[52] This has been pointed out by Carnap himself; see, for example, his (14).

[53] See Campbell (7, Chapter VI). Important parts of this chapter are reprinted in Feigl and Brodbeck (24).

[54] Campbell (7, p. 122).

[55] However, this does not show that there cannot possibly be any necessary and sufficient condition in observational terms for the theoretical sentence: the problem of proving or disproving this latter claim is subject to difficulties analogous to those discussed in section 7 in regard to definability.

[56] The intuitive notion of interpretation, as well as the conception reflected in Campbell's idea of an interpretative dictionary, would seem to call for the following additional condition: (v) Each sentence of J contains essentially terms from V_T as well as terms from V_B. However, this requirement introduces no further restriction of the concept of interpretative system; for any system J that meets conditions (i) to (iv) can be stated in an equivalent form that satisfies (v) as well. To this end, it suffices to replace the member sentences of J by their conjunction; this yields a logically equivalent interpretative system which contains only one sentence, and which satisfies (v) since J satisfies (iv).

[57] Campbell (71, p. 43).

[58] Here, and on some subsequent occasions where there is no danger of misunderstandings, logical connectives are used autonymously; the expression '$H \equiv O$', for example, represents the sentence obtained by placing the triple-bar symbol (for 'if and only if') between the sentences of which 'H' and 'O' are names.

[59] Craig's paper (16) contains the first published account of this interesting theorem. A less condensed and less technical presentation, with explicit though brief references to applications such as the one here considered, is given in Craig (17).

In application to the issue we are discussing, the result obtained by Craig may be briefly stated as follows: Let the set V_T' of primitive terms of T' and the set of postulates of T' be specified effectively, i.e., in a manner allowing anyone, given any expression, to decide in a finite number of steps whether the expression is a primitive term (or postulate) of T'. Let V_T' be divided, by an effective criterion that may otherwise be chosen at will, into two mutually exclusive vocabularies, V_T and V_B. Finally, let the rules of the logic used be such that there is an effective method of determining, for any given finite sequence of expressions, whether it is a valid deduction according to those rules.

Then there exists a general method (i.e., a method applicable in all cases meeting the conditions just outlined) of effectively constructing (i.e., characterizing the postulates and the rules of inference of) a new system T'_B whose set of primitives is V_B; and whose theorems are exactly those theorems of T' which contain no extralogical constants other than those contained in V_B.

Note that the theorem permits us to draw the dividing line between V_T and V_B wherever we please, as long as the criterion used to effect the division permits us to decide in a finite number of steps to which of the two sets a given term belongs. This condition as well as the requirement of an effective characterization of V_T' will be trivially satisfied, for example, if V_T' is finite, and its member terms as well as those of V_B and V_T are specified simply by enumerating them individually.

The further requirement of an effective characterization of the postulates and the rules of logic for T' are so liberal that no doubt any scientific theory that has yet been considered can be formalized in a manner that satisfies them—as long as the connections between theoretical and observational expressions can be assumed to be expressible in the form of definite statements. The only important case I am aware of in which this condition would be violated is that of a theory for which no definite rules of interpretation are specified—say, on the ground that the criteria of application for theoretical expressions always have to be left somewhat vague. A conception of this kind may have been intended, for example, by A. Wald's remark "In order to apply [a scientific] theory to real phenomena, we need some rules for establishing the correspondence between the idealized objects of the theory and those of the real world. These rules will always be somewhat vague and can never form a part of the theory itself." Wald (91, p. 1).

The conditions of Craig's theorem are satisfiable, however, if the vagueness here referred to is reflected in definite rules. Thus, for example, the interpretative sentences for a given theory might take the form of statistical probability statements (a possibility mentioned in Carnap (14, section 5)), or perhaps of logical probability statements (each specifying the logical probability of some theoretical sentence relative to a specified sentence in observational terms, or vice versa). Either of these procedures would yield an interpretation of a more general kind than that characterized by the definition of an interpretative system given in section 8 of the present essay. Yet even to theories which are interpreted in this wider sense, Craig's theorem can be applied.

[60] Craig (17, p. 49). It may be well to note briefly two further points which were established by Craig, in the studies here referred to: (i) A theory T′ may have a set of V_B-consequences that cannot be axiomatized by means of a *finite* set of postulates expressible in terms of V_B. (ii) There is no general method that permits an effective decision, for every theory T′, as to whether its V_B-consequences can, or cannot, be axiomatized by means of a finite set of postulates.

[61] Craig (17, p. 49). This fact does not detract in the least, of course, from the importance and interest of Craig's result as a theorem in logic.

[62] On this method, first used by von Neumann, see Carnap (9, pp. 29–30 and p. 96), where further references to the literature are given.

[63] This does not preclude the possibility of effecting the transition from the five data sentences to 'Fb' in a purely deductive fashion by means of a suitable modification of T′; in fact, just the addition, to T′, of the statement '$(x)\ ((Gx \cdot Tx \cdot Vx \cdot Ex \cdot Sx) \supset Px)$' would have the desired effect. But the modified theory thus obtained is clearly stronger than T′.

[64] Ramsey (67, pp. 212–215, 231).

[65] See Tarski (85, p. 350).

[66] On this point see, for example, MacCorquodale and Meehl (46); Lindzey (45); Feigl (21), (22); Hempel (31); Rozeboom (74).

[67] It is of interest to note here that if in addition to the conditions specified in section 8, an interpreted theory were also required to meet the criteria of significance for theoretical terms and sentences that have recently been proposed by Carnap in (14, sections 6, 7, 8), the terms and the sentences of our mock "learning theory" would be ruled out as nonsignificant.

[68] More precisely, the assertion that there exist entities of the various kinds (such as hypothetical objects and events and their various qualitative and quantitative properties and relations) postulated by an interpreted theory T′ is expressed by the Ramsey-sentence associated with T′. It is obtained by replacing all theoretical constants in the conjunction of the postulates of T′ by variables and binding all these by existential quantifiers placed before the resulting expression. The sentence thus obtained is a logical consequence of the postulates of T′; but the converse does not hold; hence strictly speaking, the assertion of the existence of the various hypothetical entities assumed in a theory is logically weaker than the theory itself.

For suggestive observations on the question of the reality of theoretical entities, see, for example, Toulmin (89, pp. 134–139) and Smart (81).

[69] The concept is used here in Quine's sense, according to which a theory is ontologically committed to those entities which must be included in the domains over which its bound variables range if the theory is to be true. Quine develops and defends this idea in several of the essays comprising his book (66).

[70] This conception of laws or theories as inferential principles has been suggested, on different grounds, by several authors, such as Schlick (who says that he owes the idea to Wittgenstein; see Schlick, 79, p. 151 and p. 155); Ramsey (see 67, p. 241),

Ryle (see 78, especially pp. 120–125), and Toulmin (see 89, Chapters III and IV). (Toulmin remarks, however, that to think of laws of nature as rules or licenses "reflects only a part of their nature" (*loc. cit.*, p. 105).) See also Braithwaite's discussion of the issue in (4, pp. 85–87). Finally, Popper's essay (64) contains several critical and constructive comments that bear on this issue and on some of the other questions discussed in the present study.

REFERENCES

1. Bergmann, Gustav. "The Logic of Psychological Concepts," *Philosophy of* Science, 18:93–110 (1951).
2. Bergmann, Gustav. "Comments on Professor Hempel's 'The Concept of Cognitive Significance,' " *Proceedings of the American Academy of Arts and Sciences*, 80(No.1):78–86 (1951). Reprinted in Gustav Bergmann, *The Metaphysics of Logical Positivism*. New York: Longmans, Green and Co., 1954.
3. Bergmann, Gustav, and Kenneth Spence. "Operationism and Theory in Psychology," *Psychological Review*, 48:1–14 (1941). Reprinted in Marx (50).
4. Braithwaite, R. B. *Scientific Explanation*. Cambridge, England: Cambridge Univ. Press, 1953.
5. Bridgman, P. W. *The Logic of Modern Physics*. New York: Macmillan, 1927.
6. Burks, Arthur W. "The Logic of Causal Propositions," *Mind*, 60:363–382 (1951).
7. Campbell, Norman R. *Physics: The Elements*. New York: Dover, 1920.
8. Carnap, Rudolf. "Testability and Meaning," *Philosophy of Science*, 3:420–468 (1936); 4:1–40 (1937). Reprinted as a monograph by Whitlock's Inc., New Haven, Conn., 1950. Excerpts reprinted in Feigl and Brodbeck (24).
9. Carnap, Rudolf. *The Logical Syntax of Language*. New York and London: Humanities, 1937.
10. Carnap, Rudolf. *Foundations of Logic and Mathematics*. Chicago: Univ. of Chicago Press, 1939.
11. Carnap, Rudolf. *Logical Foundations of Probability*. Chicago: Univ. of Chicago Press, 1950.
12. Carnap, Rudolf. "Meaning Postulates," *Philosophical Studies*, 3:65–73 (1952).
13. Carnap, Rudolf. "Meaning and Synonymy in Natural Languages," *Philosophical Studies*, 6:33–47 (1955).
14. Carnap, Rudolf. "The Methodological Character of Theoretical Concepts," in H. Feigl, and M. Scriven (eds.), *The Foundations of Science and the Concepts of Psychology and Psychoanalysis*, pp. 38–76. Minneapolis: Univ. of Minnesota Press, 1956.
15. Cohen, M. R., and E. Nagel. *Introduction to Logic and Scientific Method*. New York: Harcourt, Brace, 1934.
16. Craig, William. "On Axiomatizability within a System," *Journal of Symbolic Logic*, 18:30–32 (1953).
17. Craig, William. "Replacement of Auxiliary Expressions," *Philosophical Review*, 65:38–55 (1956).
18. Cramér, Harald. *Mathematical Methods of Statistics*. Princeton: Princeton Univ. Press, 1946.
19. Estes, W. K., S. Koch, K. MacCorquodale, P. E. Meehl, C. G. Mueller, W. S. Schoenfeld, and W. S. Verplanck. *Modern Learning Theory*. New York: Appleton-Century-Crofts, 1954.
20. Feigl, Herbert. "Some Remarks on the Meaning of Scientific Explanation," (A slightly modified version of comments first published in *Psychological Review*, 52(1948)), in Feigl and Sellars (25, pp. 510–514).
21. Feigl, Herbert. "Existential Hypotheses," *Philosophy of Science*, 17:35–62 (1950).

22. Feigl, Herbert. "Logical Reconstruction, Realism, and Pure Semiotic," *Philosophy of Science*, 17:186–195 (1950).
23. Feigl, Herbert. "Principles and Problems of Theory Construction in Psychology," in W. Dennis (ed.), *Current Trends in Psychological Theory*, pp. 179–213. Pittsburgh: Univ. of Pittsburgh Press, 1951.
24. Feigl, Herbert, and May Brodbeck (eds.). *Readings in the Philosophy of Science*. New York: Appleton-Century-Crofts, 1953.
25. Feigl, Herbert, and Wilfrid Sellars (eds.). *Readings in Philosophical Analysis*. New York: Appleton-Century-Crofts, 1949.
26. Goodman, Nelson. "On Likeness of Meaning," *Analysis*, 10:1–7 (1949). Reprinted in a revised form in Linsky (45).
27. Goodman, Nelson. *The Structure of Appearance*. Cambridge, Mass.: Harvard Univ. Press, 1951.
28. Goodman, Nelson. *Fact, Fiction, and Forecast*. Cambridge, Mass.: Harvard Univ. Press, 1955.
29. Grice, H. P., and P. F. Strawson. "In Defense of a Dogma," *Philosophical Review*, 65:141–158 (1956).
30. Hempel, Carl G. "Geometry and Empirical Science," *American Mathematical Monthly*, 52:7–17 (1945). Reprinted in Feigl and Sellars (25), in Wiener (97), and in James R. Newman (ed.), *The World of Mathematics*. New York: Simon and Schuster, 1956.
31. Hempel, Carl G. "A Note on Semantic Realism," *Philosophy of Science*, 17: 169–173 (1950).
32. Hempel, Carl G. "General System Theory and the Unity of Science," *Human Biology*, 23:313–322 (1951).
33. Hempel, Carl D. *Fundamentals of Concept Formation in Empirical Science*. Chicago: Univ. of Chicago Press, 1952.
34. Hempel, Carl G. "Implications of Carnap's Work for the Philosophy of Science," in P. A. Schilpp (ed.), *The Philosophy of Rudolf Carnap*. New York: Tudor (forthcoming).
35. Hempel, Carl G. "The Concept of Cognitive Significance: A Reconsideration," *Proceedings of the American Academy of Arts and Sciences*, 80(No. 1):61–77 (1951).
36. Hempel, Carl G., and Paul Oppenheim. "Studies in the Logic of Explanation," *Philosophy of Science*, 15:135–175 (1948). Reprinted in part in Feigl and Brodbeck (24).
37. Hermes, H. "Eine Axiomatisierung der allgemeinen Mechanik," *Forschungen zur Logik und Grundlegung der exakten Wissenschaften*. Neue Folge, Heft 3. Leipzig, 1938.
38. Hull, C. L. "The Problem of Intervening Variables in Molar Behavior Theory," *Psychological Review*, 50:273–291 (1943). Reprinted in Marx (50).
39. Hull, C. L., C. I. Hovland, R. T. Ross, M. Hall, D. T. Perkins, and F. B. Fitch. *Mathematico-Deductive Theory of Rote Learning*. New Haven: Yale Univ. Press, 1940.
40. Hutten, Ernest H. *The Language of Modern Physics: An Introduction to the Philosophy of Science*. London and New York: Macmillan, 1956.
41. Kemeny, John G. Review of Carnap, *Logical Foundations of Probability*. *Journal of Symbolic Logic*, 16:205–207 (1951).
42. Kemeny, John G. "Extension of the Methods of Inductive Logic," *Philosophical Studies*, 3:38–42 (1952).
43. Kemeny, John G., and Paul Oppenheim. "On Reduction," *Philosophical Studies*, 7:6–19 (1956).
44. Lindzey, Gardner. "Hypothetical Constructs, Conventional Constructs, and the Use of Physiological Data in Psychological Theory," *Psychiatry*, 16:27–33 (1953).

45. Linsky, Leonard (ed.). *Semantics and the Philosophy of Language*. Urbana, Ill.: Univ. of Illinois Press, 1952.
46. MacCorquodale, K., and P. Meehl. "On a Distinction between Hypothetical Constructs and Intervening Variables," *Psychological Review*, 55:95–107 (1948). Reprinted in Feigl and Brodbeck (24) and, with omissions, in Marx (50).
47. MacCorquodale, K., and P. Meehl. "Edward C. Tolman," in Estes *et al.* (19), pp. 177–266.
48. Margenau, Henry. *The Nature of Physical Reality*. New York: McGraw-Hill Book Co., 1950.
49. Martin, R. M. "On 'Analytic,'" *Philosophical Studies*, 3:42–47 (1952).
50. Marx, Melvin H. (ed.). *Psychological Theory*. New York: Macmillan, 1951.
51. Mates, Benson. "Analytic Sentences," *Philosophical Review*, 60:525–534 (1951).
52. McKinsey, J. C. C., A. C. Sugar, and P. Suppes. "Axiomatic Foundations of Classical Particle Mechanics," *Journal of Rational Mechanics and Analysis*, 2:253–272 (1953).
53. McKinsey, J. C. C., and P. Suppes. "Transformations of Systems of Classical Particle Mechanics," *Journal of Rational Mechanics and Analysis*, 2:273–289 (1953).
54. Mises, R. von. *Positivism: A Study in Human Understanding*. Cambridge, Mass.: Harvard Univ. Press, 1951.
55. Nagel, Ernest. *Principles of the Theory of Probability*. Chicago: Univ. of Chicago Press, 1939.
56. Nagel, Ernest. "The Meaning of Reduction in the Natural Sciences," in Robert C. Stauffer (ed.), *Science and Civilization*. Madison, Wis.: Univ. of Wisconsin Press, 1949. Reprinted in Wiener (97).
57. Nagel, Ernest. "Mechanistic Explanation and Organismic Biology," *Philosophy and Phenomenological Research*. 11:327–338 (1951).
58. Neumann, John von, and Oskar Morgenstern. *Theory of Games and Economic Behavior*, 2d ed. Princeton: Princeton Univ. Press, 1947.
59. Northrop, F. S. C. *The Logic of the Sciences and the Humanities*. New York: Macmillan, 1947.
60. Pap, Arthur. "Reduction Sentences and Open Concepts," *Methodos*, 5:3–28 (1953).
61. Pap, Arthur. *Analytische Erkenntnistheorie*. Wien: J. Springer, 1955.
62. Popper, Karl. *Logik der Forschung*. Wien: J. Springer, 1935.
63. Popper, Karl. *The Open Society and its Enemies*. London: G. Routledge & Sons, 1945.
64. Popper, Karl. "Three Views Concerning Human Knowledge," in H. D. Lewis (ed.), *Contemporary British Philosophy: Personal Statements*. New York: Macmillan, 1956.
65. Quine, W. V. "Two Dogmas of Empiricism," *Philosophical Review*, 60:20–43 (1951). Reprinted in Quine (66).
66. Quine, W. V. *From a Logical Point of View*. Cambridge, Mass.: Harvard Univ. Press, 1953.
67. Ramsey, Frank Plumpton. *The Foundations of Mathematics and other Logical Essays*. London and New York: Humanities, 1931.
68. Reichenbach, Hans. *Axiomatik der relativistischen Raum-Zeit-Lehre*. Braunschweig: F. Vieweg & Sohn, 1924.
69. Reichenbach, Hans. *Philosophie der Raum-Zeit-Lehre*. Berlin: W. de Gruyter & Co., 1928.
70. Reichenbach, Hans. *Philosophic Foundations of Quantum Mechanics*. Berkeley and Los Angeles: Univ. of California Press, 1944.
71. Reichenbach, Hans. *The Theory of Probability*. Berkeley: Univ. of California Press, 1949.

72. Reichenbach, Hans. *The Rise of Scientific Philosophy*. Berkeley and Los Angeles: Univ. of California Press, 1951.
73. Reichenbach, Hans. *Nomological Statements and Admissible Operations*. Amsterdam: North Holland Pub. Co., 1954.
74. Rozeboom, William W. "Mediation Variables in Scientific Theory," *Psychological Review*, 63:249–264 (1956).
75. Rubin, H., and P. Suppes. *Transformations of System of Relativistic Particle Mechanics*. Technical Report No. 2. Prepared under contract for Office of Naval Research. Stanford University, Stanford, 1953.
76. Russell, Bertrand. *Introduction to Mathematical Philosophy*. London and New York: Macmillan, 1919.
77. Russell, Bertrand. *Mysticism and Logic*. New York: W. W. Norton & Co., 1929.
78. Ryle, Gilbert. *The Concept of Mind*. London: Hutchinson's Univ. Libr., 1949.
79. Schlick, M. "Die Kausalitaet in der gegenwaertigen Physik," *Die Naturwissenschaften*, 19:145–162 (1931).
80. Skinner, B. F. *Science and Human Behavior*. New York: Macmillan, 1953.
81. Smart, J. J. C. "The Reality of Theoretical Entities," *Australasian Journal of Philosophy*, 34:1–12 (1956).
82. Spence, Kenneth W. "The Nature of Theory Construction in Contemporary Psychology," *Psychological Review*, 51:47–68 (1944). Reprinted in Marx (50).
83. Tarski, Alfred. "Einge methodologische Untersuchungen über die Definierbarkeit der Begriffe," *Erkenntnis* 5:80–100 (1935). English translation in Tarski (86).
84. Tarski, Alfred. *Introduction to Logic and to the Methodology of Deductive Sciences*. New York: Oxford University Press, 1941.
85. Tarski, Alfred. "The Semantic Conception of Truth," *Philosophy and Phenomenological Research*, 4:341–375 (1944). Reprinted in Feigl and Sellars (25) and in Linsky (45).
86. Tarski, Alfred. *Logic, Semantics, Metamathematics*. Tr. by J. H. Woodger. Oxford: The Clarendon Press, 1956.
87. Tolman, E. C. "Operational Behaviorism and Current Trends in Psychology," *Proceedings of the 25th Anniversary Celebration of the Inauguration of Graduate Study*, Los Angeles, 1936, pp. 89–103. Reprinted in Marx (50).
88. Tolman, E. C., B. F. Ritchie, and D. Kalish. "Studies in Spatial Learning. I. Orientation and the Short-Cut," *Journal of Experimental Psychology*, 36:13–24 (1946).
89. Toulmin, Stephen. *The Philosophy of Science*. London: Hutchinson's Univ. Libr., 1953.
90. Verplanck, W. S. "Burrhus F. Skinner," in Estes *et al.* (19), pp. 267–316.
91. Wald, A. *On the Principles of Statistical Inference*. Notre Dame: Univ. of Notre Dame Press, 1942.
92. Walker, A. G. "Foundations of Relativity: Parts I and II," *Proceedings of the Royal Society of Edinburgh*, 62:319–335 (1943–1949).
93. Walsh, W. H. *An Introduction to Philosophy of History*. London: Hutchinson's Univ. Libr., 1951.
94. Wang, Hao. "Notes on the Analytic-Synthetic Distinction," *Theoria*, 21:158–178 (1955).
95. White, Morton G. "The Analytic and the Synthetic: An Untenable Dualism," in S. Hook (ed.), *John Dewey: Philosopher of Science and of Freedom*. New York: Dial Press, 1950. Reprinted in Linsky (45).
96. White, Morton G. *Toward Reunion in Philosophy*. Cambridge, Mass.: Harvard Univ. Press, 1956.
97. Wiener, Philip P. (ed.). *Readings in Philosophy of Science*. New York: Scribner, 1953.

98. Woodger, J. H. *The Axiomatic Method in Biology*. Cambridge, England: Cambridge Univ. Press, 1937.
99. Woodger, J. H. *The Technique of Theory Construction*. Chicago: Univ. of Chicago Press, 1939.
100. Woodger, J. H. *Biology and Language*. Cambridge, England: Cambridge Univ. Press, 1952.

WHAT IS A SCIENTIFIC THEORY?

Patrick Suppes

Often when we ask what is a so-and-so, we expect a clear and definite answer. If, for example, someone asks me what is a rational number, I may give the simple and precise answer that a rational number is the ratio of two integers. There are other kinds of simple questions for which a precise answer can be given but for which in ordinary talk a rather vague answer is usually given and accepted. Someone reads about nectarines in a book, but has never seen a nectarine, or possibly has seen nectarines but is not familiar with their English name. He may ask me, "What is a nectarine?" and I would probably reply, "A smooth-skinned sort of peach." Certainly, this is not a very exact answer, but if my questioner knows what peaches are, it may come close to being satisfactory.

The kind of question I want to discuss fits neither one of these patterns. Scientific theories are not like rational numbers or nectarines. Certainly they are not like nectarines, for they are not simple physical objects. They are like rational numbers in not being physical objects, but they are totally unlike rational numbers in that scientific theories cannot be defined in any simple or direct way in terms of other non-physical, abstract objects.

Good examples of the kind of question we shall analyze in this chapter are provided by the familiar inquiries: "What is physics?," "What is psychology?," "What is science?" To none of these questions do we expect a simple and precise answer. On the other hand, there are many interesting things to be said about the sort of thing physics or psychology is. I shall be trying to show in this essay that this is also true of scientific theories.

THE STANDARD SKETCH

The standard sketch of scientific theories—and I emphasize the word "sketch"—runs something like the following. A scientific theory consists of two parts. One part is an abstract logical calculus. In addition to the vocabulary of logic, this calculus includes the primitive symbols of the theory, and the logical structure of the theory is fixed by stating the axioms or postulates of the theory in terms of its primitive symbols. For many theories the primitive symbols will be thought of as theoretical terms like "electron" or "particle" that are not possible to relate in any simple way to observable phenomena.

The second part of the theory is a set of rules that assign an empirical content to the logical calculus by providing what are usually called "co-ordinating definitions" or "empirical interpretations" for at least some of the primitive and defined symbols of the calculus. It is always emphasized that the first part alone is not sufficient to define a scientific theory, for without a systematic specification of the intended empirical interpretation of the theory, it is not possible in any sense to evaluate the theory as a part of science, although it can be studied simply as a piece of pure mathematics.

The most striking thing about this characterization is its highly schematic nature. Concerning the first part of a theory, the logical calculus, it is unheard of to find a substantive example of a theory actually worked out as a logical calculus in the writings of most philosophers of science. Much handwaving is indulged in to demonstrate that this working out of the logical calculus is simple in principle and only a matter of tedious detail, but concrete evidence is seldom given.

The sketch of the second part of a theory, that is, the co-ordinating definitions or empirical interpretations of some of the terms, is also highly schematic. A common defense of the relatively vague schema offered is that the variety of different empirical interpretations, for example, the many different methods of measuring mass—make a precise characterization difficult. Moreover, as we move from the precisely formulated theory on to the very loose and elliptical sort of experimental language used by almost all scientists, it is difficult to impose a definite pattern on the rules of empirical interpretation.

The view I want to support in this essay is not that this standard sketch is wrong, but rather that it is far too simple. Its very sketchiness makes it possible to omit both important properties of theories and significant distinctions that may be introduced between different theories.

MODELS VERSUS EMPIRICAL INTERPRETATIONS

To begin with, there has been a strong tendency on the part of many philosophers to speak of the first part of a theory as a logical calculus purely in syntactical terms. The co-ordinating definitions provided in the second part do not in the sense of modern logic provide an adequate semantics for the formal calculus. Quite apart from questions about direct empirical observations, it is pertinent and natural from a logical standpoint to talk about the models of the theory. These models are highly abstract, non-linguistic entities, often quite remote in their conception from empirical observations. It may well be asked what does the concept of a model have to add to the familiar discussions of empirical interpretation of theories.

I think it is true to say that most philosophers find it easier to talk about theories than about models of theories. The reasons for this are several, but perhaps the most important two are the following: In the first place, philosophers' examples of theories are usually quite simple in character, and therefore are easy to discuss in a straightforward linguistic manner. In the second place, the introduction of models of a theory inevitably introduces a stronger mathematical element into the discussion. It is a natural thing to talk about theories as linguistic entities—that is, to speak explicitly of the precisely defined set of sentences of the theory and the like—when the theories are given in what is called standard formalization.

Theories are ordinarily said to have a standard formalization when they are formulated within first-order logic. Roughly speaking, first-order logic is just the logic of sentential connectives and predicates holding for one type of object. Unfortunately, when a theory assumes more than first-order logic, it is neither natural nor simple to formalize it in this fashion. For example, if in axiomatizing geometry, we want to define lines as certain *sets* of points, we must work within a framework that already includes the ideas of set theory. To be sure, it is theoretically possible to axiomatize simultaneously geometry *and* the relevant portions of set theory, but this is awkward and unduly laborious.

Theories of more complicated structure like quantum mechanics, classical thermodynamics, or a modern quantitative version of learning theory, need to use not only general ideas of set theory but also many results concerning the real numbers. Formalization of such theories in first-order logic is utterly impractical. Theories of this sort are similar to the theories mainly studied in pure mathematics in their degree of complexity. In such contexts it is very much simpler to assert things about models of the theory rather than to talk directly and explicitly about the sentences of the theory, perhaps the main reason for this being that the notion of a sentence of the theory is not well defined when the theory is not given in standard formalization.

I would like to give just two examples in which the notion of model enters in a natural and explicit way in discussing scientific theories. The first example is concerned with the nature of measurement. The primary aim of a given theory of measurement is to show in a precise fashion how to pass from qualitative observations to the quantitative assertions needed for more elaborate theoretical stages of science. An analysis of how this passage from the qualitative to the quantitative may be performed is provided by axiomatizing appropriate algebras of experimentally realizable operations and relations. Given an axiomatized theory of measurement of some empirical quantity such as mass, distance, or force, the mathematical task is to prove a representation theorem for *models* of the theory which establishes, roughly speaking, that any empirical model is isomorphic to some numerical model of the theory. The existence of this isomorphism between models justifies the application of numbers to things.

We cannot literally take a number in our hands and apply it to a physical object. What we can do is to show that the structure of a set of phenomena under certain empirical operations is the same as the structure of some set of numbers under arithmetical operations and relations. The definition of isomorphism of models in the given context makes the intuitive idea of *same structure* precise. The great significance of finding such an isomorphism of models is that we may then use all our familiar knowledge of computational methods, as applied to the arithmetical model, to infer facts about the isomorphic empirical model. A linguistic formulation of this central notion of an empirical model of a theory of measurement being isomorphic to a numerical model is extremely awkward and tedious to formulate. But in model-theoretic terms the notion is simple and in fact represents a direct application of the very general notion of isomorphism used throughout all domains of pure mathematics.

The second example of the use of models concerns the discussion of reductionism in the philosophy of science. Many of the problems formulated in connection with the question of reducing one science to another may be formulated as a series of problems using the notion of a representation theorem for the models of a theory. For instance, the thesis that psychology may be reduced to physiology would be for many people appropriately established if one could show that for any model of a psychological theory it was possible to construct an isomorphic model within physiological theory.

The absence at the present time of any deep unitary theory within either psychology or physiology makes present attempts to settle such a question of reductionism rather hopeless. The classical example from physics is the reduction of thermodynamics to statistical mechanics. Although this reduction is usually not stated in absolutely satisfactory form from a logical standpoint, there is no doubt that it is substantially correct and represents one of the great triumphs of mathematical physics.

INTRINSIC VERSUS EXTRINSIC CHARACTERIZATION

Quite apart from the two applications just mentioned of the concept of a model of a theory, we may bring this concept to bear directly on the question of characterizing a scientific theory. The contrast I wish to draw is between intrinsic and extrinsic characterization. The formulation of a theory as a logical calculus or, to put it in terms that I prefer, as a theory with a standard formalization, gives an intrinsic characterization, but this is certainly not the only approach. For instance, a natural question to ask within the context of logic is if a certain theory *can* be axiomatized with standard formalization, that is, within first-order logic. In order to formulate such a question in a precise manner, it is necessary to have some extrinsic way of characterizing the theory. One of the simplest ways of providing such an extrinsic characterization is simply to define the intended class of models of the theory. To ask if we can axiomatize the theory is then just to ask if we can state a set of axioms such that the models of these axioms are precisely the models in the defined class.

As a very simple example of a theory formulated both extrinsically and intrinsically, consider the extrinsic formulation of the theory of simple orderings that are isomorphic to a set of real numbers under the familiar less-than relation. That is, consider the class of all binary relations isomorphic to some fragment of the less-than relation for the real numbers. The extrinsic characterization of a theory usually follows the sort given for these orderings: namely, we designate a particular model of the theory (in this case, the numerical less-than relation) and then characterize the entire class of models of the theory in relation to this distinguished model.

The problem of intrinsic characterization is now to formulate a set of axioms that will characterize this class of models without referring to the relation between models, but only to the intrinsic properties of any one model. With the present case the solution is relatively simple, although even it is not easily formulated within first-order logic. The intrinsic axioms are just those for a simple ordering plus the axiom that the ordering must contain in its domain a countable subset, dense with respect to the ordering in question.

A casual inspection of scientific theories suggests that the usual formulations are intrinsic rather than extrinsic in character, and therefore that the question of extrinsic formulations usually arises only in pure mathematics. This would also seem to be a happy result, for our philosophical intuition is surely that an intrinsic characterization is in general to be preferred to an extrinsic one.

However, the problem of intrinsic axiomatization of a scientific theory is more complicated and considerably more subtle than this remark would indicate. Fortunately, it is precisely by explicit consideration of the class of models of the theory that the problem can be put into proper perspective and formulated in a fashion that makes possible consideration of its exact solution. I shall give one simple example. The axioms for classical particle mechanics are ordinarily stated in such a way that a co-ordinate system, as a frame of reference, is tacitly assumed.

One effect of this is that relationships deducible from the axioms are not necessarily invariant with respect to Galilean transformations. We can view the tacit assumption of a frame of reference as an extrinsic aspect of the familiar characterizations of the theory. From the standpoint of the models of the theory, the difficulty in the standard axiomatizations of mechanics is that a large number of formally distinct models may be used to express the same mechanical facts. Each of these different models represents the tacit choice of a different frame of reference, but all models representing the same mechanical facts are related by Galilean transformations. It is thus fair to say that in this instance the difference between models related by Galilean transformations does not have any theoretical significance, and it may be regarded as a defect of the axioms that these trivially distinct models exist. It is important to realize that this point about models related by Galilean transformations is not the kind of point usually made under the heading of empirical interpretations of the theory.

It is a conceptual point that just as properly belongs to the theoretical side of physics. I have introduced this example here in order to provide a simple instance of how the explicit consideration of models can lead to a more subtle discussion of the nature of a scientific theory. It is certainly possible from a philosophical standpoint to maintain that particle mechanics as a scientific theory should be expressed only in terms of Galilean invariant relationships, and that the customary formulations are defective in this respect.

CO-ORDINATING DEFINITIONS AND HIERARCHY OF THEORIES

I turn now to the second part of theories mentioned above. It is true that in the foregoing discussion we have been using the word "theory" to refer only to the first part of theories—that is, to the axiomatization of the theory or the expression of the theory as a logical calculus—but as I emphasized at the beginning, the necessity of providing empirical interpretation of a theory is just as important as the development of the formal side of the theory. My central point on this aspect of theories is that the story is much more complicated than the familiar remarks about co-ordinating definitions and empirical interpretations of theories would indicate. The kind of co-ordinating definitions often described by philosophers have their place in popular philosophical expositions of theories, but in the actual practice of testing scientific theories a more elaborate and more sophisticated formal machinery for relating a theory to data is required.

The concrete experience that scientists label an experiment cannot itself be connected to a theory in any complete sense. That experience must be put through a conceptual grinder that in many cases is excessively coarse. Once the experience is passed through the grinder, often in the form of the quite fragmentary records of the complete experiment, the experimental data emerge in canonical form and constitute a model of the experiment. It is this model of the experiment rather than a model of the theory for which direct co-ordinating definitions are provided. It is also characteristic that the model of the experiment is of a relatively different logical type from that of the model of the theory. It is common for the models of a theory to contain continuous functions or infinite sequences, but for the model of the experiment to be highly discrete and finitistic in character.

The assessment of the relation between the model of the experiment and some designated model of the theory is a characteristic fundamental problem of modern statistical methodology. What is important about this methodology for present purposes is that, in the first place, it is itself formal and theoretical in nature; secondly, it has been a typical function of this methodology to develop an elaborate theory of experimentation that intercedes between any fundamental scientific theory and raw experimental experience.

It is not possible in the confines of this essay to spell out the rather elaborate hierarchy of theories that are customarily interposed between the fundamental scientific theory and the experiments presumed to support it. My only point here is to make explicit the existence of this hierarchy and to point out that there is no simple procedure for giving co-ordinating definitions for a theory. It is even a bowdlerization of the facts to say that co-ordinating definitions are given to establish the proper connections between models of the theory and models of the experiment in the sense of the canonical form of the data just mentioned. The elaborate methods, for example, for estimating theoretical parameters in the model of the theory from models of the experiment are not adequately covered by a reference to co-ordinating definitions.

If someone asks, "What is a scientific theory?" it seems to me there is no simple response to be given. Are we to include as part of the theory the well-worked-out statistical methodology for testing the theory? If we are to take seriously the standard claims that the co-ordinating definitions are part of the theory, then it would seem inevitable that we must also include in a more detailed description of theories a methodology for designing experiments, estimating parameters and testing goodness-of-fit of the models of the theory. It does not seem to me important to give precise definitions of the form: *X* is a scientific theory if, and only if, so-and-so. What is important is to recognize that the existence of a hierarchy of theories arising from the methodology of experimentation for testing the fundamental theory is an essential ingredient of any sophisticated scientific discipline.

INSTRUMENTAL VIEW OF THEORIES

There is one view of scientific theories which is undoubtedly of considerable importance and which I have not yet mentioned. This is the view that theories are to be looked at from an instrumental viewpoint. The most important function of a theory, according to this view, is not to organize or assert statements that are true or false but to furnish material principles of inference that may be used in inferring one set of facts from another. Thus, in the familiar syllogism "all men are mortal; Socrates is a man; therefore, Socrates is mortal," the major premise "all men are mortal," according to this instrumental viewpoint, is converted into a principle of inference. And the syllogism now has only the minor premise "Socrates is a man."

From a logical standpoint it is clear that this is a fairly trivial move, and the question naturally arises if there is anything more substantial to be said about the instrumental viewpoint. Probably the most interesting argument for claiming that there is more than a verbal difference between these two ways of looking at theories or laws is the argument that when theories are regarded as principles of inference rather than as major premises, we are no longer concerned directly to establish their truth or falsity but to evaluate their *usefulness* in inferring new statements of fact. It is characteristic of discussions in this vein by philosophers that no genuinely original formal notions have arisen out of these discussions to displace the classical semantical notions of truth and validity. To talk, for instance, about laws having different jobs than statements of fact is trivial unless some systematic semantical notions are introduced to replace the standard analysis.

From another direction there has been one concerted serious effort to provide a formal framework for the evaluation of theories which replaces the classical concept of truth. What I have in mind is modern statistical decision theory. It is typical of statistical decision theory to talk about actions rather than statements. Once the focus is shifted from statements to actions, it seems quite natural to replace the concept of truth by that of expected loss or risk. It is appropriate to ask if a statement is true, but it does not make much sense to ask if it is risky. On the other hand, it is reasonable to ask how risky an action is, but not to ask if it is true. It is apparent that statistical decision theory, when taken literally, projects a more radical instrumental view of theories than does the view already sketched.

Theories are not regarded even as principles of inference but as methods of organizing evidence to decide which one of several actions to take. When theories are regarded as principles of inference, it is a straightforward matter to return to the classical view and to connect a theory as a principle of inference with the concept of a theory as a true major premise in an argument. The connection between the classical view and the view of theories as instruments leading to the taking of an action is certainly more remote and indirect.

Although many examples of applications of the ideas of statistical decision theory have been worked out in recent literature on the foundations of statistics, these examples in no case deal with complicated scientific theories, and I have seen no serious discussion of the treatment of scientific theories from the standpoint of statistical decision theory. Again, it is fair to say that when we want to talk about the evaluation of a sophisticated scientific theory, disciplines like statistical decision theory have not yet offered any genuine alternative to the semantical notions of truth and validity. In fact, even a casual inspection of the literature of statistical decision theory shows that in spite of the instrumental orientation of the fundamental ideas, formal development of the theory is wholly dependent on the standard semantical notions and in no sense replaces them.

What I mean by this is that in concentrating on the taking of an action as the terminal state of an inquiry the decision theorists have found it necessary to use standard semantical notions in describing evidence, their own theory, and so forth. For instance,

I cannot recall a single discussion by decision theorists in which particular observation statements are treated in terms of utility rather than in terms of their truth or falsity.

It seems apparent that statistical decision theory does not at the present time offer a genuinely coherent or deeply original new view of scientific theories. Perhaps future developments of decision theory will proceed in this direction. Be that as it may, there is one still more radical instrumental view that I would like to discuss as the final point to be covered in this essay. As I have already noted, it is characteristic of many instrumental analyses to distinguish the status of theories from the status of particular assertions of fact. It is the point of a more radical instrumental, behavioristic view of the use of language to challenge this distinction and to look at the entire use of language, including the statement of theories as well as of particular matters of fact, from a behavioristic viewpoint.

According to this view of the matter, all uses of language are to be analyzed with strong emphasis on the language users. It is claimed that the semantical analysis of modern logic gives a very inadequate account even of the cognitive uses of language, because it does not explicitly consider the production and reception of linguistic stimuli by speakers, writers, listeners, and readers. It is plain that for the behaviorist an ultimately meaningful answer to the question "What is a scientific theory?" cannot be given in terms of the kinds of concepts considered earlier. An adequate and complete answer can be given only in terms of an explicit and detailed consideration of both the producers and consumers of the theory. There is much that is attractive in this behaviorist way of looking at theories or language in general. What it lacks at present, however, is sufficient scientific depth and definiteness to serve as a genuine alternative to the precise notions of modern logic and semantics. Moreover, much of the language of models and theories discussed earlier in this chapter is surely so approximately correct that any behaviorist revision of our way of looking at theories must yield the ordinary talk about models and theories as a first approximation. It is a matter for the future to see whether or not the behaviorist's approach will deepen our understanding of the nature of scientific theories.

In current perspective, the methods and concepts of modern logic provide a satisfactory and powerful set of tools for analyzing the detailed structure of scientific theories. What would seem to be needed for the present is deeper and more detailed application of these tools to the job of analysis. I have tried to indicate what I think are some of the more fruitful directions for future investigation.

The Semantic Approach to Scientific Theories

BAS C. VAN FRAASSEN
Princeton University

The purpose of this paper is not to be new or original, but to provide a concise exposition of a certain approach in philosophy of science, at the hands of a loosely associated group of contemporary philosophers.* I do not want anyone to suffer from guilt by asociation, so I should emphasize that any two authors mentioned here may agree on only a few aspects of the discussed approach, and disagree on others.

I locate the real beginning of the semantic approach, the place where it became consciously distinct, in E. W. Beth's *Natuurphilosophie* (1948), and summarized rather cryptically in two short articles in English and French in 1948/49. I read this book in 1965 and I gave an exposition of his views in my 'On the Extension of Beth's Semantics of Physical Theories' (1970). As I will explain below, some of the main ideas I found there were already in some way 'in the air' in North America. Since his view could be sloganized by saying that *the analysis of quantum logic provides the paradigm for the semantic analysis of physical theory*, it is clear that his view also had antecedents in the thirties and forties, flourishing perhaps entirely unconsciously of the implied opposition to the approach of the logical positivists. A good example is Herman Weyl's article 'The Ghost of Modality' (1940).

My own particular variant of the semantic approach is strongly colored by my attempts to become a more thoroughgoing empiricist, while such other advocates as Frederick Suppe and Ronald Giere are outspoken scientific realists. My exposition could not easily remain unbiased on this issue, but I hope that the essential neutrality thereon, of the semantic view, will be clear nevertheless.

1. The aim of science

Philosophy of science attempts to answer the question *What is science?* in just the sense in which philosophy of art, philosophy of religion, and the like answer the

* The main content of this paper is accordingly a combination of several previous ones; mainly my 'Aim and Structure of Scientific Theories' and 'Theory Construction and Experiment'; see bibliography. A more complete defense of my own particular variant of the approach will be found in *Images of Science*, edited by P. M. Churchland and C. A. Hooker, Chicago: University of Chicago Press, 1985.

Nancy J. Nersessian (editor), The Process of Science. ISBN 90-247-3425-8.

similar question about their subject. For better or for worse our tradition has focussed on the scientific theory rather than on scientific activity itself (on the product, rather than on the aim, conditions, and process of production, to draw an analogy, which is already one that points in its terminology to the product as most salient feature). Yet all aspects of scientific activity must be illumined if the whole is to become intelligible. I shall therefore devote a preliminary section to the aim of science, and to the proper form of epistemic or doxastic attitudes toward scientific theories, before entering upon a description of their structure.

The activity of constructing, testing, and refining of scientific theories – that is, the production of theories to be accepted within the scientific community and offered to the public – what is the aim of this activity?

I do not refer here either to the motives of individual scientists for participating, or the motives of the body civic for granting funds and otherwise supporting the activity. Nor do I ask for some theoretically postulated 'fundamental project' which would explain this activity. It is part of the straightforward description of any activity, communal or individual, large-scale or small, to describe the end that is pursued as one of its defining conditions. In the most general terms, the end pursued is success, and the question is what counts as success, what are the criteria of success in this particular case?

We cannot answer our particular questions here without some reflection on what sort of thing this product, the scientific theory, is. A scientific theory must be the sort of thing that we can accept or reject, and believe or disbelieve: accepting a theory implies the opinion that it is successful: science aims to give us acceptable theories. To put it more generally, a theory is an object for epistemic or at least doxastic attitudes. A typical object for such attitudes is a proposition, or a set of propositions, or more generally a body of putative information about what the world is like, what the facts are. If anyone wishes to be an instrumentalist, he has to deny the appearances which I have just described. An instrumentalist would have to say that the apparent expression of a doxastic attitude toward a theory is elliptical; 'to believe theory T' he would have to construe as 'to believe that theory T has certain qualities'. I shall not follow that path. Let me state here at once, as a first assumption, that the theory itself is what is believed or disbelieved.

At this point we readily can see that there is a very simple possible answer to all our questions, the answer we call *scientific realism*. This philosophy says that a theory is the sort of thing which is either true or false; and that the criterion of success is truth. As corollaries we have that acceptance of a theory as successful is, or involves the belief that it is true; and that the aim of science is to give us (literally) true theories about what the world is like.

That answer must of course be qualified in various ways to allow for our epistemic finitude and the consequent tentativeness of reasonable doxastic attitudes. Thus we add that although it cannot generally be *known* whether or not the criterion of success has been met, we may reasonably have a high degree of belief that it has been, or that it is met approximately (i.e., met exactly by one member of a set of 'small variants' of the theory), and this imparts similar qualifications to acceptance in practice. And we add furthermore of course that empiricism precludes dogmatism, that is, *whatever* doxastic attitude we adopt, we stand ready to revise in the face of further evidence. These are all qualifications of a sort that anyone must acknowledge, and should therefore really go without saying. They do not detract from the appealing and as it were pristine clarity of the scientific realist position.

I did not want to discuss the structure of theories before bringing this position into the open, and confronting it with alternatives. For it is very important, to my mind, to see that an analysis of theories – even one that is quite traditional with respect to what theories are – does not presuppose it. Let us keep assuming with the scientific realist, that theories are the sort of thing which can be true or false, that they say what the world is like. What they say may be true or false, but it is nevertheless literally meaningful information, in the neutral sense in which the truth value is 'bracketed'.

There are a number of reasons why I advocate an alternative to scientific realism. One point is that reasons for acceptance include many which, *ceteris paribus*, detract from the likelihood of truth. In constructing and evaluating theories, we follow our desires for information as well as our desire for truth. For belief, however, all but the desire for truth must be 'ulterior motives'. Since therefore there are reasons for acceptance which are not reasons for belief, I conclude that acceptance is not belief. It is to me an elementary logical point that a more informative theory can not be more likely to be true – and attempts to describe inductive or evidential support through features that require information (such as 'inference to the best explanation') must either contradict themselves or equivocate.

It is still a long way from this point to a concrete alternative to scientific realism. Once we have driven the wedge between acceptance and belief, however, we can reconsider possible ways to make sense of science. Let me just end these preliminary remarks now by stating my own position, which I call *constructive empiricism*. It says that the aim of science is not truth as such but only *empirical adequacy*, that is, truth with respect to the observable phenomena. Acceptance of a theory involves as belief only that the theory is empirically adequate – but it involves more than belief.

While truth as such is therefore, according to me, irrelevant to success for theories, it is still a category that applies to scientific theories. Indeed, the *content* of a theory is what it says the world is like; and this is either true or false. The applicability of this notion of truth value remains here, as everywhere, the basis of all logical analysis. When we come to a specific theory, there is an immediate philosophical question, which concerns the content alone: *how could the world possibly be the way this theory says it is?*

This is for me the foundational question *par excellence*. And it is a question whose discussion presupposes no adherence to scientific realism, nor a choice between its alternatives. This is the area in philosophy of science where realists and anti-realists can meet and speak with perfect neutrality.

2. Theory structure – models and their logical space

In this section I shall present a view of theories which makes language largely irrelevant to the subject. Of course, to present a theory, we must present it in and by language. That is a trivial point, for any effective communication proceeds by language, except in those rare cases in which information can be conveyed by the immediate display of an object or happening. In addition, both because of our own history – the history of philosophy of science which became intensely language oriented during the first half of this century – and because of its intrinsic importance, we cannot ignore the language of science. Hence I shall make it the subject of the last section.

In what is now called the received view, a theory was conceived of as an axiomatic theory. That means, as a set of sentences, defined as the class of logical consequences of a smaller set, the axioms of that theory. A distinction was drawn: since the class of axioms was normally taken to be effectively presentable, and hence syntactically describable, the theory could be thought of as in itself uninterpreted. The distinction is then that scientific theories have an associated interpretation, which links their terms with their intended domain. We all know the story of misery and pitfalls that followed. Only two varieties of this view of scientific theories as a special sort of interpreted theories, emerged as anywhere near tenable. The first variety insists on the formal character of the theory as such, and links it to the world by a partial interpretation. Of this variety the most appealing to me is still Reichenbach's, which said that the theoretical relations have *physical correlates*. Their partial characters stand out when we look at the paradigm example: light rays provide the physical correlate for straight lines. It will be immediately clear that not every line is the path of an actual light ray, so the language-world link is partial. The second variety, which came to maturity in Hempel's later writings, hinges for its success on treating the axioms as already stated in natural language. The interpretative principles have evolved into axioms among axioms. This means that the class of axioms may be divided into those which are purely theoretical, in which all non-logical terms are ones specially introduced to write the theory, and those which are mixed, in which non-theoretical terms also appear. It will be readily appreciated that in both these developments, despite lip service to the contrary, the so-called problem of interpretation was left behind. We do not have the option of interpreting theoretical terms – we only have the choice of regarding them as either (a) terms we do not fully understand but know how to use in our reasoning, without detriment to the success of science, or (b) terms which are now part of natural language, and no less well understood than its other parts. The choice, the correct view about the meaning and understanding of newly introduced terms, makes no practical difference to philosophy of science, as far as one can tell. It is a good problem to pose to philosophers of language, and to leave them to it.

In any tragedy,[1] we suspect that some crucial mistake was made at the very beginning. The mistake, I think, was to confuse a theory with the formulation of a theory in a particular language. The first to turn the tide was Patrick Suppes, with his well-known slogan: the correct tool for philosophy of science is mathematics, *not* metamathematics. This happened in the fifties – bewitched by the wonders of logic and the theory of meaning, few wanted to listen. Suppes' idea was simple: *to present a theory, we define the class of its models directly,* without paying any attention to questions of axiomatizability, in any special language, however relevant or simple or logically interesting that might be. And if the theory as such, is to be identified with anything at all – if theories are to be reified – then a theory should be identified with its class of models.[2]

This procedure is in any case common in modern mathematics, where Suppes had found his inspiration. In a modern presentation of geometry we find not the axioms of Euclidian geometry, but the definition of a Euclidean space. Similarly Suppes and his collaborators sought to reformulate the foundations of Newtonian mechanics, by replacing Newton's axioms with the definition of a Newtonian mechanical system. This gives us, by example, a *format* for scientific theories. In Ronald Giere's recent encapsulation, a theory consists of (a) the *theoretical definition,* which defines a certain class of systems; (b) a *theoretical hypothesis,* which asserts that certain (sorts of) real systems belong to that class.

This is a step forward in the direction of less shallow analysis of the structure of a scientific theory. The first level of analysis addresses the notion of theory *überhaupt,* but we do not want to stop there. We can go still a bit further by making a division between relativistic and non-relativistic theories. In the latter, the systems are physical entities developing in time. They have accordingly a space of possible states, which they take on and change during this development. This introduces the idea of a cluster of models united by a common *state-space;* each has in addition a domain of objects plus a 'history function' which assigns to each object a history, i.e., a trajectory in that space. A real theory will have many such clusters of models, each with its state-space. So the presentation of the theory must proceed by describing a class of *state-space types.*

In the case of relativistic theories, early formulations can be described roughly as relativistically invariant descriptions of objects developing in time – say in their proper time, or in the universal time of a special cosmological model (e.g., Robertson-Noonan models). A more general approach, developed by Glymour and Michael Friedman, takes space-times themselves as the systems. Presentation of a space-time theory T may then proceed as follows: a *(T-)space-time* is a four dimensional differentiable manifold M, with certain geometrical objects (defined on M) required to satisfy the *field equations* (of T), and a special class of curves (the possible trajectories of a certain class of physical particles) singled out by the *equations of motion* (of T).

Clearly we can further differentiate both sorts of theories in other general ways, for example with respect to the stochastic or deterministic character of imposed laws. (It must be noted however that except in such special cases as the flat space-time of special relativity – its curvature independent of the matter-energy distribution – there are serious conceptual obstacles to the introduction of indeterminism into the space-time picture.)

I must leave aside for now the details of foundational research in the sciences. But I want to point out that the point of view which I have been outlining – the *semantic view* as opposed to the received view – is much closer to practice there. The scientific literature on a theory makes it relatively easy to identify and isolate classes of structures to be included in the class of theoretical models. It is on the contrary usually quite hard to find laws which could be used as axioms for the theory as a whole. Apparent laws which frequently appear are often partial descriptions of special subclasses of models. their generalization being left vague and often shading off into logical vacuity. Let me give two examples. The first is from quantum mechanics: *Schrödinger's equation.* This is perhaps its best known and most pervasively employed law – but it cannot very well be an axiom of the theory since it holds only for conservative systems. If we look into the general case. we find that we can prove the equation to hold. for some constant Hamiltonian. under certain conditions – but this is a metamathematical fact, hence empirically vacuous. The second is the Hardy-Weinberg law in population genetics. Again. it appears in any foundational discussion of the subject. But it could hardly be an axiom of the theory. since it holds only under certain special conditions. If we look into the general case. we find a logical fact: that certain assumptions imply that it describes an equilibrium which can be reached in a single generation. and maintained. The assumptions are very special. and more complex variants of the law can be deduced for more realistic assumptions – in an open and indefinite sequence of sophistications.

What we have found. in this approach. is a way to describe relevant structures in ways that are also directly relevant, and seen to be relevant. to our subject matter. The scholastically logistical distinctions that the logical positivist tradition produced – observational and theoretical vocabulary. Craig reductions, Ramsey sentences. first-order axiomatizable theories, and also projectible predicates. reduction sentences. disposition terms. and all the unholy rest of it – had moved us *mille milles de toute habitation scientifique*, isolated in our own abstract dreams. Since Suppes' call to return to a non-linguistic orientation. now about thirty years ago. we have slowly regained contact.

3. Theory structure – relation to the world

Above I mentioned Giere's elegant capsule formulation of the semantic view: a theory is presented by giving the definition of a certain kind (or kinds) of systems plus one or more hypotheses to the effect that certain real (kinds of) systems belong to the defined class(es). We speak then of the *theoretical definition* and the *theoreti-*

cal hypothesis which together constitute the given formulation (in. so to say, canonical form) of the theory. A 'little' theory might for example define the class of Newtonian mechanical systems and assert that our solar system belongs to this class.

Truth and falsity offer no *special* perplexities in this context. The theory is true if those real systems in the world really do belong to the indicated defined classes. From a logical. or more generally semantic point of view. we may consider as implicitly given models of the world as a whole. which are as the theoretical hypotheses say it is. There is of course a very large class of models of the world as a whole. in which our solar system is a Newtonian mechanical system. In one such model. nothing except this solar system exists at all: in another the fixed stars also exist. and in a third. the solar system exists and dolphins are its only rational inhabitants. Now the world must be one way or another; so the theory is true if the real world itself is (or is isomorphic to) one of these models. This is equivalent to either of two familiar sorts of formulations of the same point: the theory is true exactly if (a) one of the possible worlds allowed by the theory is the real world: or (b) all real things are the way the theory says they are.

But while the subject of truth yields no special conceptual difficulties in this context. I do not believe that it marks the relation to the world. which science pursues in its theories. This. as you will recall. is the point at issue between scientific realism and empiricism. Leaving the issue itself aside. I think that even scientific realists need to be acutely interested in a much closer. more empirical relation of theory to world. I call this relationship *empirical adequacy*.

The logical positivist tradition gave us a formulation of such a concept which was not only woefully inadequate but created a whole cluster of 'artifact problems' (by this I mean. problems which are artifacts of the philosophical approach. and not inherent in its subject). In rough terms. the empirical content of a theory was identified with a set of sentences. the consequences of that theory in a certain 'observational' vocabulary. In my own studies. I first came across formulations of more adequate concepts in the work of certain Polish writers (Przeleweski 1969, Wojcicki 1974). of Dalla Chiara and Toraldo di Francia (1973 and 1977) and finally of course in Patrick Suppes' own writings on what he calls empirical algebras and data models (1967. 1969). While some of these formulations were still more language-oriented than I liked. the similarity in their approach was clear: certain parts of the models were to be identified as *empirical substructures*. and these were the candidates for representation of the observable phenomena which science can confront within our experience.

At this point I perceived that the relationship thus explicated corresponds exactly to the one Reichenbach attempted to identify through this concept of coordinative definitions. once we abstract from the linguistic element. Thus in a space-time the geodesics are the candidates for the paths of light rays and particles in free fall. More generally. the identified spatio-temporal relations provide candidates for the relational structures constituted by actual genidentity and signal connections. These actual physical structures are to be embeddable in certain substructures of space-time, which allows however for many different possibilities, of which the actual is, so to say, some arbitrary fragment.

Thus we see that the empirical structures in the world are the parts which are at once *actual* and *observable;* and empirical adequacy consists in the embeddability of all these parts in some single model of the world allowed by the theory.

Patrick Suppes has very carefully investigated the construction of data models, and the empirical constraint they place on theoretical models. Thought of as concerned with exactly this topic, much apparently 'a prioristic' theorizing on the foundations of physics takes on a new intelligibility. A reflection on the possible forms of structures definable from joint experimental outcomes yields constraints on the general form of the models of the theories 'from below' which can then be narrowed down by the imposition of postulated general laws, symmetry constraints, and the like, 'from above'.

4. Theorizing: data models and theoretical models

New theories are constructed under the pressure of new phenomena, whether actually encountered or imagined. By 'new' I mean here that there is no room for these phenomena in the models provided by the accepted theory. There is no room for a mutable quantity with a discrete set of possible values in the models of a theory which says that all change is continuous. In such a case the old theory does not allow for the phenomenon's description, let alone its prediction.

I take it also that the response to such pressure has two stages, logically if not chronologically distinguishable. First the existing theoretical framework is widened so as to allow the possibility of those newly envisaged phenomena. And then it is narrowed again, to exclude a large class of the thereby admitted possibilities. The first move is meant to ensure empirical adequacy, to provide room for all actual phenomena, the rock-bottom necessary condition of success. The second move is meant to regain empirical import, informativeness, predictive power.

It need hardly be added that the moves are not made under logical compulsion. When a new phenomenon, say X, is described it is no doubt possible to react with the assertion that if it looked as if X occurred, one would only conclude that familiar fact or event Y had occurred. A discrete quantity can be approximated by a continuous one, and an underlying continuous change can be postulated. From a purely logical point of view, it will always be up to the scientists to take a newly described phenomenon seriously or to dismiss it. Logic knows no bounds to *ad hoc* postulation. This also brings out the fact of creativity in the process that brings us the phenomena to be saved. Ian Hacking put this to me in graphic terms when he described the quark hunters as seeking to create new phenomena. It also makes the point long emphasized by Patrick Suppes that theory is not confronted with raw data but with models of the data, and that the construction of these data models is a sophisticated and creative process. To these models of data, the dress in which the debutante phenomena make their debut, I shall return shortly.

In any case, the process of new theory construction starts when described (actual or imagined) phenomena are taken seriously as described. At that point there certainly is logical compulsion, dimly felt and, usually much later, demonstrated. Today Bell's Inequality argument makes the point that certain quantum mechanical phenomena cannot be accommodated by theories which begin with certain classical assumptions. This vindicates, a half century after the fact, the physicists' intuition that a radical departure was needed in physical theory.

Of the two aspects of theorizing, the widening of the theoretical framework, and it's narrowing to restore predictive power, I wish here to discuss the former only. There we see first of all a procedure so general and common that we recognize it readily as a primary problem-solving method in the mathematical and social as well as the natural sciences, any place where theories are constructed, including such diverse areas familiar to philosophers as logic and semantics. This method may be described in two ways: as *introducing hidden structure*, or 'dually' as *embedding*. Here is one example.

Cartesian mechanics hoped to restrict its basic quantities to ones definable from the notions of space and time alone, the so-called kinematic quantities. Success of the mechanics required that later values of the basic quantities depend functionally on the earlier values. There exists no such function. Functionality in the picture of nature was regained by Newton, who introduced the additional quantities of mass and force. Behold the introduction of hidden parameters.

The world 'hidden' in 'hidden parameters' does not refer to lack of experimental access. It signifies that we see parameters in the solution which do not appear in the statement of the problem.

We can 'dually' describe the solution as follows: the kinematic relational structures are embedded in structures which are much larger – larger in the sense that there are additional parameters (whether relations or quantities or entities). *The phenomena are small but chaotic; they are treated as fragments of a 'whole' that is much larger but orderly and simple.* This point could, I believe, be illustrated by examples from every stage of the history of science. When a point has such generality one assumes that it must be banal, and carry little insight. In such a general inquiry as ours, however, perspective is all; and we need general clues to find a general perspective. (That this particular point may be productive of more specific insights is in any case not an unreasonable hope. The most spectacular recent theoretical development may well be the deduction of Maxwell electrodynamics, Einstein geometrodynamics, and the Yang-Mills quark-binding field dynamics from the requirement of embeddability in space-time, by Hojman, Kuckar and Teitelboim.)

5. Theorizing illustrated

In order to illustrate the general view of what theorizing is like, which I have just presented, and to try and persuade you that it is a reasonable one, I shall now describe some recent activities in the foundations of quantum mechanics.

The whole point of having theoretical models is that they should fit the phenomena, that is, fit the models of data. So we need to look at what the latter must be like in general. Hence the development of Randall and Foulis', 'empirical logic'; Mackey, Jauch, and Piron's preliminary discussions of experimental questions; Ludwig and Mielnik's filters; and so forth. These authors write sometimes as if their program is one of transcendental deduction: study what the data models must be like, deduce what structure theoretical models must have if the data models are to be deducible, demonstrate the basic axioms of quantum mechanics as corollaries to this deduction. Since the theory has clear empirical content, success can be, at most, partial; but it is astonishing how much can be achieved in this way. Moreover the very fact that success must necessarily be partial is what gives the approach its value for the future: it brings within our ken alternatives to the extant theory that can rival it.

In Suppes' description, the experimentalist brings to the theoretician a small relational structure, constructed carefully from selected data. The examples Suppes mentions are specific and the little structures are algebras; hence he calls them 'empirical algebras'. The authors in quantum mechanics point to such small structures that represent data, and they are not always algebras; they are most generally partial algebras, or just partially ordered sets with some operations. Let us see how this happens.

In the typical sort of experiment discussed in connection with the Einstein-Podolski-Rosen paradox and Bell's Inequalities, we have two apparatus, L (for *'left')* and R. Each has (say) three settings or orientations; let, for instance, L1 be the proposition that L has been given the first setting. The experiments have each (say) two distinct possible outcomes, which we may represent by the numbers zero and one. Let, for instance, L30 be the proposition that L has the third setting and outcome zero.

When we carry out a particular run on this dual apparatus, we can give a score of T (for 'true') or F to some of these propositions. For example, the first time we do it, each apparatus was placed in the first setting; L had outcome 1 and R had outcome 0. An experimental report looks, in part, as follows:

Proposition	Score
L1	T
L2	F
R1	T
L10	F
L20	no score
R10	T

We note that L20 received no score. It could have been give F, simply on the basis that L had not been given the second setting. But this is useless information. and does not appear in the experimental report.

This single report is not likely to come to the theoretician's desk. What reaches him rather are reports of the form:

A) With initial preparation X, the probability of outcome Lia. given setting Li. equals r.

B) For all initial preparations. the probability of (Lia & Ria). given settings Li and Ri, equals zero.

There was an extrapolation before these conclusions were reached: the extrapolation from found relative frequencies to probabilities. This is regarded as no different from the extrapolation of data points on a graph, to a smoothed curve.

But the report that comes in forms (A) and (B) leads us to a mathematical structure that may properly be called the *data model.* The important relation stated between Lia and Ria in (B) is that when they can receive an informative score at all (i.e., when the preconditions Li and Ri obtain), they cannot both receive the score T. We then call Lia and Ria *orthogonal.* It also means that Ri1 must receive score T when Li0 does (modulo probability zero). and again when the informative scoring conditions obtain; and we call that *implication.* The latter is a partial ordering. and so we have here a partially ordered set *(poset)* with an orthogonality relation.

Reflection on this form of representation leads to assertions of the form: all data models can take the form . . . A popular way to fill in the dots is to say 'ortho-poset' (i.e., poset with orthocomplement). A. R. Marlow. whose work I am about to take as a special example, used the more general characterization *dual poset;* that is, a partially ordered set with zero element and equipped with a single operation. *duality* ($x \neq x'$; $x = x''$; *and x implies* y only if y' *implies* x').

In the world of mathematical entities there are many dual posets. Widening our theoretical framework will consist in the provision of models that can have very strange dual posets of experimental propositions embedded. But the embedding must be good, that is, we must be able to see in the theoretical model all the significant features. I mentioned parts of the experimental report labelled (A). They start 'With initial preparation X. . . .' and then they mention probabilities. These probabilities characterize what is called the *state* prepared by procedure X. And these states (they look like fragments of ordinary probability functions, in that they assign probabilities only to propositions for which the informative scoring conditions obtain) must be 'visible' in a certain sense in the computational structure of the theoretical model.

Here is Marlow's theorem. It requires two preliminary definitions. A *probability function* on a dual poset is any function f with the properties $f(0) = 0$, $f(x') = 1 - f(x)$, and $f(x) \leq f(y)$ if x implies y. A *base* for the dual poset is any set B of elements which does not contain the zero element, nor does it contain two orthogonal elements (here defined by the relation (x *implies* y')), but does contain either x or x' for each x. (Remark: every set with the first two properties can be extended to one that has all three. Note also that any set of elements that have all received score T on a particular occasion is intuitively required to have the first two properties.) The theorem says now that if we have a dual poset and a base, we can embed the poset in the algebra of projection operators on a Hilbert space, in such a way that duality becomes ortho-complementation, the partial ordering is preserved for elements within the base, and each probability function on the poset can be associated with a vector and becomes calculable by means of the familiar trace computation used in quantum mechanics.

This is an extraordinarily general result. Marlow takes the result as justification for his project to write space-time theories in Hilbert space formalism. The result provides good reason, after all, to write all physical theory in that mathematical framework. Of course he realizes that in some ways the theorem is less than totally general (the implication order is preserved only within the base!) and in some ways less than informative (there are enormous Hilbert spaces with room to embed almost anything) but the postulates he adds, and intends to add, will narrow down the *embarras de richesse* to recover empirical content.

Let us look, however, as second main illustration, at a line of thought born from dissatisfaction with the way phenomenal structures are embeddable in the Hilbert space framework. I refer to *operational quantum mechanics*, associated with Ludwig, Mielnik, Davies, and Edwards.

To explain how data models can take the form of ortho-posets, or more generally dual posets, I already gave a brief sketch of a quite typical experimental set-up, of an intermediate degree of complexity. Let me now start an alternative sketch; the two will not be incompatible. In a typical simple test, a system is *prepared* in a certain way; an *operation* is performed on it; and a *question* is asked. In the simplest case the question has *yes* or *no* as possible answers. We may keep count, as we repeat the text, of how often we receive *yes* as answer. We visualize the situation by imagining a *source* which sends out a beam of particles, that encounter a *barrier*, and a *counter* on the other side of the barrier that clicks every time a particle reaches it. This is a good picture for it has all the general features indicated; we appear to lose no generality if we focus attention on it.

The barrier affects the intensity of the beam; for example, if the barrier were not there, the counter would be clicking twice as rapidly. The sort of barrier determines the sort of question being asked. When the situation, so conceived, is embedded in the mathematical apparatus of Hilbert space, each source has an associated statistical operator W (representing the *prepared state),* and each barrier an associated projection operator P (representing the *question asked);* the probability of a single particle passing the barrier (the factor by which the beam intensity is diminished) is calculated by the Born rule as p = Tr(PW).

Many situations of the general sort described can indeed be modelled in this way. But others have to be treated in more round-about fashion. Intuitively we say that there is an *observable* in such a set-up. Whether that is so in the sense now instilled in us by quantum theory, has a simple criterion: there must be a state such that the *yes* answer becomes certain, that is, receives probability one. That state is then called an eigenstate of the observable being measured.

It is certainly possible to find examples where it looks intuitively as if we are measuring something, but the criterion is not met. Suppose we place an atom of a radioactive substance near a Geiger counter, and ask *Will the counter click within four to five minutes from now?* Is that not a simple form of measurement, of something we could call, say, the decay time of the atom? Yet we cannot prepare the atom in a state so as to make the *yes* answer certain.

A clearer example, first introduced into the literature, I think, by Shimony, and discussed especially by the Dutch authors de Muynk, Cooke, and Hilgevoort, is pictured in Figure 1. We have a battery of three Stern-Gerlach apparatus, testing for spin. They are so arranged that particles exiting from the first, along the top channel, encounter the second, while those exiting in the first top channel will exit along the second top channel, and so forth. We may choose the orientations so that those transition probabilities equal 1/2. As is easily seen in the diagram, there is no initial state which makes certain an exit along the second top channel: for the probability of

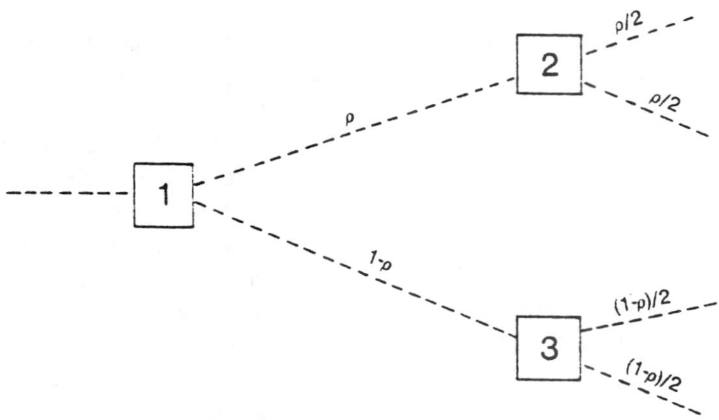

Fig. 1.

that exit equal p/2, which has maximum value 1/2. Here we have a question for which the *yes* answer *cannot* be certain. We can at most represent it by means of a sequence of questions, of the directly representable sort. Perhaps we could say that a 'derived observable' has been measured.

Operational quantum mechanics may be thought of as enlarging the theoretical models so as to allow the embedding of all such empirical structures in the same way. The questions asked by our battery of apparatus are treated on a par with those asked by a single apparatus. This enlargement proceeds as follows. We represent the (ensemble prepared by the) source by means of two parameters: the statistical operator W and a number t which represents the production intensity. The barrier then affects both state W and intensity t, changing W to some other state W', and reducing the intensity t to rt (with $0 \leq r \leq 1$). We may set the initial intensity equal to 1, so that t itself must belong to the interval (0, 1). Referring to the statistical operators W – positive definite Hermitean operators with unit trace – as the 'old states', we represent the new states by their multiples rW. These are not in general themselves statistical operators, for if W has unit trace then rW has trace r.

Mathematically the enlargement proceeds in several steps. We begin with the 'ball' S of old states: generate the real cone

$$B+ = \{rW : W \varepsilon S \text{ and } 0 \leq r \leq 1\}$$

and then enlarge that cone by closing it under the usual linear operations, thus forming a real vector space B, a Banach space with the trace as norm. The physical operations can now be represented uniformly, by considering their effects on 'old state' and trace/intensity.

These two longish examples I take to reveal philosophically especially significant aspects of scientific theorizing. I turn to experimentation.

6. Experimentation: as test and as means of inquiry

Theory construction I have described as being ideally divisible into two stages: the construction of sufficiently rich models to allow for the possibility of described phenomena, and the narrowing down of the family of models so as to give the theory greater empirical content. There must be a constant interplay between the theoretician's desk and the experimenter's laboratory. Here too I wish to distinguish two aspects, of which I shall discuss one at greater length than the other.

The well-known function of experimentation is hypothesis testing. The experimenter reads over the theoretician's shoulder, and designs experiments to test whether the narrowing down has not gone too far and made the theory empirically inadequate. This characterization is simple and appealing; it is unfortunately over-simple. It overlooks first of all the fact that hypothesis testing is in general comparative, and ends not with support or refutation of a single hypothesis but with support of one hypothesis against another. It overlooks secondly that testability has to do with informativeness; a theory may be empirically adequate – or at least adequate with respect to a certain class of phenomena – but not sufficiently informative about them to allow the design of a test which could bring it support. Hence when we try to evaluate theories on the basis of their record even in the most assiduous testing, the ranking will reflect not only adequacy or truth, but also informativeness. Certainly these are both virtues, and both epistemic values to be pursued (as has been cogently argued by Isaac Levi), but it means that it is a mistake to think in terms of pure confirmation. To give an analogue: the number of votes a candidate receives is a measure of voter support for his platform but it is also a function of media exposure, and so it is not a pure measure of voter support. These reflections clearly bear on Glymour's theory of testing and relevant evidence, and his use of this important and original theory in arguments concerning scientific realism; but I have pursued them elsewhere.

There is another function of experimentation which is less often discussed and in the present context more interesting.

This second function is the one we describe in the language of discovery. Chadwick discovered the neutron, Millikan the charge of the electron, and Livingstone the Zambesi river. Millikan's case is a good illustration. He observed oil droplets drifting down in the air between two plates, which he could connect and disconnect with a battery. By friction with the air, the droplets could acquire an electrostatic charge; and Millikan observed their drifting behavior, calculating their apparent charges from their motion. Thus he found the largest number of which all apparent charges were integral multiples, and concluded that number to be the charge of a single electron. That number was *discovered*. The number of such charges per Faraday equals Avogadro's number! No one could have predicted that! You would think that empiricists would be especially bothered by such a scientific press release. For it appears that by carefully designed experiment we can discover facts about the unobservable entities behind the phenomena.

We must note first however that the division of experiments into means of testing and means of discovery is a division neither by experimental procedure nor by experimental apparatus. The set-up and operations performed would have been just the same if Millikan had made bold conjectures beforehand about that number, and had set out to test these conjectures. The division is rather by function *vis à vis* the ongoing process of theory construction.

The theory is written, so to say, step by step. At some point, the principles laid down so far imply that the electron has a negative charge. A blank is left for the magnitude of the charge. If we wish to continue now, we can go two ways. We could certainly proceed by trial and error, hypothesizing a value, testing it, offering a second guess, testing again. But alternatively, we can let the experimental apparatus write a number in the blank. What I mean is: in this case the experiment shows that unless a certain number (or a number not outside a certain interval) is written in the blank. the theory will become empirically inadequate. For the experiment has shown by actual example, that no other number will do: that is the sense in which it has filled in the blank. So regarded, *experimentation is the continuation of theory construction by other means*.

Recalling the similar saying about war and diplomacy, I should like to call this view the 'Clausewitz doctrine of experimentation'. It makes the language of construction, rather than of discovery, appropriate for experimentation as much as for theorizing.

7. The language(s) of science

We arrive now finally at the subject which I have mostly tried to banish from our discussion: language. I must admit that I too was, to begin, overly impressed by certain successes of modern logic. Thus one reviewer (John Worrall) of my book *The Scientific Image*, was able to quote the remark in my first paper on the subject, that the interrelations between the syntactic and semantic characterizations of a theory 'make implausible any claim of philosophical superiority for either approach'. The interrelations referred to are of course those described by the generalized completeness proof. I have long since changed my mind about its significance, both theoretical and practical.

To begin with the theoretical point, when a theory is presented by defining the class of its models, that class of structures cannot generally be identified with an elementary class of models of any first-order language.[5] The reason is found in the limitative meta-theorems, which brought to light the dark side of completeness. To take only the most elementary example, if a scientist describes a class of models, the mathematical object he is most likely to include is the real number continuum. There is no elementary class of models of a denumerable first-order language each of which includes the real numbers. As soon as we go from mathematics to metamathematics, we reach a level of formalization where many mathematical distinctions cannot be captured – except of course by *fiat,* as when we speak of 'standard' or 'intended' models. The moment we do so, we are using a method of description not accessible to the syntactic mode.

On the practical side we must mention the enormous distance between actual research on the foundation of science, and syntactically capturable axiomatics. While this disparity will not affect philosophical points which hinge only on what is possible 'in principle', it may certainly affect the real possibility of understanding and clarification.

Given this initial appreciation of the situation, shall we address ourselves to language at all? The answer I think is *yes,* not on any general grounds, but for a number of specific reasons.

Before detailing those specific reasons, let us look for a moment at language and the study of language in a general way. Russell made familiar to us the idea of an underlying *ideal language*. This is the skeleton, natural language being the complete living organic body built on this skeleton, the flesh being of course rather accidental, idiosyncratic, and molded by the local ecology. The skeleton, finally, is the language of logic; and for Russell's contemporaries the question was only whether *Principia Mathematica* needed to be augmented with some extra symbols to fully describe the skeleton.

Against this we must advance the conception of natural language as not being constituted by any one realization of any such logical skeleton. Logic has now provided us with a great many skeletons. Linguists have uncovered fragments of language in use for which no constructed logical skeleton yet provides any satisfactory model. Natural language consists in the resources we have for playing many different possible language games. Languages studied in logic texts are models, rather shallow models, of some of these specific language games, some of these fragments. To think that there must in principle exist a language in the sense of the objects described by logic, which is an adequate model for natural language taken as a whole, may be strictly analogous to the idea that there must exist a set which is the universe of set theory.

So if we now apply our logical methods in the philosophy of science we should, as elsewhere, set ourselves the task of modelling interesting fragments of language specially relevant to scientific discourse. These fragments may be large or small.

In my opinion, choosing the task of describing a language in which a given theory can be formulated, is a poor choice. The reason is that descriptions of structure in terms of satisfaction of sentences is, as far as I can see, generally less informative and less illuminating than direct mathematical (instead of meta-mathematical) description. It is the choice, explicit or implicit, to be formed in almost all linguistically oriented philosophical studies of science. It was the implicit choice behind, certainly, almost all logical positivist philosophy of science.

At the other extreme, we may choose a very small fragment, such as what I have called the fragment of *elementary statements*. Originally I characterized these as statements which attribute some value to a measurable physical magnitude. The syntactic form was therefore trivial – it is always something like 'm has value r' – and therefore the semantic study alone has some significance. Under pressure of various problems in the foundations of quantum mechanics, I broadened my conception of elementary statements in two ways. First, I admitted as possibly logically distinct the attributions of ranges (or Borel sets) of values. Second, I admitted as possibly distinct the attribution of states of certain sorts, on the one hand, from that of values to measurable magnitudes. (It should be added however that I soon found it much more advantageous to concentrate on the propositions expressible by elementary statements, rather than on the statements themselves. At that point there is not even a bow in the direction of syntactic description.)

There are points between these two extremes. I would point here especially to certain forms of natural discourse that are prevalent in the informal presentation of scientific theory, but which have a long history of philosophical perplexities. The main examples are causality and physical modality. From an empirical point of view, there are besides relations among actual matters of fact, only relations among words and ideas. Yet causal and modal locutions appear to introduce relations among possibilities, relations of the actual to the possible. Since irreducible probability is now a fact of life in physics, and probability is such a modality, there is no escaping this problem. Yet, if we wish to be empiricists, we have nowhere to turn besides thought and language for the locus of possibility. In other words, an empiricist position must entail that the philosophical exploration of modality, even where it occurs in science, is to be part of the theory of meaning.

In sections 1 and 3 I already made clear one important point in the empiricist view of scientific models. They may, without detriment to their function, contain much structure which corresponds to no elements of reality. The part of the model which represents reality includes the representation of actual observable phenomena, and *perhaps* something more, but is explicitly allowed to be only a proper part of the whole model.

This gives us I think the required leeway for a program in the theory of meaning. If the link between language and reality is mediated by models, it may be a very incomplete link – without depriving the language of a complete semantic structure. The idea is that the interpretation of language is not simply an association of a real denotata with grammatical expressions. Instead the interpretation proceeds in two steps. First, certain expressions are assigned values in the family of models and their logical relations derive from relations among these values. Next, reference or denotation is gained indirectly because those model elements *may* correspond to elements of reality. The exploration of modal discourse may then draw largely on structure in the models which outstrips their representation of reality.

A graphic, if somewhat inaccurate way to put this would be: causal and modal discourse describes features of our models, not features of the world. The view of language presented here – that discourse is guided by models or pictures, and that the logic of discourse is constituted by this guidance – I recommend as a general empiricist approach for a theory of meaning without metaphysics.[4]

Notes

1. I use the word deliberately: it was a tragedy for philosophers of science to go off on these logico-linguistic tangles, which contributed nothing to the understanding of either science or logic or language. It is still unfortunately necessary to speak polemically about this, because so much philosophy of science is still couched in terminology based on a mistake.
2. The impact of Suppes' innovation is lost if models are defined, as in many standard logic texts, to be partially linguistic entities, each yoked to a particular syntax. Here the models are mathematical structures, called models of a given theory only by virtue of belonging to the class defined to be the models of that theory. See section 7.

3. This answers a question posed in another review, the one by Michael Friedman. Unfortunately Friedman assumed the contrary answer, and built part of his critique on that conjecture.
4. For the background of this paper I wish to refer the reader to pp. 21–30 of the Introduction to F. Suppe (ed.) *The Structure of Scientific Theories* (Urbana: University of Illinois Press, 1974); Ch. 5 'Theories' of R. Giere, *Understanding Scientific Reasoning* (New York: Holt, Rinehart, Winston, 1979); Ch. 3 section 4 of my *An Introduction to the Philosophy of Time and Space* (New York: Random House, 1970); my book *Scientific Image* (Oxford: Oxford University Press, 1980): Patrick Suppes, 'What is a scientific theory?' pp/55–67 in S. Morgenbesser (ed.) *Philosophy of Science Today* (New York: Basic Books, 1967).

Bibliography

Bell, J.S. (1964). 'On the Einstein Podolsky Rosen Paradox'. *Physics* 1: 195–200.

Bell, J.S. (1966). 'On the Problems of Hidden Variables in Quantum Mechanics'. *Reviews of Modern Physics* 38: 447–452.

Bell, J.S. (1971). 'Introduction to the Hidden Variable Question', in *Foundations of Quantum Mechanics*. Edited by B. d'Espagnat. New York: Academic Press. Pages 171–181.

Beltrametti, E. and van Fraassen B.C. (eds.). (1981), *Current Issues in Quantum Logic*. New York: Plenum.

Cooke, R.M. and Hilgevoord, J. (1980). 'The Algebra of Physical Magnitudes'. *Foundations of Physics* 10: 363–373.

Dalla Chiari, M.L. and Toraldo di Francia G. 'A Logical Analysis of Physical Theories', *Rivista di Nuovo Cimento*, Serie 2, 3 (1973): 1–20.

Dalla Chiara, M.L. 'Formal Analysis of Physical Theories', in G. Toraldo di Francia (ed.) *Problems in the Foundations of Physics*. Amsterdam: North-Holland, 1979).

Davies, E.B. (1970), 'On the Repeated Measurements of Continuous Observables in Quantum Mechanics'. *Journal of Functional Analysis* 6: 318–346.

Davies, E.B. and Lewis J.T. (1970), 'An Operational Approach to Quantum Probability'. *Communications in Mathematical Physics* 17: 239–260.

Edwards, C.M. (1970), 'The Operational Approach to Algebraic Quantum Theory I.' *Communications in Mathematical Physics* 16: 207–230.

Edwards, C.M. (1971), 'Classes of Operations in Quantum Theory'. *Communications in Mathematical Physics* 20: 26–56.

Einstein, A., Podolsky B., and Rosen N. (1935), 'Can Quantum Mechanical Description of Reality be Considered Complete?' *The Physical Review* 47: 777–780.

Foulis, D.J. and Randall C.H. (1974), 'Empirical Logic and Quantum Mechanics'. *Synthese* 29: 81–111.

Glymour, C. (1980), *Theory and Evidence*. Princeton University Press.

Hojman, S., Kuchar K. and Teitelboim C. (1973), 'New Approach to General Relativity'. *Nature of Physical Science* 245: 97–98.

Hojman, S. (1976), 'Geometrodynamics Regained'. *Annals of Physics* 96: 88–135.

Jauch, M.M. (1968), *Foundations of Quantum Mechanics*. New York: Addison-Wesley.

Ludwig, G. (1967), 'Attempt of an Axiomatic Foundation of Quantum Mechanics and More General Theories II'. *Communications in Mathematical Physics*. 4: 331–348.

Marlow, A.R. (1978), 'Quantum Theory and Hilbert Space'. *Journal of Mahematical Physics* 19: 1841–1846.

Marlow, A.R. (1980), 'An Extended Quantum Mechanical Embedding Theorem' in *Quantum Theory and Gravitation*. A. R. Marlow (ed.) New York: Academic Press. pp 71–77.

Millikan, Robert A. (1917), *The Electron*. J. W. M. DuMond. (ed.) Chicago: University of Chicago Press.

Peacocke, C. (1981), 'Not Real But Observable'. Review of van Fraassen (1980). *Times Literary Supplement* 30: 121.

Przelewski, M., *The Logic of Empirical Theories* London: Routledge and Kegan Paul, 1969.

Suppe, F., 'Theories, Their Formulations and the Operational Imperative.' *Synthese* 25 (1972), pp. 129–159.

Suppes, P., 'What is a Scientific Theory?' in *Philosophy of Science Today*, S. Morgenbesser (ed.) New York: Basic Books, pp. 55–67.

Suppes, P. (1974), 'The Structure of Theories and the Analysis of Data'. in *The Structure of Scientific Theories*, F. Suppe (ed.) Urbana: University of Illinois Press, pp. 266–283.

Suppes, P., *Studies in Methodology and Foundations of Science* (Dordrecht: Reidel, 1969).

Van Fraassen, B.C., 'The Charybdis of Realism: Epistemological Implications of Bell's Inequality' *Synthese* 52 (1982), pp. 25–38.

Van Fraassen, B.C., 'A Formal Approach to the Philosophy of Science.' pp. 303–366 in R. Colodny (ed.) *Paradigms and Paradoxes: The Challenge of the Quantum Domain*. Pittsburgh: University of Pittsburgh Press, 1972.

Van Fraassen, B.C. (1980). *The Scientific Image*. Oxford and New York: Oxford University Press.

Van Fraassen, B.C., 'Theory Construction and Experiment: An Empiricist View.' pp. 663–678 in P. Asquith and R. Giere (eds.) *PSA 1980* Vol 2 East Lansing, Mich.: Philosophy of Science Association, 1981.

Weyl, H., 'The Ghost of Modality' in M. Farber (ed.) *Philosophical Essays in Memory of Edmund Husserl*. Cambridge, Mass. 1940.

Wojcicki, R., 'Set Theoretic Representations of Empirical Phenomena'. *Journal of Philosophical Logic* 3 (1947), 337–343.

Worrall, J., 'An Unreal Image'. Review of van Fraassen (1980). *British Journal for the Philosophy of Science*. 35 (1984), pp. 65–80.

TO SAVE THE PHENOMENA *

AFTER the demise of logical positivism, scientific realism has once more returned as a major philosophical position. I shall not try here to criticize that position, but rather attempt to outline a comprehensive alternative.[1]

I

What exactly is scientific realism? Naively stated, it is the view that the picture science gives us of the world is true, and the entities postulated really exist. (Historically, it added that there are real necessities in nature; I shall ignore that aspect here.[2]) But that statement is too naive; it attributes to the scientific realist the belief that today's scientific theories are (essentially) right.

The correct statement, it seems to me, must indeed be in terms of epistemic attitudes, but not so directly. The aim of science is to give us *a literally true story of what the world is like;* and the proper form of acceptance of a theory is to believe that it is true. This is the statement of scientific realism: "To have good reason to accept a theory is to have good reason to believe that the entities it postulates are real," as Wilfrid Sellars has expressed it. Accordingly, an anti-realism is a position according to which the aims of science can well be served without giving such a literally true story,

* To be presented in an APA symposium on Scientific Realism, December 28, 1976. Richard N. Boyd and Clark Glymour will comment; see this JOURNAL, this issue, pp. 633–635 and 635–637, respectively.

The research for this paper was supported by Canada Council Grant S74-0590. An earlier version was presented at the Western Division of the Canadian Philosophical Association (Calgary, October 1975).

I want to acknowledge my debt to Clark Glymour, Princeton University, for the challenge of his critiques of conventionalism in his dissertation and unpublished manuscripts.

1 For some criticisms, see my "Theoretical Entities: The Five Ways" *Philosophia* 4 (1974): 95–109, and "Wilfrid Sellars on Scientific Realism" *Dialogue,* XIV, 4 (December 1975): 606–616.

2 Cf. my "The Only Necessity Is Verbal Necessity," forthcoming in this JOURNAL, LXXIV, 2 (February 1977).

and acceptance of a theory may properly involve something less (or other) than belief that it is true.

The idea of a literally true account has two aspects: the language is to be literally construed; and, so construed, the account is true. This divides the anti-realists into two sorts. The first sort holds that science is or aims to be true, properly (but not literally) construed. The second holds that the language of science should be literally construed, but its theories need not be true to be good. The anti-realism I advocate belongs to the second sort.

II

When Newton wrote his *Mathematical Principles of Natural Philosophy* and *System of the World,* he carefully distinguished the phenomena to be saved from the reality he postulated. He distinguished the "absolute magnitudes" that appear in his axioms from their "sensible measures" which are determined experimentally. He discussed carefully the ways in which, and extent to which, "the true motions of particular bodies [may be determined] from the apparent," via the assertion that "the apparent motions . . . are the differences of true motions."[3]

The "apparent motions" form relational structures defined by measuring relative distances, time intervals, and angles of separation. For brevity, let us call these relational structures *appearances.* In the mathematical model provided by Newton's theory, bodies are located in Absolute Space, in which they have real or absolute motions. But within these models we can define structures that are meant to be exact reflections of those appearances and are, as Newton says, identifiable as differences between true motions. These structures, defined in terms of the relevant relations between absolute locations and absolute times, which are the appropriate parts of Newton's models, I shall call *motions,* borrowing Simon's term.[4]

When Newton claims empirical adequacy for his theory he is claiming that his theory has some model such that *all actual appearances are identifiable with* (*isomorphic to*) motions in that model.

Newton's theory goes a great deal further than this. It is part of his theory that there is such a thing as Absolute Space, that absolute motion is motion relative to Absolute Space, that absolute acceleration causes certain stresses and strains and thereby deformations in

[3] F. Cajori, ed. *Sir Isaac Newton's Mathematical Principles of Natural Philosophy and His System of the World* (Berkeley: University of California Press, 1960), p. 12.

[4] Herbert A. Simon, "The Axiomatization of Classical Mechanics," *Philosophy of Science,* XXI, 4 (October 1954): 340–343.

the appearances, and so on. He offered, in addition, the hypothesis (his term) that the center of gravity of the solar system is at rest in Absolute Space. But, as he himself noted, the appearances would be no different if that center were in any other state of constant absolute motion.

Let us call Newton's theory (mechanics and gravitation) *TN*, and *TN*(v) the theory *TN* plus the postulate that the center of gravity of the solar system has constant absolute velocity. By Newton's own account, he claims empirical adequacy for *TN*(0); and also claims that, if *TN*(0) is empirically adequate, then so are all the theories *TN*(v).

Recalling what it was to claim empirical adequacy, we see that all the theories *TN*(v) are empirically equivalent exactly *if all the motions in a model of TN*(v) *are isomorphic to motions in a model TN*($v + w$), for all constant velocities v and w. For now, let us agree that these theories are empirically equivalent, referring objections to a later section.

III

What exactly is the "empirical import" of *TN*(0)? Let us focus on a fictitious and anachronistic philosopher Leibniz*, whose only quarrel with Newton's theory is that he does not believe in the existence of Absolute Space. As a corollary, of course, he can attach no "physical significance" to statements about absolute motion. Leibniz* believes, like Newton, that *TN*(0) is empirically adequate; but not that it is true. For the sake of brevity, let us say that Leibniz* *accepts* the theory but that he does not *believe* it; when confusion threatens we may expand that idiom to say that he *accepts the theory as empirically adequate,* but does not *believe it to be true.* What does Leibniz* believe, then?

Leibniz* believes that *TN*(0) is empirically adequate, and hence, equivalently, that all the theories *TN*(v) are empirically adequate. Yet we cannot identify the theory that Leibniz* holds about the world—call it *TNE*—with the common part of all the theories *TN*(v). For each of the theories *TN*(v) has such consequences as that the earth has *some* absolute velocity, and that Absolute Space exists. In each model of each theory *TN*(v) there is to be found something other than motions, and there is the rub.

To believe a theory is to believe that one of its models correctly represents the world. A theory may have isomorphic models; that redundancy is easily removed. If it has been removed, then to believe the theory is to believe that exactly one of its models correctly represents the world. Therefore, if we believe of a family of theories that all are empirically adequate, but each goes beyond the phenomena, then we are still free to believe that each is false, and hence their common part is false. For that common part is phrasable as: one of the models of one of those theories correctly represents the world.

IV

It may be objected that theories will seem empirically equivalent only so long as we do not consider their possible extensions.[5] The equivalence may generally, or always, disappear when we consider their implications for some further domain of application. The usual example is Brownian motion; but this is imperfect, for it was known that phenomenological and statistical thermodynamics disagreed even on macroscopic phenomena over sufficiently long periods of time. But there is a good, *fictional* example: the combination of electromagnetism with mechanics, if we ignore the unexpected null results that led to the replacement of classical mechanics.

Maxwell's theory was not developed as part of mechanics, but it did have mechanical models. This follows from a result of Koenig, as detailed by Poincaré in the preface of his *Electricité et Optique* and elsewhere. But the theory had the strange new feature that velocity itself, not just its derivative, appears in the equations. A spate of thought experiments was designed to measure absolute velocity, the simplest perhaps that of Poincaré:

> Consider two electrified bodies; though they seem to us at rest, they are both carried along by the motion of the earth; . . . therefore, equivalent to two parallel currents of the same sense and these two currents should attract each other. In measuring this attraction, we shall measure the velocity of the earth; not its velocity in relation to the sun or the fixed stars, but its absolute velocity.[6]

The null outcome of all experiments of this sort led to the replacement of classical by relativistic mechanics. But let us imagine that values *were* found for the absolute velocities; specifically for that of the center of the solar system. Then, surely, one of the theories $TN(v)$ would be confirmed and the others falsified?

This reasoning is spurious. Newton made the distinction between true and apparent motions without presupposing more than that basic mechanics in which Maxwell's theories has models. Each mo-

[5] See, for example, Richard N. Boyd, "Realism, Undetermination, and a Causal Theory of Evidence," *Noûs*, VII, 1 (March 1973): 1–12.

[6] Henri Poincaré, *The Value of Science*, B. Halsted, tr. (New York: Dover, 1958), page 98.

tion in a model of $TN(v)$ is isomorphic to one in some model of $TN(v + w)$, for all constant velocities v and w. Could this assertion of empirical equivalence possibly be controverted by those nineteenth-century reflections? The answer is *no*. The thought experiment, we may imagine, confirmed the theory that added to TN the hypothesis:

HO. The center of gravity of the solar system is at absolute rest.
EO. Two electrified bodies moving with absolute velocity v attract each other with force $F(v)$.

This theory has a consequence strictly about appearances:

CON. Two electrified bodies moving with velocity v relative to the center of gravity of the solar system, attract each other with force $F(v)$.

However, that same consequence can be had by adding to TN the two alternative hypotheses:

Hw. The center of gravity of the solar system has absolute velocity w.
Ew. Two electrified bodies moving with absolute velocity $v + w$ attract each other with force $F(v)$.

More generally, for each theory $TN(v)$ there is an electromagnetic theory $E(v)$ such that $E(0)$ is Maxwell's and all the combined theories $TN(v)$ plus $E(v)$ are empirically equivalent.

There is no originality in this observation, of which Poincaré discusses the equivalent immediately after the passage I cited above. Only familiar examples, but rightly stated, are needed, it seems, to show the feasibility of concepts of empirical adequacy and equivalence. In the remainder of this paper I shall try to generalize these considerations, while showing that the attempts to explicate those concepts *syntactically* had to reduce them to absurdity.

V

The idea that theories may have hidden virtues by allowing successful extensions to new kinds of phenomena, is too pretty to be left. Nor is it a very new idea. In the first lecture of his *Cours de philosophie positive,* Comte referred to Fourier's theory of heat as showing the emptiness of the debate between partisans of calorific matter and kinetic theory. The illustrations of empirical equivalence have that regrettable tendency to date; calorifics lost. Federico Enriques seemed to place his finger on the exact reason when he wrote: "The hypotheses which are indifferent in the limited sphere of the actual theories acquire significance from the point of view of their possible extension." [7] To evaluate this suggestion, we must ask what exactly is an extension of a theory.

Suppose that experiments really had confirmed the combined theory $TN(0)$ plus $E(0)$. In that case mechanics would have won a *victory*. The claim that $TN(0)$ was empirically adequate would have been borne out by the facts. But such victorious extensions could never count for a theory as against one of its empirical equivalents.

Therefore, if Enriques' idea is to be correct, there must be another sort of extension, which is really a defeat—but qualified. For a theory T may have an easy or obvious modification which is empirically adequate, while another theory empirically equivalent to T does not. One example may be the superiority of Newton's celestial mechanics over the variant produced by Brian Ellis; Ellis himself seems to be of this opinion.[8] This is a *pragmatic* superiority and cannot suggest that theories, empirically equivalent in the sense explained, can nevertheless have different empirical import.

VI

We still need a general account of empirical adequacy and equivalence. It is here that the syntactic approach has most conspicuously failed. A theory was conceived as identifiable with the set of its theorems in a specified language. This language has a vocabulary, divided into two classes—the observational and theoretical terms. Let the first class be E; then the empirical import of theory T was said to be its subtheory T/E—those theorems expressible in that subvocabulary. T and T' were declared empirically equivalent if T/E was the same as T'/E.

Obvious questions were raised and settled. Craig showed that, under suitable conditions, T/E is axiomatizable in the vocabulary E. Logicians attached importance to questions about restricted vocabularies, and this was apparently enough to make philosophers think them important too. The distinction between observational and theoretical terms was more debatable, and some changed the division into "old" and "newly introduced" terms.[9] But all this is mistaken. Empirical import cannot be isolated in this syntactic

[7] *Historical Development of Logic*, J. Rosenthal, tr. (New York: Holt, 1929), p. 230.

[8] "The Origins and Nature of Newton's Laws of Motion," in R. Colodny, ed., *Beyond the Edge of Certainty* (Englewood Cliffs, N.J.: Prentice-Hall, 1965), pp. 29–68.

[9] For example, David Lewis, "How to Define Theoretical Terms," this JOURNAL, LXVII, 13 (July 9, 1970): 427–446. This paper is not subject to my criticisms here; on the contrary, it provides independent reasons to conclude that the empirical import of a theory cannot be syntactically isolated.

fashion. If that could be done, then T/E would say exactly what T says about what is observable, and nothing else. But consider: the quantum theory, Copenhagen version, says that there are things which sometimes have a position in space and sometimes do not. This consequence I have just stated without using theoretical terms. Newton's theory TN implies that there is something (to wit, Absolute Space) which neither has a position nor occupies volume. As long as unobservable entities differ systematically from observable entities with respect to observable characteristics, T/E will say that there are such things if T does.

The reduced theory T/E is not a description of the observable part of the world of T; rather, it is a hobbled and hamstrung version of T's description of everything. Empirical equivalence fares as badly. In section II, $TN(0)$ and TNE *must* be empirically equivalent, but the above remark about TN shows that $TN(0)/E$ is not TNE/E. To eliminate such embarrassments, extensions of theories were considered in attempts to redefine empirical equivalence.[10] But these have similar absurd consequences.

The worst consequence of the syntactic approach was surely the way it focused philosophical attention on irrelevant technical questions. The expressions 'theoretical object' and 'observational predicate' mark category mistakes. Terms may be theoretical, but 'observable' classifies putative entities. Hence there cannot be a "theoretical/observable distinction." It is true surely that elimination of all theory-laden terms would leave no usable language; also that 'observable' is as vague as 'bald'. But these facts imply not at all that 'observable' marks an unreal distinction. It refers quite clearly to our limitations, the limits of observation, which are not incapacitating, but also not negligible.

VII

The phenomena are saved when they are exhibited as fragments of a larger unity. For that very reason it would be strange if scientific theories described the phenomena, the observable part, in different terms from the rest of the world they describe. And so an attempt to draw the conceptual line between phenomena and the transphenomenal by means of a distinction of vocabulary, must always have looked too simple to be good.

Not all philosophers who discussed unobservables, by any means, did so in terms of vocabulary. But there was a common assumption:

[10] See fn 5 above. We could say that Boyd's paper, like Lewis's, provides independent evidence that empirical import cannot be syntactically isolated. But Boyd concludes also that there can be no distinction between truth and empirical adequacy for scientific theories.

that the distinction marked is philosophical. Hence it must be drawn, if at all, by philosophical analysis and, if attacked, by philosophical arguments. This attitude needs a Grand Reversal. If there are limits to observation, these are empirical, and must be described by empirical science. The classification marked by 'observable' must be of entities in the world of science. And science, in giving content to the distinction, will reveal how much we believe when we accept it as empirically adequate.

A future Unified Science may detail the limits of observation exactly; meanwhile, extant theories are not silent on them. We saw Newton's delineation; for relativity theory, we have two revealing studies by Clark Glymour. The first shows that local (hence, I should think, measurable) quantities do not uniquely determine global features of space-time.[11] The second shows that these features also are not uniquely determined by structures each lying wholly within some absolute past cone—hence, I should think, by observable structures. It is the theory of relativity itself, after all, that places *an absolute* limit on the information we can gather, through the limiting function of the speed of light.

In the foundations of quantum mechanics much more attention has been given to measurement. Much of the discussion is about necessary limitations: the role of noise in amplification, the distinction between macro- and micro-observables.[12] Yet we have no such clarity as Glymour gave us for relativity theory, concerning the extent to which macro-structure determines micro-structure. The debate over scientific realism may at least have the virtue of directing attention to such questions.

Science itself distinguishes the observable that it postulates from the whole it postulates. The distinction, being in part a function of the limits science discloses on human observation, is anthropocentric. But, since science places human observers among the physical systems it means to describe, it also gives itself the task of describing anthropocentric distinctions. It is in this way that even the scientific realist must observe a distinction between the phenomena and the transphenomenal in the scientific world picture.

VIII

I have laid some philosophical misfortunes at the door of a mistaken orientation toward syntax. The alternative is to say that the-

[11] "Cosmology, Convention, and the Closed Universe," *Synthese*, XXIV, 1/2 (July/August 1972): 195–218; discussed in my "Earman on the Causal Theory of Time," *op. cit.*, pp. 87–95 (referred to therein by an earlier title).

[12] See for example N. D. Cartwright, "Superposition and Macroscopic Observation," *Synthese*, XXIX (December 1974): 229–242, and references therein.

ories are presented directly by describing their models. But does this really introduce a new element? When you give the theorems of T, you give the set of models of T—namely, all those structures which satisfy the theorems. And, if you give the models, you give at least the set of theorems of T—namely all those sentences which are satisfied in all the models. Does it not follow that we can as advantageously identify T with its theorems as with its models?

But there is an ellipsis in the argument. It is being assumed that there is a specific language L which is the one language that belongs to T. And indeed, the theorems of T in L determine and are determined by the set of model structures of L (that is, structures in which L is interpreted) in which those theorems are satisfied. However, the assumption that there is a language L which plays this role for T places important restrictions on what the set of models of T can be like.

A theory provides, among other things, a specification (more or less complete) of the parts of its models that are to be direct images of the structures described in measurement reports. In the case of Newton's mechanics, I called those parts *motions;* in general, let us call them *empirical substructures*. The structures described in measurement reports we may continue to call *appearances*. A theory is *empirically adequate* exactly if all appearances are isomorphic to empirical substructures in at least one of its models. Theory T is *empirically no stronger* than theory T' exactly if, for each model M of T, there is a model M' of T' such that all empirical substructures of M are isomorphic to empirical substructures of M'. Theories T and T' are *empirically equivalent* exactly if neither is empirically stronger than the other. In that case, as an easy corollary, each is empirically adequate if and only if the other is.

In section v, I distinguished two kinds of extensions, the first a sort of victory and the second a sort of defeat. Let us call the first a *proper extension:* this simply narrows the class of models. We may call a theory *empirically minimal* if it is not empirically equivalent to any of its proper extensions. Glymour has convincingly argued, in the work cited above, that General Relativity is not empirically minimal. The reason is, in my present terms, that only local properties of space-time enter the descriptions of the appearances, but models may differ in global properties. This is a further nontrivial example of empirical equivalence.

The second sort of extension I shall not try to define precisely. The idea is that models of the theory may differ in structure other than that of the empirical substructures. In that case the theory is not empirically minimal, but this may put it in the advantageous position of offering modeling possibilities when radically new phenomena come to light. An example may yet be offered by hidden-variable theories in quantum mechanics.[13]

In terms of the concepts now at our disposal, and the examples given, we can conclude that there are indeed nontrivial cases of empirical equivalence, non-uniqueness, and extendability, both proper and improper. Such cases are now seen to be quite possible *even if the formulation of the theory has not a single term that cannot be called observational, in some way.* And now it should be possible to state the issue of scientific realism, which concerns our epistemic attitude toward theories rather than their internal structure.

All the results of measurements are not in; they are never all in. Therefore we cannot know what all the appearances are. We can say that a theory is empirically adequate, that all the appearances will fit (the empirical substructures of) its models. Though we cannot know this with certainty, we can reasonably believe it. All this is the case not only for empirical adequacy but for truth as well. Yet there are two distinct epistemic attitudes that can be taken: we can *accept* a theory (accept it as empirically adequate) or *believe* the theory (believe it to be true). We can take it to be the aim of science to produce a literally true story about the world, or simply to produce accounts that are empirically adequate. This is the issue of scientific realism versus its (divided) opposition. The intrascientific distinction between the observable and the unobservable is an anthropocentric distinction; but it is reasonable that the distinction should be drawn in terms of *us,* when it is a question of *our* attitudes toward theories.

BAS C. VAN FRAASSEN
University of Toronto and University of Southern California

[13] See Stanley Gudder, "Hidden Variables in Quantum Mechanics Reconsidered," *Review of Modern Physics,* XL (1968): 229–231; and section III of my "Semantic Analysis of Quantum Logic," in C. A. Hooker, ed., *Contemporary Research in the Foundations and Philosophy of Quantum Theory* (Dordrecht: Reidel, 1973), pp. 80–113.

HOW TO DEFINE THEORETICAL TERMS

MOST philosophers of science agree that, when a newly proposed scientific theory introduces new terms, we usually cannot define the new terms using only the old terms we understood beforehand. On the contrary, I contend that there is a general method for defining the newly introduced theoretical terms.

Most philosophers of science also agree that, in order to reduce one scientific theory to another, we need to posit bridge laws: new laws, independent of the reducing theory, which serve to identify phenomena described in terms of the reduced theory with phe nomena described in terms of the reducing theory. On the contrary, I deny that the bridge laws must be posited independently. They may follow from the reducing theory, via the definitions of the theoretical terms of the reduced theory. In such cases it would be wrong to think that theoretical reduction is done voluntarily, for the sake of parsimony, when the reduced and reducing theories are such as to permit it. Sometimes reduction is not only possible but unavoidable.

F. P. Ramsey proposed that theoretical terms might be eliminated in favor of existentially quantified bound variables. Rudolf Carnap, in his most recent discussions of theoretical terms, has used Ramsey's method to split any term-introducing theory into two parts: an analytic postulate to partially interpret the theoretical terms, and a synthetic postulate in which these terms do not occur. My proposal will be in the spirit of Ramsey's and Carnap's.

My proposal could be called an elimination of theoretical terms, if you insist; for to define them is to show how to do without them. But it is better called a vindication of theoretical terms; for to define them is to show that there is no good reason to want to do without them. They are no less fully interpreted and no less well understood than the old terms we had beforehand.

I am not planning to define theoretical terms within an observation language, whatever that is. Some statements report observations, and some do not; but I do not know of any special compartment of our language that is reserved for the reporting of observations. I do not understand what it is just to be a theoretical term, not of any theory in particular, as opposed to being an observational term (or a logical or mathematical term).[1] I believe I do understand what it is to be a *T-term:* that is, a theoretical term introduced by a given theory *T* at a given stage in the history of science. If so, then I also understand what it is to be an *O-term:* that is, any *other* term, one of our *original* terms, an *old* term we already understood before the new theory *T* with its new *T*-terms was proposed. An *O*-term can have any epistemic origin and priority you please. It can belong to any semantic or syntactic category you please. Any old term can be an *O*-term, provided we have somehow come to understand it. And by understand I mean "understand"—not "know how to analyze."

I am also not planning to "dispense with theoretical entities." Quite the opposite. The defining of theoretical terms serves the cause of scientific realism. A term correctly defined by means of other terms that admittedly have sense and denotation can scarcely be regarded as a mere bead on a formal abacus. If it purports to name something, then if the theory that introduced it is true it does name something. I suppose a theoretical entity is something we believe in only because its existence, occurrence, etc. is posited by some theory—especially some recent, esoteric, not-yet-well-established scientific theory. Theoretical entities might better be called (as they sometimes are called) *hypothetical* entities. Theoretical terms need not name theoretical entities: consider 'H_2O'. Theoretical entities need not be named by theoretical terms: consider "living creature too small to see." Theoretical entities need not be invisible, intangible, etc.: consider the dark companions of stars. Theoretical entities are not entities of a special category, but entities we know of (at present) in a special way.

THE POSTULATE OF *T*

Suppose the best scientific explanation we can devise for some body of data includes a new theory *T*, formulated by means of a postulate in which there occur some new terms $\tau_1 \ldots \tau_n$, terms we have never used before. We shall call these newly introduced terms the *theoretical terms of T*, or just *T-terms;* and we shall call all the other

[1] My reasons are more or less those discussed by Hilary Putnam in "What Theories Are Not," *Logic, Methodology and Philosophy of Science* (Stanford: University Press, 1962), ed. by E. Nagel, P. Suppes, and A. Tarski.

terms of our scientific vocabulary *O-terms*. We shall accordingly call any sentence of our language that is free of *T*-terms an *O-sentence*. Let us assume that the *O*-terms have conventionally established standard interpretations, well known to us. The *T*-terms, on the other hand, are unfamiliar. Our only clue to their meaning is the postulate of *T* that introduced them. We are accustomed to say that it implicitly defines them; but we would prefer explicit definitions.

We may stipulate that the postulate of *T* is a single sentence; if it was a set of sentences, take their conjunction. If it was a finite set, we can take their conjunction within ordinary logic. If it was a decidably infinite set, we must introduce devices for infinite conjunction—to do so would be bothersome, but not problematic.

We may stipulate that our *T*-terms are names, not predicates or functors. No generality is lost, since names can purport to name entities of any kind: individuals, species, states, properties, substances, magnitudes, classes, relations, or what not. Instead of a *T*-predicate '*F* __', for instance, we can use '__ has *F*-hood'; '*F*-hood' is a *T*-name purporting to name a property, and '__ has __' is an *O*-predicate. It is automatic to reformulate all *T*-terms as names, under the safe assumption that our *O*-vocabulary provides the needed copulas:

__ has the property __
__ is in the state __ at time __
__ has __ to degree __

and the like. We will later replace *T*-terms by bound variables; by making the *T*-terms grammatically uniform, we avoid the need to introduce variables of diverse types.

We must assume that all occurrences of *T*-terms in the postulate of *T* are purely referential, open to existential generalization and to substitution by Leibniz's law. We need not assume, however, that the language of *T* is an extensional language. Among the *O*-terms there may be nonextensional operators, for instance 'it is a law that __'; nonextensional connectives, for instance '__ because __'; and so on.

We must assume, finally, that the postulate of *T* will be false in case any of the *T*-terms is denotationless. This is not a legitimate assumption for sentences in general: "There is no such substance as phlogiston" is true just because 'phlogiston' is denotationless. However, it does seem to be a legitimate assumption for the postulate of a term-introducing theory. The postulate of *T* will therefore imply, for each *T*-term τ_i, the sentence '$(\exists x)(x=\tau_i)$', which says that τ_i names something.

Such sentences sometimes count as logical truths. Many systems of logic avoid the difficulties of denotationless names by stipulating that an otherwise denotationless name is deemed artificially to name some arbitrarily chosen "null" individual. We must be able to take seriously the possibility of denotationless *T*-terms; it is worth the trouble to use a system of logic designed to tolerate them. Such a system has been given by Dana Scott[2]; its salient features are as follows.

(1) Improper descriptions and other denotationless terms are *really* denotationless: they name nothing in the domain of discourse. (The domain itself serves as a null individual for technical convenience; but it is not *in* the domain, and no term literally names it.)

(2) Atomic sentences containing denotationless terms are either true or false, depending on the predicate and other terms involved; we might, but need not, stipulate that they are always false.

(3) Identities containing denotationless terms on both sides are true; identities containing a denotationless term on one side only are false.

(4) Denotationless terms are interchangeable *salve veritate* in extensional contexts; necessarily denotationless names are interchangeable *salve veritate* even in intensional contexts.

THE RAMSEY AND CARNAP SENTENCES OF *T*

Let us write the postulate of our theory *T* in a way that exhibits the occurrences of *T*-terms therein: '$\mathrm{T}[\tau_1 \ldots \tau_n]$'.

If we replace the *T*-terms uniformly by variables $x_1 \ldots x_n$ respectively (distinct variables that do not occur there already), we get a formula which we may call the *realization formula* of *T*: '$\mathrm{T}[x_1 \ldots x_n]$'.

Any *n*-tuple of entities that satisfies this formula, under the fixed standard interpretations of its *O*-terms, may be said to *realize*, or to be a *realization* of, the theory *T*.

Therefore we recognize the postulate of *T* as the sentence that says that *T* is realized by the *n*-tuple of entities denoted, respectively, by the *T*-terms $\tau_1 \ldots \tau_n$. If so, then *a fortiori* *T* is realized. We can write another sentence, called the *Ramsey sentence* of *T*, which says only that *T* is realized: '$\exists x_1 \ldots x_n \mathrm{T}[x_1 \ldots x_n]$'.

We can write a third sentence, called the *Carnap sentence* of *T*, which is neutral as to whether *T* is realized, but says that if *T* is realized, then the *n*-tuple of entities named respectively by $\tau_1 \ldots \tau_n$ is one realization of *T*. The Carnap sentence is the conditional of the Ramsey sentence and the postulate:

$$\exists x_1 \ldots x_n \mathrm{T}[x_1 \ldots x_n] \supset \mathrm{T}[\tau_1 \ldots \tau_n]$$

[2] "Existence and Description in Formal Logic," in *Bertrand Russell: Philosopher of the Century* (London: Allen & Unwin, 1967), ed. by Ralph Schoenman.

Our three sentences are related as follows: (1) The postulate is logically equivalent to the conjunction of the Ramsey sentence and the Carnap sentence. (2) The Ramsey sentence and the postulate logically imply exactly the same O-sentences. (3) The Carnap sentence logically implies no O-sentences except logical truths.

Therefore, insofar as the theory T serves as a device for systematizing O-sentences, the Ramsey sentence of T will do the job as well as the postulate itself. That was Ramsey's observation.[3] The Ramsey sentence can obviously do nothing to help interpret the T-terms, since they do not occur in it. The Carnap sentence, on the other hand, does nothing to help systematize O-sentences; but it does contain the T-terms, and it does seem to do as much toward interpreting them as the postulate itself does. And the Ramsey and Carnap sentences between them do exactly what the postulate does.

Accordingly, Carnap proposes[4] to take the Ramsey sentence as the synthetic postulate of T and the Carnap sentence as the analytic postulate of T. They divide the labor of the original postulate, which both systematized O-sentences and partially interpreted the T-terms. (Here and henceforth, when I speak of Carnap's proposal it should be understood that I mean Carnap's proposal *minus* Carnap's stipulation that the O-terms belong to an observation language.)

THE INTERPRETATION OF T-TERMS

Let us see whether we want to agree that the Carnap sentence does specify the appropriate interpretations of the T-terms, insofar as appropriate interpretations can be specified at all. Put aside, for the time being, Carnap's idea that the Ramsey and Carnap sentences in partnership should be a perfect substitute for the original postulate.

It is important to separate three cases. T may have precisely one realization, or no realization, or more than one realization.

In case T is uniquely realized, the Carnap sentence clearly gives exactly the right specification. It says that the T-terms name the entities in the n-tuple that is the unique realization of T. The first T-term, τ_1, names the first component of the unique realization of T; τ_2 names the second component; and so on.

[3] "Theories," *The Foundations of Mathematics* (London: Routledge & Kegan Paul, 1931), ed. by R. B. Braithwaite.

[4] *Philosophical Foundations of Physics* (New York: Basic Books, 1966), ed. by Martin Gardner, ch. 28; "Replies and Systematic Expositions," in *The Philosophy of Rudolph Carnap* (La Salle, Ill.: Open Court, 1963), ed. by P. A. Schilpp, section 24D; "On the Use of Hilbert's ε-Operator in Scientific Theories," in *Essays on the Foundations of Mathematics, Dedicated to A. A. Fraenkel on his Seventieth Anniversary* (Jerusalem: Magnes Press, 1961), ed. by Y. Bar-Hillel *et al.*; "Beobachtungssprache und theoretische Sprache," *Logica: Studia Paul Bernays dedicta* (Neuchâtel: Griffon, 1959).

In case T is not realized, the Carnap sentence says nothing about the denotation of the T-terms. But this modesty seems to be uncalled for. The T-terms were introduced on the assumption that T was realized, in order to name components of a realization of T. There is no realization of T. Therefore they should not name anything. 'Phlogiston' presumably is a theoretical term of an unrealized theory; we say without hesistation that there is no such thing as phlogiston. What else could we possibly say? Should we say that phlogiston is something or other, but (unless phlogiston theory turns out to be true after all) we have no hope of finding out what?

Let us say, then, that the theoretical terms of unrealized theories do not name anything. That will do very well, at least in the case of a theory like phlogiston theory which comes nowhere near being realized. It will not do so well in the case of an unrealized theory with a (unique) *near*-realization: that is, an n-tuple that does not realize the original theory, but does realize some theory obtained from it by a slight weakening or a slight correction. We might want to say that the theoretical terms name the components of whichever n-tuple comes nearest to realizing the theory, if it comes near enough. We will ignore this complication, in part for the sake of simplicity and in part because we might hope to handle it as follows. Given a theory T, we might find a slightly weaker T', implied by but not implying T, such that an n-tuple is a realization of T' if and only if it is a near-realization of T. Then we could say that T', not T, is the real term-introducing theory; everything we have been saying about T really ought to be taken as applying to T' instead. T itself may be recovered as the conjunction of T' with further hypotheses containing the theoretical terms already introduced by T'.

There remains the case in which T is multiply realized. In this case, the Carnap sentence tells us that the T-terms name the components of some realization or other. But it does not tell us which; and there seems to be no nonarbitrary way to choose one of the realizations. So either the T-terms do not name anything, or they name the components of an arbitrarily chosen one of the realizations of T. Either of these alternatives concedes too much to the instrumentalist view of a theory as a mere formal abacus. Neither does justice to our naive impression that we understand the theoretical terms of a true theory, and without making any arbitrary choice among realizations. We should not accept Carnap's treatment in this case if we can help it. Can we?

We might say instead that the theoretical terms of multiply realized theories do not name anything. If multiple realization is a defect that theorists can reasonably hope to avoid, then we can afford to treat multiply realized theories as failures: call them false, and call their theoretical terms denotationless. But if multiple realization is inevitable, we cannot afford to disdain multiply realized theories. We can have denotations arbitrarily chosen, or no denotations at all.

A uniquely realized theory is, other things being equal, certainly more satisfactory than a multiply realized theory. We should insist on unique realization as a standard of correctness unless it is a standard too high to be met. Is there any reason to think that we must settle for multiply realized theories? I know of nothing in the way scientists propose theories which suggests that they do not hope for unique realization. And I know of no good reason why they should not hope for unique realization. Therefore I contend that we ought to say that the theoretical terms of a multiply realized theories are denotationless.

Many philosophers do seem to think that unique realization is an extravagant hope, unlikely in scientific practice or even impossible in principle. Partly, this is professional skepticism; partly it is skepticism derived from confusions that I shall try to forestall.

In the first place, I am not claiming that scientific theories are formulated in such a way that they could not possibly be multiply realized. I am claiming only that it is reasonable to hope that a good theory will not in fact be multiply realized.

In the second place, I am not claiming that there is only one way in which a given theory *could* be realized; just that we can reasonably hope that there is only one way in which it *is* realized.

Finally, I should say again that we are talking only about realizations that make T true under a fixed interpretation of all of its O-vocabulary. And this O-vocabulary may be as miscellaneous as you please; in practice it is likely to be very miscellaneous indeed. An O-term is *any* term, of any character, which we already understood before the new theory T came along. It does not have to belong to an observation language. If anyone hopes to adapt my proposals to the task of interpreting theoretical terms using only an observation language—if there is any such thing—I would not be at all surprised if he ran into trouble with multiple realizations. But his project and his troubles are not mine.

John Winnie has announced a proof that scientific theories cannot be uniquely realized.[5] Though his proof is sound, it goes against nothing I want to say. Most of Winnie's multiple realizations of a given theory—all but one, perhaps—are not what I call realizations of the theory. I am concerned only with realizations

[5] "The Implicit Definition of Theoretical Terms," *British Journal for the Philosophy of Science*, XVIII (1967): 223–229.

under a fixed interpretation of the O-vocabulary; whereas Winnie permits variation in the interpretation of certain O-terms from one realization to another, provided that the variation is confined to theoretical entities. For instance, he would permit variation in the extension of the O-predicate '__ is bigger than __' so long as the extension among observational entities remained fixed. Winnie's proof does not show that a theory is multiply realized in my sense unless the postulate of the theory is free of "mixed" O-terms: O-predicates whose extension includes theoretical entities, and the like. I would claim that mixed O-terms are omnipresent, and that there is no reason not to grant them a fixed interpretation even as applied to theoretical entities.

Perhaps another reason to think that theories cannot be uniquely realized comes from the idea that theoretical terms are somehow partially interpreted. It seems that the stronger a theory is, the better its theoretical terms are interpreted. If the postulate of the theory is a tautology, for instance, the theoretical terms are not interpreted at all. It is tempting to explain this by saying that the stronger theory has fewer realizations. Since no consistent theory interprets its terms so well that it could not have done better if it had been still stronger, it seems that unique realization—perfect interpretation—is a limit we can never reach. But this is a mistake. The stronger theory *may* have fewer actual realizations or it may not; but it *must* have less *risk* of multiple realization, and that is enough to explain why strength seems to make for better interpretation. On the other hand, the stronger theory must also have more risk of nonrealization.

Let us conclude, therefore, that the T-terms ought to name the components of the unique realization of T if there is one, and ought not to name anything otherwise. We can record our conclusion by laying down three meaning postulates. The first

$$\exists y_1 \ldots y_n \forall x_1 \ldots x_n (\mathrm{T}[x_1 \ldots x_n] \equiv . y_1 = x_1 \& \ldots \& y_n = x_n) \supset \mathrm{T}[\tau_1 \ldots \tau_n]$$

says that if T is uniquely realized, then it is realized by the entities named, respectively, by $\tau_1 \ldots \tau_n$. It is logically implied by the Carnap sentence of T. The second

$$\sim\exists x_1 \ldots x_n \mathrm{T}[x_1 \ldots x_n] \supset . \sim\exists x (x = \tau_1) \& \ldots \& \sim\exists x (x = \tau_n)$$

says that, if T is not realized, then $\tau_1 \ldots \tau_n$ do not name anything. It is logically independent of the Carnap sentence. The third

$$\exists x_1 \ldots x_n \mathrm{T}[x_1 \ldots x_n] \& \sim\exists y_1 \ldots y_n \forall x_1 \ldots x_n (\mathrm{T}[x_1 \ldots x_n] \equiv . y_1 = x_1 \& \ldots \& y_n = x_n) . \supset . \sim\exists x (x = \tau_1) \& \ldots \& \sim\exists x (x = \tau_n)$$

says that, if T is multiply realized, then, again, $\tau_1 \ldots \tau_n$ do not name anything. It disagrees with the Carnap sentence, inasmuch as the third postulate and the Carnap sentence together imply that T has at most one realization, but that conclusion ought not to follow from meaning postulates.

Now we have specified the denotations of the T-terms. What about their senses? We have specified their senses already. For we have specified their denotations in any possible world, not just here in our actual world. In any possible world, they are to name the components of whatever uniquely realizes T in that world, and they are to name nothing in that world unless T is uniquely realized there. We know what it is for an n-tuple of entities to realize T in an arbitrary possible world w: namely, the n-tuple satisfies the realization formula of T in the model determined jointly by the state of affairs in the possible world w and by the fixed interpretation of the O-vocabulary.

Here I rely on, but do not argue for, a doctrine of senses due originally to Carnap.[6] I am supposing that the sense of a name—at least, of a name that cannot be decomposed into constituents—is given in full by specifying what (if anything) it names in each possible world. If we like, we can say that a sense *is* a function from (some or all) possible worlds to named entities. The most important objection to this doctrine is that possible worlds and their unactualized inhabitants are occult; I have argued elsewhere[7] that they are no more occult than the infinite sets we have learned to live with, and just as useful in systematic philosophy.

Putnam, arguing against Carnap's notation of partial interpretation, has objected that "theories with false observational consequences have *no* interpretation (since they have no model that is 'standard' with respect to the observation terms). This certainly flies in the face of our usual notion of an interpretation, according to which such a theory is *wrong*, not *senseless*."[8] The objection is mistaken. The theoretical terms of such a theory are not senseless, just *denotationless*. Their sense is given by their denotation in those possible worlds in which the theory *is* uniquely realized and, therefore, does *not* have false consequences. They have just as much sense as the denotationless term 'Santa Claus'.

A *logically determinate* name is one which names the same thing in every possible world. Its sense is a constant function. Numerals, for instance, seem to be logically determinate names of numbers.

[6] *Meaning and Necessity*, 2nd ed. (Chicago: University Press, 1956), pp. 181–182.

[7] *Convention* (Cambridge, Mass.: Harvard, 1969), p. 208.

[8] "What Theories Are Not," p. 247.

But 'the number of solar planets' is a logically indeterminate name of a number. Here in our actual world it happens to name the number 9. Elsewhere it may name other numbers, or nothing at all. *Anything* can have a logically indeterminate name—even a property. (This should have been obvious, but wasn't.) Take the name 'the physical property detected by means of the instrument with catalog number 12345 in so-and-so catalog' (filling in the name of a catalog). This happens, perhaps, to name the property of fluorescing in the ultra-violet. But if the catalog numbers had been different, it would have named some other property, or none.

We should notice that T-terms purporting to name properties will normally turn out to be logically indeterminate names of properties. Likewise for other T-terms; but let us stick to the special case. Suppose τ_1 is a T-term purporting to name a property. More precisely: T is formulated in such a way that the postulate of T cannot be true unless τ_1 names a property. It follows that if any n-tuple of entities uniquely realizes T in any possible world, the first component of that n-tuple is a property. But unless T is a very special theory, different n-tuples with different first components will uniquely realize T in different possible worlds. The sense of τ_1 will not be a constant function.

The logical indeterminacy of τ_1 makes for subtle equivocation in any context in which possible worlds other than our actual world are under discussion, either explicitly or implicitly. For instance, there will be trouble whenever τ_1 occurs in the scope of a modal operator, in the scope of a nomological operator, in the scope of an epistemic operator, or in a subjunctive conditional. We have to keep track of when τ_1 names the first component of the unique realization of T (if any) in our actual world, and when τ_1 instead names the first component of the unique realization of T (if any) in some other possible world under discussion.

For instance, someone might plausibly object that, on my account, it is impossible for T to have a unique realization but for the first component thereof not to be the property named by τ_1. But that does seem to be possible. Just consider the property named by τ_1. Fix your attention on that property, whatever it is. Now surely T could have a unique realization in which *that* property was not the first component! So what is impossible according to me is really possible. I reply that the objection commits a fallacy of equivocation. What I assert is this: for no possible world w is it the case that T has a unique realization in w but that the first component thereof is not the property named by τ_1 in the world w. What the objector properly denies is this: for no possible world w is it the case that

T has a unique realization in w but that the first component thereof is not the property named by τ_1 in our actual world. Unfortunately, what I assert and what he properly denies can both be expressed by this ambiguous sentence: "It is impossible for T to have a unique realization but for the first component thereof not to be the property named by τ_1."

A similar difficulty arises over the name 'the property of having τ_1'. This name purports to name a property. We would suppose offhand that it names the same property as τ_1 itself. Indeed, we would suppose it has the same sense as τ_1. But at least on one reading it and τ_1 do not name the same property, neither in our actual world nor in any other possible world.

I take it that a property is identified when, and only when, we have specified exactly which things have it in every possible world. And I take it that a name of the form 'the property of doing so-and-so' names the property that belongs, in any world w, to exactly those things which, in the world w, do so-and-so. For instance, 'the property of having τ_1' names the property that belongs, in any world w, to exactly those things which, in the world w, have the property named by τ_1.

Now we can see the problem. Do we mean: (1) the property that belongs, in any world w, to exactly those things which, in the world w, have the property named by τ_1 *in our actual world?* That, of course, is just the same property which is named by τ_1 in our actual world. On this first reading, 'the property of having τ_1' and τ_1 do both name the same property.

Or do we rather mean: (2) the property that belongs, in any world w, to exactly those things which, in the world w, have the property named by τ_1 *in the world w?* On this second—and, I believe, better—reading, 'the property of having τ_1' is a logically determinate name of a certain property, which we may call the *diagonalized sense* of τ_1. The sense of τ_1 may be represented by a function $\|\tau_1\|$ which assigns to any world w a property $\|\tau_1\|_w$. A property in turn may be represented by a function P which assigns to any world w the set P_w of things which, in the world w, have the property. Then the diagonalized sense of τ_1 is the property whose representing function assigns to any world w the set of things $(\|\tau_1\|_w)_w$. It is not named by τ_1 in any world, unless T is a very peculiar theory. Neither is it the sense of τ_1; that is not a property at all, but rather a function from worlds to properties.

THE DEFINITIONS OF T-TERMS

Given our conclusion that the T-terms $\tau_1 \ldots \tau_n$ should denote the components of the unique realization of T if there is one, and should not denote anything otherwise, it is natural to define the T-terms by means of definite descriptions as follows.

$$\tau_1 = \imath y_1 \exists y_2 \ldots y_n \forall x_1 \ldots x_n (\mathrm{T}[x_1 \ldots x_n] \equiv . y_1 = x_1 \& \ldots \& y_n = x_n)$$

$$\ldots$$

$$\tau_n = \imath y_n \exists y_1 \ldots y_{n-1} \forall x_1 \ldots x_n (\mathrm{T}[x_1 \ldots x_n] \equiv . y_1 = x_1 \& \ldots \& y_n = x_n)$$

These are to be our *definition sentences* for the theory T. The first, for instance, says that τ_1 names that entity which, followed by some $n-1$ entities, comprises an n-tuple identical with all and only n-tuples that realize T. That is to say, τ_1 names the first component of the unique realization of T.

These definitions work properly, under the treatment of denotationless terms we have chosen. They are valid: true in any model in which the T-terms are interpreted as specified in the previous section, whether or not T is uniquely realized therein. They are jointly equivalent to the set of meaning postulates we put forth to replace the Carnap sentence of T. They do not imply any O-sentences except logical truths.

We can see now why it was worth the trouble to adopt Dana Scott's treatment of denotationless names rather than its more familiar alternatives.[9] Under Russell's theory of descriptions, the definition sentences are disguised existential quantifications, false unless T is uniquely realized. Under the truth-value-gap theory of Frege and Strawson, atomic contexts of denotationless names or descriptions have no truth value; so the definition sentences are neither true nor false unless T is uniquely realized. Either way, the definition sentences would not be valid identities, true whether or not T is uniquely realized. We would be forced to regard them as metalinguistic assertions of synonymy, or something of the sort. Under the chosen-individual theory of Frege and Carnap, improper descriptions name some arbitrarily chosen individual. So do the theoretical terms of unrealized or multiply realized theories, if the definition sentences are to be valid. But suppose T is multiply realized, and suppose the n-tuple consisting of the chosen individual taken n times over happens to be one of the realizations of T. In this case, by accident, the postulate of T turns out to be true, contrary to our decision that multiply realized theories should be false.

THE EXPANDED POSTULATE OF T

Given our definitions, we can eliminate T-terms in favor of the definite descriptions whereby we have defined them. Replacing each

[9] For a survey of the alternatives, see David Kaplan, "What Is Russell's Theory of Descriptions?" *Physics, Logic, and History* (New York: Plenum, 1969), ed. by Wolfgang Yourgrau.

T-term by its definiens throughout the postulate of T, we obtain an O-sentence which we may call the *(definitionally) expanded postulate* of T. It says T is realized by the n-tuple consisting of the first, second, ..., nth components of the unique realization of T. The expanded postulate is, of course, definitionally equivalent to the postulate. That is, the postulate together with the definition sentences is logically equivalent to the expanded postulate together with the definition sentences.

The expanded postulate says that T is uniquely realized. It is logically equivalent to a shorter O-sentence that says so in a more straightforward way; we may call this the *unique-realization sentence* of T:

$$\exists y_1 \ldots y_n \forall x_1 \ldots x_n (\mathrm{T}[x_1 \ldots x_n] \equiv . y_1 = x_1 \& \ \ldots \ \& y_n = x_n)$$

Thus the postulate is definitionally equivalent to the unique-realization sentence. That is what we should expect, given our decision—*contra* Carnap—to interpret the T-terms in such a way that the postulate is true if and only if T is uniquely realized.

Still we may have misgivings. The expanded postulate is an O-sentence stronger than the Ramsey sentence of T, which said merely that T had *at least* one realization. Yet if the definition sentences are part of T, the expanded postulate is an O-theorem of T. So the definitions are giving us O-theorems that could not have been derived without them. That means that the definitions themselves, unlike the Carnap sentence, are not logically implied by the postulate.

Therefore, if I want to contend that the definition sentences of T are correct definitions, I must give up the idea that the theorems of T are all and only the logical consequences of the postulate of T. I am quite willing to give up that idea. I contend that the theorist who proposed T by asserting the postulate of T explicitly, labeling it as the postulate of a term-introducing theory, has also implicitly asserted the definition sentences of T. I contend that his audience *will* take him to have implicitly asserted the definition sentences of T. That is an empirical hypothesis about the conventional semantics of our language. To test it, we should find out what would happen if the audience came to think that T was multiply realized.

If they would thereupon call T a false theory, that confirms my account of the interpretation and definitions of the T-terms. If they would call T a true theory, that refutes my account and confirms Carnap's. What if they would be unable to decide which to say? Then my account and Carnap's are rival proposals for rational reconstruction within the range of free choice left open by our conventions. My proposal has the advantage that it permits theoretical terms to be fully interpreted and explicitly defined. Carnap's proposal has the advantage that it lets us live with multiply realized theories. But I see no reason why we need to be able to live with multiply realized theories.

The expanded postulate (and its logical equivalent, the unique-realization sentence) are not by any means the only interesting definitional equivalents of the postulate of T. For any T-term τ_i, consider the sentence that says τ_i names something: '$(\exists x)(x=\tau_i)$'. This sentence is definitionally equivalent to the postulate of T and to the expanded postulate of T (but logically equivalent to neither). That is what we should expect: T is uniquely realized if and only if there is something which is the ith component of the unique realization of T.

Therefore the postulate and expanded postulate of T are definitionally implied by any sentence which contains T-terms and which comes out false in case any of its T-terms are denotationless. There are many such sentences. Let τ_1, τ_2, and τ_3 be T-terms purporting to name, respectively, a property, a relation, and a function; and let ζ and η be O-terms purporting to name individuals. Let ν be an O-term naming a real number. Now consider such sentences as: 'ζ has the property τ_1', 'ζ bears the relation τ_2 to η', or 'the value of the function τ_3 for the argument ζ is ν'. Clearly each of these sentences should be false if its T-term is denotationless. Therefore each of them definitionally implies the postulate and expanded postulate of T.

We would probably say, at first sight, that sentences such as these purport to state particular facts. Yet the theoretical postulates they imply—or some conjuncts thereof—will usually purport to state laws of nature, perhaps even laws of unrestricted universality. Therefore it is possible that a particular fact might be explained by a covering-law explanatory argument in which the explanans contain no explicit statement of any law! The required covering laws would be available by implication from certain particular-fact premises in the explanans. It seems that, if my account of theoretical terms is correct, the covering-law analysis of scientific explanation is in need of a slight reformulation. Take one of Hempel's definitions of deductive-nomological explanation:

> ...deductive arguments whose conclusion is the explandum sentence...; and whose premiss-set, the explanans, consists of general laws..., and of other statements..., which make assertions about particular facts.[10]

[10] *Philosophy of Natural Science* (Englewood Cliffs, N.J.: Prentice-Hall 1966), p. 51.

I contend we ought rather to say:

> ...deductive arguments whose conclusion is the explanandum sentence; and whose premiss-set, the explanans, consists of statements which make assertions about particular facts and *perhaps* also of statements which assert general laws; and whose premiss-set implies statements which assert general laws.

I am confident that my objection goes against the letter, not the spirit, of Hempel's covering-law analysis.

DERIVED BRIDGE LAWS FOR T

We can safely assume that there is a period, after the new theory T has been proposed, during which the T-terms retain the interpretations they received at the time of their introduction. At least for a while, our definition sentences for T remain valid. Suppose that, during this period, T is reduced by means of some other accepted scientific theory T^*. Let this be a reduction in which T survives intact; not, what is more common, a simultaneous partial reduction and partial falsification of T by T^*. We shall consider how this reduction might take place.

The reducing theory T^* need not be what we would naturally call a single theory; it may be a combination of several theories, perhaps belonging to different sciences. Parts of T^* may be miscellaneous unsystematized hypotheses which we accept, and which are not properly called theories at all. Different parts of T^* may have been proposed or accepted at different times, either before or after T itself was proposed.

The most interesting case, however, is that in which T^* is well systematized, and at least part of T^* is newer than T. It is in that case that the reduction of T by means of T^* is likely to be an important advance toward the systematization of all empirical knowledge.

T^*, or parts of T^*, may introduce theoretical terms; if so, let us assume that these T^*-terms have been introduced by means of the same O-vocabulary which was used to introduce the theoretical terms of T. This is possible regardless of the order in which T and T^* were proposed. Any term that is either an O-term or a T^*-term may be called an O^*-term; so at the time T is reduced, the relevant part of our scientific vocabulary is divided into the T-vocabulary and the O^*-vocabulary.

Suppose the following O^*-sentence is a theorem of T^*; we may call it a *reduction premise* for T: '$\mathrm{T}[\rho_1 \ldots \rho_n]$'. The terms $\rho_1 \ldots \rho_n$ are to be names belonging to the O^*-vocabulary. They may be elementary expressions and belong to the O^*-vocabulary in their own right, or they may be compound expressions—for instance, definite descriptions—whose ultimate constituents belong to the O^*-vocabulary. The reduction premise says that T is realized by an n-tuple of entities named, respectively, by the O^*-terms $\rho_1 \ldots \rho_n$. Notice that it cannot be true if any of those O^*-names are denotationless.

A reduction premise for T does not imply the postulate of T either logically or definitionally; nor conversely. But the postulate does follow logically from the reduction premise together with a set of *bridge laws* for T, as follows:

$$\rho_1 = \tau_1$$
$$\ldots$$
$$\rho_n = \tau_n$$

The bridge laws serve to identify phenomena described in terms of the reduced theory T with phenomena described in terms of the reducing theory T^* (including the O-vocabulary); and via these identifications, T can be derived from T^*.

But where do we get the bridge laws? The usual view[11] is that they are separate empirical hypotheses, independent of the reducing theory T^*. When T^* yields a reduction premise for T, we have the opportunity to choose between two rival bodies of theory. We may choose T^* augmented with bridge laws; if so, we can derive T, so we do not have to posit it. Or we may choose T^* without bridge laws, in which case we will have to posit T separately. Given this choice, we take whichever body of theory is better—more systematized, parsimonious, simple, credible, or what have you. We must decide whether the gain in systematization from reducing T is worth the loss in systematication from adding bridge laws. If it is, we choose to accept the bridge laws and perform the reduction.

The usual view assumes that it is impossible to derive the bridge laws from the unaugmented reducing theory T^*, since the bridge laws contain essential occurrences of the T-terms and these do not occur in T^*. But consider the *definitionally expanded bridge laws:* the sentences obtained from the bridge laws when we replace the T-terms by definite descriptions, according to our definition sentences for T.

$$\rho_1 = \iota y_1 \exists y_2 \ldots y_n \forall x_1 \ldots x_n (\mathrm{T}[x_1 \ldots x_n] \equiv . y_1 = x_1 \& \ldots \& y_n = x_n)$$
$$\ldots$$
$$\rho_n = \iota y_n \exists y_1 \ldots y_{n-1} \forall x_1 \ldots x_n (\mathrm{T}[x_1 \ldots x_n] \equiv . y_1 = x_1 \& \ldots \& y_n = x_n)$$

The definitionally expanded bridge laws are O^*-sentences. No incompatibility of vocabulary prevents them from being theorems of

[11] Reviewed in the first two sections of John Kemeny and Paul Oppenheim, "On Reduction," *Philosophical Studies*, VII (1956): 6–13. Since Kemeny and Oppenheim take the theoretical terms as predicates, their bridge laws are universally closed biconditionals rather than identities.

T^*. Yet they are definitional equivalents of the bridge laws; so if they are theorems of T^*, then T^* definitionally implies the bridge laws. There is no need for any empirical hypothesis other than theorems of T^*.

If T^* yields as theorems a reduction premise for T, and also a suitable set of definitionally expanded bridge laws for T, then T^*—without the aid of any other empirical hypothesis—reduces T. For T^* definitionally implies the postulate of T, as well as a set of bridge laws. Once T^* is accepted, there is no choice whether or not to reduce T. The reduction of T does not need to be justified by considerations of parsimony (or whatever) over and above the considerations of parsimony that led us to accept T^* in the first place.

It is useful to observe that the set of definitionally expanded bridge laws is logically implied by the following O^*-sentence, which we may call an *auxiliary reduction premise* for T:

$$\forall x_1 \ldots x_n (\mathrm{T}[x_1 \ldots x_n] \equiv . \rho_1 = x_1 \& \ldots \& \rho_n = x_n)$$

It says that, unless one of the terms $\rho_1 \ldots \rho_n$ is denotationless, T is uniquely realized by an n-tuple of entities named, respectively, by $\rho_1 \ldots \rho_n$. Hence the reduction premise and the auxiliary reduction premise together say just this: T is uniquely realized by an n-tuple of entities named, respectively, by $\rho_1 \ldots \rho_n$. That is what T^* must imply in order to reduce T by means of derived bridge laws. In that case, indeed, we can by-pass the bridge laws. T^* definitionally implies T by an alternate route. Since T^* guarantees that T is uniquely realized, T^* logically implies the unique realization sentence of T, which is logically equivalent to the expanded postulate of T and definitionally equivalent to the postulate of T.

Let us briefly examine two examples of reduction by means of derived bridge laws.

(1) Let T be a theory explaining the operation of a machine by means of transition laws of the form: when the machine is in state τ_i, so-and-so input causes it to go into state τ_j and produce such-and-such output. The T-terms are $\tau_1 \ldots \tau_n$, purporting to name states of the machine. Later we obtain T^*: an account, purporting to be complete, of the internal structure of the machine and the principles on which it works. Then, unless T^* falsifies T, we should be able to form state-names $\rho_1 \ldots \rho_n$ in the mechanical vocabulary of T^*, such that T^* implies that the states thus named are related to one another, and to the appropriate inputs and outputs, by the given transition laws. Thus T^* would yield a reduction premise for T. But also, since T^* purports to be a complete account of the working of the machine, we can reasonably expect T^* to leave no room for a second realization of T. T^* could imply that every

n-tuple of states describable in its mechanical vocabulary, except for the n-tuple named by $\rho_1 \ldots \rho_n$, disobeys the given transition laws; and according to the completeness claimed by T^*, no states not so describable need be considered in explaining the input-output relations of the machine. (For instance, what we think we know about ordinary electromechanical machines seems to imply that the only states causally responsible for outputs are states describable in terms of positions and momenta of moving parts, currents, and voltages.) If so, T^* yields an auxiliary reduction premise for T; hence T^* together with the definitions of the state-names $\tau_1 \ldots \tau_n$ suffices to imply the bridge laws: '$\tau_i = \rho_i$'.

(2) Let T be a theory explaining the regulation of certain biological processes by positing hormones $\tau_1 \ldots \tau_n$: chemical substances of unspecified composition, secreted by specified cells under specified conditions and regulating the rates of specified chemical reactions in a specified way. The T-terms $\tau_1 \ldots \tau_n$, in this case, purport to name substances. Let T^* comprise our body of biochemical knowledge at some later time; T^* might imply that certain substances named by chemical formulas $\rho_1 \ldots \rho_n$ realized T, and that they alone did so. To exclude multiple realization of T, T^* would have to contain the information that, e.g., a certain gland secretes *nothing but* the substance with formula ρ_1; but we often do have such knowledge.

Not only is it possible for a theory to be reduced by means of derived bridge laws; we can even regard all possible reduction of theories as working in this way. Instead of saying that in some cases the bridge laws are posited independently of the reducing theory, we may rather say that in some cases the reducing theory must be strengthened ad hoc before it yields the bridge laws we want.

Suppose we want to reduce our theory T; and we accept a theory T^* which yields as a theorem some reduction premise for T, but does not yield the corresponding set of definitionally expanded bridge laws. Let T^{**} be the theory obtained from T^* by providing these definitionally expanded bridge laws. T^* does not reduce T by means of derived bridge laws, but only by means of bridge laws posited independently of T^*. T^{**}, a theory obtained by suitable strengthening of T^*, does reduce T by means of derived bridge laws. How shall we describe what happens when we add bridge laws to T^* in order to reduce T? If we say that T^* is the reducing theory, we must say that T^* reduces T by means of independently posited bridge laws. But we could just as well say that T^{**} was the reducing theory, and that we strengthened T^* to T^{**} in order that T^{**} might reduce T by means of derived bridge laws. Of course, it does not matter very much which we say. Either way, the bridge laws have been posited ad hoc in return for the reduction of *T;* for T^{**} is definitionally equivalent to T^* plus the bridge laws.

Still, I think it best to say that the strengthening of the reducing theory is a precondition, not a part, of the reduction. For that will remind us that strengthening is not necessarily needed. A body of theory already accepted before we thought of using it to reduce *T* may already be strong enough, without any more strengthening ad hoc, to reduce *T* by means of derived bridge laws. I do not know whether most cases of theoretical reduction are of this sort; but we should at least leave the possibility open, as the usual account of reduction does not.

LATER REVISIONS OF *T*

So far, we have discussed the interpretation of *T*-terms only at the time of their introduction, the time when their parent theory *T* is first proposed. It remains to ask what happens later when *T* is amended and extended. This matters especially in connection with reduction, since theories do not often survive reduction intact. More often the original theory is falsified while a corrected version is reduced. If *T* is thus partially reduced and partially falsified, or revised for any other reason, do the *T*-terms retain their meanings?

We might say that the *T*-terms should always be defined using the currently accepted version of *T*. As *T* is corrected, modified, extended, or perhaps even when we accept miscellaneous hypotheses that contain *T*-terms but do not belong integrally to any version of *T*, the *T*-terms gradually change their meaning. In particular, they change their meanings in the revisions preparatory to partial reduction of *T*.[12] But these are very peculiar changes of meaning—so peculiar that this position seems to change the meaning of 'change the meaning of'. They occur continually, unnoticed, without impeding communication. We might try saying that a *small* enough change of meaning is not really a change of meaning, but that would imply that enough nonchanges in meaning could add up to a change in meaning. The position has other problems. How are we to define the theoretical terms of defunct or never-accepted theories? Should we use the best-known version, or what? What if different scientists accept slightly different versions of *T*, disagreeing, say, on the exact value of some physical constant? We ought not to say they give the *T*-terms different meanings. What do the *T*-terms mean if we have suspended judgment between two slightly different versions of *T*?

[12] As suggested by Paul Feyerabend, "Explanation, Reduction, and Empiricism," *Minnesota Studies in the Philosophy of Science* III (Minneapolis: University of Minnesota Press, 1962), ed. by H. Feigl and G. Maxwell.

We might therefore prefer to say that the T-terms keep the meanings they received at their first introduction. They should still be defined using the original version of T even after it has been superseded by revised versions.

This position will work only if we permit the T-terms to name components of the nearest near-realization of T, even if it is not a realization of T itself. For after T has been corrected, no matter how slightly, we will believe that the original version of T is unrealized. We will want the T-terms to name components of the unique realization (if any) of the corrected version of T. They can do so without change of meaning if a realization of the corrected version is also a near-realization of the original version.

According to this position, we may be unable to discover the meanings of theoretical terms at a given time just by looking into the minds of the most competent language-users at that time. We will need to look at the past episodes of theory-proposing in which those terms were first introduced into their language. The working physicist is the expert on electrons; but the historian of physics knows more than he about the meaning of 'electron', and hence about which things could truly have been called electrons if the facts had been different. If we were ignorant of history, we could all be ignorant or mistaken about the meanings of words in common use among us. This situation is surprising, but it has precedent: a parallel doctrine about proper names has recently been defended.[13] To know what 'Moses' means among us it is not enough to look into our minds; you must look at the man who stands at the beginning of the causal chain leading to our use of the word 'Moses'.

I do not wish to decide between these alternatives. Either seems defensible at some cost. I hope the truth lies in between, but I do not know what an intermediate position would look like.

DAVID LEWIS

Princeton University

[13] See David Kaplan, "Quantifying In," *Synthese*, XIX (1968): 178–214.

*Saving the Noumena**

LAWRENCE SKLAR
The University of Michigan

> I want to consider another defense which I find most unsatisfactory. It might be called the *positivist defense;* and some philosophers seem to think it is the only defense available to someone who wants to claim that the apparent conflict between two theories is merely verbal. According to the positivist defense, *whenever* we have two theories that have all the same observational consequences, any apparent disagreements between the two theories are merely verbal ones. Call this the *positivist principle*. . . . For instance Sklar's discussion of conventionalism about geometry seems to presuppose this view. In all the standard cases of alternate geometries (plus compensating adjustments elsewhere in the physical theories), the geometric objects of one theory are definable out of the geometric objects of the other. Sklar obscures this fact by comparing the conflict between alternative geometries to the conflict between the normal world-view and Descarte's "evil demon" hypothesis; but this latter example is one where the objects of one theory are clearly not definable in terms of the objects of the other, so only by some form of the positivist principle could one claim that the conflict between those theories was purely verbal.
>
> Hartry Field[1]

I

Most, but not all, philosophers would agree that surface differences between two theory presentations might mask underlying identity of theory. Sometimes the verbal presentations of a theory may differ in a way which does not even suggest incompatibility, for example in the case of the presentation of a theory in two different natural languages. Few would want to argue that Newton's *Principia* in Latin does not present the same theory as that work in English. More interesting are those cases where only one natural language is involved, and where, at least on the surface, some sort of incompatibility or other seems present, but where many, if not all, would agree that this surface incompatibility is deceptive.

Are our theories of the world really incompatible if they differ simply in that you talk of temporally extended physical objects whereas I deal with time-slices as my basic particulars? More radically, even if it is a case of a thing theory versus a theory which

*My thanks to Steve White for the title and to Tim McCarthy for many useful discussions.

takes spatio–temporal regions as primitive and treats things as substantival predicates of such regions, many would certainly argue that we have here a case of theoretical equivalence. Hardly anyone would deny that the Schrödinger and Heisenberg "representations" are, indeed, representations of one and the same theory, despite the fact that in the former the state function varies with time and the operators do not and in the latter the reverse is the case.

A more controversial case from physics is the famous pair, curved spacetime vs. flat spacetime with compensating metric and force fields. Here some (Eddington, Schlick, Reichenbach) assure us that we have a classic case of theoretical equivalence, while others (at various times Putnam and Glymour, for example) think, rather, that the "theories" are not equivalent at all and that, indeed, the former is much preferable to the latter as a plausible account of the data. We will have more to say about this example later. Finally we have the positivist philosophers' paradigm of equivalence, material world vs. malevolent demon or brain–in–a–vat, an alleged equivalence which, as a consequence of positivism, is taken by many (including Field, above) as a *reductio ad absurdam* of the positivist position.

The positivist position on theoretical equivalence, whatever one thinks of its adequacy or ultimate tenability, is at least clear. It is especially designed to handle those cases where we want to claim that surface incompatibility masks underlying commonality of import. The line is familiar. One discriminates among the consequences of the theory a special class of sentences which allegedly contain among them all the consequences of the theory open to empirical determination of truth independently of assuming the truth of the theory in question. If two theories agree on this observational set then they are alleged equivalent, no matter how radical the apparent incompatibility which resides in the remaining, theoretical, level of the two theories. The vices of the positivist program are well known, but here I would like to remind us of some of its virtues.

However fraught with untenable dualisms, incoherent foundationalist notions, etc., positivism may be, it at least offers an account of theoretical equivalence which brings into a harmonious whole a theory of meaning, of evidence, of ontology, of truth, of explanation and of equivalence itself. Observationally equivalent theories do not differ in meaning since the total meaning of the theory is exhausted by its set of observational consequences. Since the consequences of the theory exhaust the totality of possible evidence for or against a theory, evidentially equivalent theories are automatically genuinely equivalent, blocking any opening for skepticism. One need not take the apparently conflicting ontological claims of the theories seriously, since the ontology we are committed to by apparently referring terms at the theoretical level is only *façon de parler* in any case. Nor need we be concerned about which theory is true, since, once again, all the theories really say is what their observational consequences say, and at that level they say the same thing. Finally we are offered an account of explanation, basically the subsumption of particular correlated observable events under general rules of constant conjunction, which makes it clear that despite apparent differences theories equivalent in the positivist sense offer the same explanations of events, if explanation too is positivistically understood. I am, of course, laying all this out here grotesquely sketchily and without anything like the necessary attention to either detail or to variance and controversy within the positivist camp, but it will have to do for the time being.

Nor will I be concerned here with the manifold, familiar objections to the positivist program, replete with the well-known proofs of its alleged impossibility and absurdity. I will focus, rather, on one particular kind of realist alternative to positivism. In particular I will explore one way a realist might try to offer an alternative to the positivistic account of theoretical equivalence and will explore the problems encountered in such a program which an anti-realist might put forward as reasons for skepticism regarding the viability of a notion of equivalence differing from his own. I make no claim that the realist account of equivalence I will offer is the only one a realist might offer, nor that this particular account of equivalence could not possibly overcome the difficulties I will raise in its path. I do think, however, that the account I offer does capture the instincts many realists have had regarding theoretical equivalence, and the problems I lay out do capture the skeptical doubts regarding such a realist program latent in many recalcitrant positivist's skeptical remarks.

What is wrong with the positivist notion of equivalence according to the realist? And how is the central deficiency of this account of equivalence to be remedied? What is wrong with the positivistic notion, according to the realist I have in mind, is this: Designed to handle plausible cases of theoretical equivalence, say the Schrödinger and Heisenberg formulations of quantum mechanics, the positivistic notion of equivalence misconstrues just what it is about these two theory presentations that makes them merely two expressions of one and the same theory. His misconstrual leads him, fallaciously, to believe that any reasonable commitment to the equivalence of theories must force one by an irresistible slippery–slope argument to such absurd conclusions as the equivalence of a realistic account of the world with an extended–dream account or a brain–in–a–vat account. The source of this misconstrual, says my realist, is the positivist's failure to take sufficient account of the interrelationships theories can bear to one another at the theoretical level. Once one sees that in all the plausible cases of theoretical equivalence there is a far stronger relationship between the two theory presentations than merely a common ability to save the phenomena, a relationship at the theoretical level, one will realize that it is this stronger relationship which warrants the justified claim of theoretical equivalence. But then we will realize that we are by no means forced from the acceptance of such genuine cases of equivalence into any assertions of equivalence of the kind devastating to the realist program.

So what I wish to explore is this: What is the structural relationship between theories sufficient for a realist to attribute equivalence to them? In particular, how does this additional requirement go beyond the weak positivist constraint of merely having all observational consequences in common? I will argue that the realist will find, unless he is very careful, that his program all too often seems to lead him into assuming far more in common with the positivist than he might at first think. And, I will argue, it is incumbent upon the realist to offer us not just a notion of equivalence more stringent than that of the positivist, but an integrated theory of meaning, truth, ontology, confirmation and explanation into which his notion of equivalence will naturally fit. I certainly will not maintain that this is impossible for the realist, but only insist that profound difficulties must be faced up to by any realist who takes this program seriously enough to counter the positivist's claim that, dubious as his account may be, it at least offers us a fully integrated, coherent account of all these meta–theoretical aspects of theories.

II

The realist I have in mind argues that it is not sufficient for theoretical equivalence that the two theories in question save all the same phenomena. A more profound and tighter structural interrelationship between the theories, an interrelationship which takes account of the relationship of the structure of one theory at the theoretical level to the structure of the other theory at its theoretical level, is required for genuine equivalence. What could this additional structural component be?

One thing we can be sure of. Whatever this structural "isomorphism" is to be, it cannot be a purely formal notion. It cannot be, that is, an interrelationship which can be determined to hold solely on the basis of the logical form of the theories in question. Why not? We will consider the strongest such possible formal interrelationship one can imagine. Let us suppose that the theories are term by term "inter–translatable," that is, that each can be obtained from the other merely by a substitution of terms of one theory for terms of the other. Would that be enough to show the theories equivalent?

Surely not. Let the two theories be 'All lions have stripes', and 'All tigers have stripes', with all the words in both theories taking on their usual meanings. The theories are inter–translatable in the purely formal sense. They are exactly alike in logical form and one can be obtained from the other by a simple term for term substitution. But they are most assuredly not equivalent. I am not denying that if we found a speaker who persistently asserted that lions had stripes and tigers didn't that we would probably take him to mean by 'tiger' what we mean by 'lion' and vice versa. Nor am I denying that it is just the question of the meaning of the theoretical terms of a theory that is at issue in many questions about the necessary and sufficient conditions for theoretical equivalence. We will return to that issue in detail. All I am claiming here is that mere commonality of logical form, even of a total theory when compared with another total theory, is certainly not by itself sufficient for theoretical equivalence. The meanings of the terms in the theories, however construed, are crucial to questions of equivalence.

But "translatability" of one theory to another, in some sense, is just the additional constraint, over and above saving the phenomena, that some realists want to demand as necessary for theoretical equivalence. So if that isn't the purely formal notion of commonality of logical form, what is it?

Two options suggest themselves. If we have some grasp of the meanings of the terms of the theories which comes to us from outside the role these terms play in the theories, and if on the basis of this knowledge of meanings we can affirm the logical equivalence of the theories, then, of course, the theories are equivalent. Straightforward translations of a theory phrased in one language to a version of the theory in some other language, where the theories are small fragments of the totality of beliefs of speakers of the languages and where the full vocabulary of the theories appears embedded in a far broader context than the particular theory in question and in such a way that we are inclined to say that we grasp the meanings of the terms independently of the role they play in the theories in question, are of this sort. Thus we won't have any trouble affirming that the theory that 'Salt is white', is equivalent to the theory that 'Salz ist weiss'. But this notion of equivalence will hardly be of help to us in the most interesting and crucial cases. Our conviction that Schrödinger quantum mechanics is equivalent to Heisenberg quantum mechanics hardly comes from some outside grasp of the meanings of the respective psi–functions which informs us that one simply means the time transform of the other.

So what does assure us over and above mere formal similarity, which we saw was never enough, that in this and similar cases the theories are genuinely equivalent to one another? I think the answer is clear. We believe that we have no antecedent understanding of the meanings of the theoretical terms (psi–functions and operators in the example of quantum mechanics) other than the role they play in the theories in question. We have no external grasp on the meanings of these terms that we could go to to determine whether the existence of a formal mapping from one theory to the other did or did not constitute a genuine translation, and hence a genuine demonstration of theoretical equivalence. Rather, all we have is this: Relative to whatever understanding we have of the notion of observational consequence of a theory, we are satisfied that the two theories in question really do have all their observational consequences in common. Over and above this we can demonstrate the existence of a formal mapping of some appropriate kind between the theories at the theoretical level. For the purpose of our argument it really won't matter very much just exactly what this mapping, necessary to show appropriate commonality of theoretical structure, amounts to. Together these *two* features of the theory, commonality of observational consequences and the existence of the appropriate structural mapping at the theoretical level, are taken to be enough to establish theoretical equivalence.

The intuition behind the position is clear. The meaning of the theoretical terms is fixed entirely by the role they play in their respective theories, in the holistic account of the meaning of theoretical terms now so familiar to us. The combined force of observational commonality and structural correspondence at the theoretical level is taken to be sufficient to establish commonality of meaning for the theoretical terms, their having, as they do, their meaning entirely determined by the place they occupy in the theoretical structure which generates the observational consequences. The details of the argument will depend on one's detailed account of the holistic structure of theories and on what one takes to be a sufficient structural interrelationship for theoretical equivalence. But consideration of such simple cases as theories which have common observational consequences and which can be obtained from one another by such simple operations as inter-changing of terms for terms ("You just are using 'positive charge' to mean what we meant by 'negative charge' and vice versa") will be enough to give us the general idea of the motivations and arguments of the program.

Seen from this point of view a realist conception of theoretical equivalence should not be viewed as a total rejection of the positivist account. Rather, it amounts to the claim that the positivist constraint of commonality in saving the phenomena, while necessary for theoretical equivalence, is simply not sufficient. Again the intuition is clear. Two theories might have all the same observational predictions but be so radically different in their structure at the theoretical level that one ought to take them as attributing (realistically) quite different explanatory structures to the world. Only commonality of structure at the level of the theoretical ontology introduced by the theories to explain the commonly predicted observational results is enough for us to say that the two accounts are genuinely, realistically, equivalent to one another.

Notice how much of the positivist account the realist, construed this way, must first accept. The project of characterizing a notion of theoretical equivalence in this way at least presupposes that the notion of commonality of observational consequences is a coherent one. The argument which takes Schrödinger and Heisenberg to be offering us the same theory must at least assume that some coherent limits to the notion of observationality can be given. If the psi-function itself could be considered, under any circumstances, an observational quantity, then the existence of the appropriate mathematical transformation from one representation to another would no more constitute a demonstration of equivalence than does the trivial formal transformation from 'Lions have stripes' to 'Tigers have stripes' demonstrate the genuine equivalence of those assertions. Of course the realist need not lay down where observationality ends and theoreticity begins. He need only be assured that the consequences common to the two theories (in the example chosen probabilities of outcomes of measurements in the quantum mechanical sense) *include* anything which could be called observational. He must, that is, be sure that the apparent incompatibilities between the theories, incompatibilities he will demonstrate to be only apparent by means of his equivalence–establishing interrelationship shown to hold among them, are all firmly "trapped" at the non–observational level. At least that much of the presuppositions of positivism is presupposed by this kind of realist as well.

To summarize, what my realist asserts is this: The point of theories is to introduce theoretically posited structures to explain the observable phenomena. Merely predicting the same phenomena is not enough for two theories to be equivalent, for they may explain these phenomena in radically different ways. But if the theories share both commonality of predicted phenomena *and* an, appropriately characterized, commonality of structure at the theoretical level, as demonstrated by the existence of the appropriate structural mapping between them at the theoretical level, then the theories plainly are equivalent. Any residual appearance of incompatibility must be due to merely verbal equivocation at the theoretical level. But realizing as we do, on the basis of the theory of the meaning of theoretical terms which tells us that the totality of their meaning is the place they play in the theoretical structure in question, that the apparent disagreement between the two theories is a mere superficiality of alternative verbal designations for common structural elements, we will not be deceived but will realize that, properly speaking, we have only one theory expressed in two misleadingly different ways. But it is essential for genuine equivalence that two conditions be met. On the observational level the theories must be identical. On the theoretical level they must bear the appropriate structural similarity.

III

What interrelationship at the theoretical level should the realist demand, over and above commonality in saving the phenomena, before he admits two theories to be equivalent? I hope I don't have to say, for each specific proposal be it term–by–term translation, common definitional extensions, or whatever, would require its own careful analysis. I hope that I will be able to say all I wish to say without pinning the imagined realist or myself down to a specific proposal. But then how can we say anything interesting about potential pitfalls in his realist path?

One obvious problem for the realist is getting his principles to coincide with his intuitions. He may, in the so-familiar manner, find that his proposal is too weak or too strong in its demands to coincide with what he takes intuitively to be genuine cases of equivalent theories and what he takes to be cases of non-equivalent theories which happen to save the same phenomena.

Reichenbach, for example, wanted to hold, along with Eddington and Schlick, that curved spacetime and flat spacetime with universal forces were equivalent theories. But on a model of equivalence suggested by *some* of his remarks, remarks to the effect that equivalent theories simply called the same theoretical entities by alternative names, the model of equivalence he should have in mind would require a strict term-by-term interdefinability of the two theories. This won't be the case, however, since the flat spacetime theory, being "ontologically otiose" compared to the curved spacetime theory, can't have its parameters defined from those of the ontologically more parsimonious theory. That is, given a flat spacetime metric and the "universal forces" we can determine spacetime curvature in the curved spacetime theory. But a full specification of curvature is not enough to *uniquely* specify the proper flat spacetime and universal forces. Crudely this is related to the nineteenth-century observation that uniform gravitational fields are empirically undetectable. (I will have more to say about this case later.) So one will either have to give up his stringent model of equivalence or renege on his intuition that these two theories are, indeed, equivalent. In the case of Reichenbach matters are more complex, of course. Actually what he does is to suggest in other, dominating places that commonality of observational consequences is sufficient for equivalence, which certainly saves the equivalence of curved spacetime and flat spacetime plus universal forces. Naturally, though, adopting this notion of equivalence plays havoc with his alleged scientific realism.[2]

One can see this problem of a conflict of intuition with principles arise in the other direction as well. Even quite stringent realist notions of equivalence may not be strong enough to exclude as equivalent pairs of theories the realist does not wish to think of as equivalent.

Surely the realist, although he wants Schrödinger and Heisenberg to say the same thing, does not want the usual wild Cartesian possibilities to count as theories equivalent to our ordinary scheme of the external material world. Now some Cartesian alternatives, say the one that tells us that there is simply nothing but our private experience and nothing "out there" to explain it, will be rejected as inequivalent to our ordinary theory because of a lack of structural similarity to our ordinary theory on the theoretical level. But other Cartesian fantasies, say the one in which all our private experience is caused by an appropriate signal fed into our tank–immersed brain along a cord (or caused by the appropriate state of the cybernetic machine at the other end of the cord) look as though they have the possibility of being made to look as structurally similar to our ordinary "external world" theory as we like. Indeed, what seems to differentiate the real cases of discovered scientific equivalence from these spurious Cartesian cases is not interdefinability in the former cases and not in the latter. Rather, what makes the scientific cases interesting scientifically and the Cartesian cases not is that it is a significant scientific task to demonstrate interdefinability in the scientific cases, whereas interdefinability is just too trivial to be interesting in the kind of Cartesian cases I have in mind.

Here quite a bit of caution is called for, though. There is the kind of realist who rejects the holistic–role–in–a–theory account of the meaning of theoretical terms. For him there is no problem in asserting that the Cartesian alternatives, no matter how structurally similar to our ordinary external–world account, are not equivalent to the ordinary account. But then just what account of theoretical equivalence he will offer, at least what account which will handle the cases we are interested in where we think equivalence is clear despite surface incompatibility, isn't clear. The realist we have in mind, espousing as he does the doctrine of meaning which gives to theoretical terms only the meaning they possess from their holistic role in the theories in question, will find it difficult, I think, to avoid attributing equivalence of the suitably constructed Cartesian alternatives to our ordinary account. But then his account of theoretical equivalence is sliding perilously close to the positivist account. For if the brain–in–a–vat account of the world is really equivalent to the ordinary material object world account, so long as the brain–in–a–vat account is suitably formally structured, then have we really gotten very far from merely saving the phenomena as sufficient for theoretical equivalence? Once again the tension between an "instrumentalistic" account of the meaning of theoretical terms and a realist account of theoretical ontology is clear. But our realist can just bite the bullet and affirm that, appearances to the contrary, such cleverly tidied–up Cartesian alternatives are simply equivalent to our ordinary world–view. Contrary to the view expressed in the quote of Field at the beginning of this paper, all "evil demon" hypotheses would not then be equal. Some would be genuine alternatives to our ordinary world–view, presumably to be rejected as unacceptable on some epistemic ground or other. Other "evil demon" hypotheses would just be our ordinary world–view dressed out in peculiarly misleading terminology. It all depends on whether or not our hypothesis about the doings of the evil demon has him failing or succeeding in structurally duplicating our ordinary world in his alternative causal structure of our private experiences.

IV

Let us suppose that our realist is satisfied that his formal notion of theoretical equivalence is consonant with his intuitions about just what pairs are pairs of genuinely equivalent theories. What further concerns must then arise?

One is fairly obvious given a principle component of the motivation behind the positivist position on equivalence to which our realist is opposed. Surely one of the major thrusts behind the positivistic notion is its ability to limit the possibilities of skepticism. Faced with alternative theories all equally compatible with all possible empirical data, we are at a loss as to how to decide which of these we ought to believe. Taking the positivist line and adjudicating all of these theories as saying the same thing, there is no longer a decision to be made. Hence there is less room for skepticism to intrude. We still are faced with the inference problem from observed empirical phenomena to full generalizations over all possible empirical phenomena, of course the problem of induction in its most general form, but at least we need no longer fret about alternatives at the level of the in–principle unobservable haunting us.

Once we insist on stricter conditions for equivalence than mere commonality at the empirical level, the problem of rationalizing the choice of alternatives to be believed is reinstated. Nor, from a realistic point of view, with its insistence on truth as the goal of inquiry and on a correspondence theory of truth (whatever that means) even at the theoretical level, do such ways out as allowing rationality to be permissive ("You are rational if you believe any one of the non–equivalent theories which saves the phenomena equally well") appeal. What we have a right to expect from the realist is a systematic account of what rationalizes the choice of one theory as true over any of the non–equivalent but empirically indiscriminable alternatives to it.

The realist can, of course, simply refuse the challenge and accept the skeptical consequences. But suppose he doesn't. What we will be offered is an account of the rationality of theory choice which will allow us to go beyond the conformity of a theory with all possible empirical data in deciding its worthiness for our belief. One problem will be, though, that we will need to be assured that the powers of this model of confirmation are sufficient to bring us up to the boundaries of equivalence. To avoid the intrusion of skepticism, that is, we will need assurance that for any two non–equivalent theories one will be preferred epistemically to the other. In other words there will have to be a close internal harmony between the notion of theoretical equivalence offered and the notion of the epistemic worthiness of theories.

Without a specific model of theoretical equivalence and a specific, allegedly harmonized, model of confirmation, I can only illustrate this problem by briefly considering in a qualitative way the issues I have in mind. Typically one can generate alternative theories saving the same phenomena by some process which introduces into a theory otiose elements whose place in the theory "cancels out." Most interesting, of course, are the historical cases where the theory with the otiose elements came first and where it was an important scientific discovery that one could eliminate them by a conceptual revision. For example one could replace the aether theories used to account for the null round–trip experiment results with the special relativity, thereby replacing a manifold of theories with a "cancelled out" element (the velocity of the observer relative to the aether) with a theory without that vacuous element. Again Einstein showed us that one could replace earlier gravitational theories with their otiose elements of "real inertial frames" and "real gravitational fields" by curved spacetime with its merely local inertial frames.

While the positivist seems to be committed to the view that a theory and its excessively otiose counterparts are all equivalent, the realist will, in general, claim that they are not. What we can expect from him is some attempt at a systematization of the notion of ontological simplicity which will credit the less otiose theory with greater simplicity and the simpler theory with greater warrant for belief. Obviously this will require a notion of confirmation which differs from that which associates confirmation with having the right observational consequences, and attempts in this direction have been made.[3]

But what if we have a pair of theories such that each member of the pair is preferable to the other in ontological simplicity in some respect? We can easily imagine a situation where *A* is preferable to *B* in terms of ontological parsimony for one aspect of theoretical structure, and *B* preferred to *A* along another component. Here one will probably hope to establish that relative to this pair of theories there is some third alternative, *C*, which is ontologically more parsimonious than either *A* or *B*, and, hence, preferable to both of them. What we will expect from the realist, though, if he takes this approach, is some reason to believe that relative to the notion of theoretical equivalence which he espouses, in each possible case there will always result a single unique "maximally parsimonious" theory. Or, rather, that relative to his notions of theoretical equivalence and ontological parsimony, any two theories which, relative to a given set of observational consequences, are both adjudged maximally parsimonious will turn out, on his criterion, to be theoretically equivalent. Unless he can show us that there is a most parsimonious theory, and that this is unique modulo his notion of theoretical equivalence, we will once again be faced with room for the intrusion of skepticism.

But there is a more profound problem of an epistemic sort than this for the realist. Taking equivalence to demand more than commonality of observational consequences the realist is faced with the threat of skepticism which arises when he tolerates inequivalent theories having all their observational consequences in common. So his confirmation theory is supplemented with elements going beyond reflection on the observational consequences of a theory in evaluating its merit for belief. Whether this be ontological simplicity, as I have chosen for the example above, or conservatism with respect to antecedent theory which has frequently been proposed, or some other possible feature of theories altogether, is immaterial. The first problem the realist faces, unless he is simply willing to tolerate states of irresolvable suspension of belief, is to try to demonstrate to us that his combination of a notion of theoretical equivalence and a notion of warranted believability always produces as most believable only a single theory or a class of theories all equivalent to one another relative to his criterion of theoretical equivalence.

But even if he does this should we then be fully satisfied? I think not. For the question will always arise: Why should we accept *this* notion of confirmation or epistemic preferability? Why should we believe that the simpler theory, or the theory which deviates minimally from previously accepted theory, is the theory which is, in any sense, most likely to be true? The point is hardly new. Insofar as one places in one's structure for warranted belief any considerations that go beyond mere conformity with the observational data, it is hard to see why one ought to take these additional espoused "marks of believability" as genuine "..narks of truth."

Of course there are the familiar options open of adopting as a priori principles that simpler theories are more likely to be true, etc. I do not wish to pursue these here. We ought to note that one familiar approach is not open to the realist we have in mind. There is a familiar "justification" for such principles of warranted belief of a pragmatist sort which claims that the normative principles of believability rest ultimately on our actual practice of belief, and that the connection between truth and believability rests not on showing that some process for belief leads (more likely than not) to the truth, but rather on understanding the very notion of truth as ideal warranted believability. However persuasive such a tack may be in trying to rationalize our actual reliance on principles of simplicity, conservatism, etc., in science, the realist we have in mind should be loath to undercut the possibility of skepticism by this means. For if there is anything realistic to his realism, it should include at least the rejection of such a pragmatist line on truth and on justification. Whatever realism about theories means, it should include a claim that truth is correspondence to objective reality (in the metaphysical sense in which this is a controversial thesis), and this brings with it strong pressure for an account of just *why* we ought to accept anything over and above conformity with the observational data as relevant to the believability of a theory. But some such invocation of principles over and above conformity with observational data is necessary for anyone who wishes to impose a constraint on equivalence which outruns commonality in saving the phenomena and who does not wish to accede to the possible skeptical consequences of that demand.

There is, of course, a familiar response the realist can give. Unfortunately, it is of the "You're no better" form with the usual unsatisfactoriness of the *tu quoque* reply. This response is the familiar one of the realist to the positivist that even eschewing postulation of unobservables still leaves us with the problem of leaping beyond present data to fully general theory of the world. And, the reply will go on, the problem of purely inductive inference is as infected with the invocation of notions like "simplest hypothesis" etc., as is the discussion of inference from the observable to the unobservable in principle. Perhaps so. Still we shall want some rationalization from the realist for his invocation of whatever principles he chooses to suggest limit the range of skepticism inevitably introduced by his belief in an ontology of in-principle unobservables, and by his invocation of a criterion of theoretical equivalence which demands that in at least some cases theories be inequivalent even when they save all the same phenomena in all possible worlds.

V

If the kind of realist notion of theoretical equivalence we have been discussing runs into problems in the upward direction of confirmation, there are difficulties it must deal with in the other, downward, relation of theory to data as well. We take it that our theories are explanatory. How must the realist's notions of explanation and of theoretical equivalence mesh?

First there will be the obvious requirement that his account of explanation satisfy at least some version of an equivalence principle. If two theories are genuinely equivalent, then, one hopes, in at least some sense they ought to offer the "same explanation" of the phenomena. And if the theories are inequivalent on the realist's criterion, then there ought to be a discrimination on the level of explanation which tells us why they are not, even if they are predictively equivalent, explaining the predicted phenomena in the same way. Actually the case here is subtle, for there might be alleged to be a kind of intensionality which allows theories to be equivalent in the sense the realist has in mind yet not fully equivalent in some "finer grained" explanatory sense. The individuation of "kind of explanation" might, for example, depend upon alleged modes of expression so that one and the same theory might in two equivalent versions be said in some sense to offer different explanations due solely to the mere difference in the manner in which the theory is expressed in the two versions. Someone who finds the notion of explanation loaded with "pragmatic" aspects might indeed hold to this. In the same vein it would, I suppose, be possible to maintain that two equivalent versions of a theory received differential confirmation from the same data, despite the obvious persuasiveness of a principle of equivalence as a criterion of adequacy of an acceptable notion of confirmation.

What is certain, I think is that an exponent of a realistic notion of equivalence must needs be quite wary of allowing equivalent theories to either offer inequivalent explanations of the data or to be inequivalently confirmed by the same evidential base. If he allows both of these, his notion of equivalence is beginning to drain into a bloodless ghost of that notion as pre-analytically understood. At the very least we would expect from him at least coarse-grained notions of explanation and confirmation which would mesh with his notion of theoretical equivalence in the natural way, and an account as to why the finer-grained notions, which allowed equivalent theories to offer inequivalent explanations or to be inequivalently confirmed, did not vitiate the real sense of theoretical equivalence he originally had in mind.

Once again, a detailed critique of a realistic attempt to "mesh" notions of theoretical equivalence and explanation will be impossible for us here, for that would require what I am trying to avoid, the presentation of a specific principle of equivalence and specific accounts of explanation and confirmation. Instead I will focus on what to the positivist seem to be certain fundamental difficulties with any such realistic program. What the positivist will allege is familiar. Suppose we have available that minimal, best confirmed theory compatible with the data (and, of course, all of those equivalent members of the equivalence class of theories of which this particular one is representative). Presumably this theory is the best possible explanatory account of the data. Indeed, any reasonable notion of confirmation and explanation should lead us in at least some weak sense to take it that the best confirmed, most plausible account is the best explanatory account, a relationship which becomes, of course, trivial if we adopt notions of confirmation as inference to the best explanation.

One thing had better be the case if the realist account is to hold onto even a skeletal version of its fundamentals. That theory which is the Craigian reduction of the theory in question had better not appear among the set of equivalent best alternative theories. For if it does the notion of theoretical equivalence in mind would seem automatically to reduce to the positivist notion. But how is the realist to avoid the introduction of the Craigian reduction, if not as equivalent to his best theoretical account, then as a simpler (hence, superior) alternative account to it?

Relying on the syntactical complexity (or, indeed, infinitistic nature) of the Craigian alternative to exclude it seems rather "un-realistic" in nature, a move more suitable to a frank pragmatist notion of theoretical preferability. Surely the right way out for the realist is to argue that the Craigian alternative is to be rejected for its explanatory inadequacy. According to the realist, the best realistic theory and its equivalent alternatives explain the observational phenomena in a way in which the Craigian reduction does not. Indeed, the Craigian surrogate is alleged to be devoid of explanatory power at all. Theoretical postulation, we are told, does not merely summarize observational generality nor accommodate it in a compact syntactical form. Theoretical postulation is taken to *explain* the observational generality. Hence the mere statement of that generality is, in a fundamental sense, totally devoid of any real explanatory power.

But what is this notion of explanation which makes the best realist theory explanatory in a way the Craigian reduction of it is not? Here we must once again focus on two quite distinct kinds of realism, one the kind of realism which takes a view of the meaning of theoretical terms as given by their holistic role in the theories in which they appear, and the other which allows the attribution of "excess" meaning to theoretical terms, by semantic analogy or otherwise. Thus, for example, for the first kind of realist molecules exist, but the meaning of 'molecule' is given solely by the complex role it plays in the sum total of molecular theories. For the other kind of realist, we understand 'molecule' by means of our understanding of such notions as 'particle', notions whose meaning is originally accrued in their use to refer to observables, but whose meaning is preserved intact when the term is used in the quite different contexts of reference to "particles too small to see" or even "particles unobservable in principle." The first kind of realist, the one with the idea of the meaning of theoretical terms as being solely their role in the holistic theory, is the one we have been emphasizing. Why? Because, I have claimed, it is only for such a realist that the notion of theoretical equivalence as being structural isomorphism super-added to observational equivalence is plausible. Without that theory of the meaning of theoretical terms, it is hard to see why even the most strict commonality of structure on the theoretical level would be enough to guarantee equivalence. But, I want to argue here, it is from the point of view of the second kind of realist that the notion of realistic explanation as something over and above subsumption of observed facts into generalizations about observables makes its most plausible sense. In other words, I want to argue that there is a fundamental tension between the realist's desire to posit a notion of explanation over and above that adopted by the positivist, and his desire to maintain a theory of the meaning of theoretical terms which allows for the notion of theoretical equivalence we have been discussing.[4]

Suppose we have a generalization over observable phenomena.

What does the realist demand over and above that? Presumably the postulation of a theoretical entity or property which explains the observational generality. But what additional explanatory power does this theoretical postulation give us? A frequent answer is "unification." Positing the theoretical structure provides, somehow, a *unified* account of the phenomena.[5]

Let us look at some cases to see why I think there are difficulties for the realist here. The observational facts of dynamics require us to single out a set of preferred states of motion, the local inertial motions. Additionally optical phenomena pick out as distinguished these same reference frames. The realist accounts for all of this by positing the spacetime structure itself as existing over and above the (for this purpose) observational moving systems and light waves. The dynamic and optical distinctions between inertial and non–inertial states are then accounted for by the relation of the physical systems to the underlying spacetime structure.

But, replies the positivist, what kind of additional explanatory force does positing "spacetime itself" give us? Whereas before we had the irreducibly inexplicable basic distinction between inertial systems and non–inertial, now we have a new, and trivially introduced, "deeper" explanation in terms of the underlying spacetime and the relations of the physical systems to it. But have we done anything more than added an otiose layer to what was already a perfectly adequate theory, a layer which only gives the spurious appearance of further "unification" or further "explanation" to the phenomena?

Many positivists would argue like this: The appearance of additional explanatory value in the substantivalist spacetime account, over the purely relationist theory, is due to the fact that while the reference to spacetime itself is really introduced only in the holistic–place–in–the–theory manner of the realist we have been emphasizing, the naive picture of spacetime as a kind of "ghostly" substance, sort of like a thin "rigid" extended material thing, gives us the impression that we are getting an explanation like that which the other kind of realist would offer. If we really understood what we were talking about when we posit the theoretical apparatus, independently of positing its explanatory role in the situation in question, then we could attribute further explanatory force to the substantival theory over the relational in the manner familiar from some older realists (reducing the unfamiliar to the familiar, demonstrating "mechanism" in the Newtonian sense, etc.). But we have no such grasp of what the theoretical apparatus is over and above its place in the theory in question. As a consequence, its putative explanatory force is void.

To see this, the positivist continues, consider another case. Study of the symmetries of the interactions of elementary particles gives us a systematization of these particles in terms of various symmetry groups. One can capture this symmetry structure by positing the existence of various "charges" for the particles (strangeness, charm, etc.) and associating with the posited charges various appropriate conservation rules. But in positing these charges and their conservation, are we *explaining* the symmetries involved? Most theoretical physicists would argue, I think rightly, that the invocation of the charges is simply another way of stating what the various symmetries are. There is a redundancy of the "theoretical structure" on the "observational data" to be explained. If anything would count as genuinely explanatory of the symmetries, they would argue, it would be the subsumption of these symmetries into some higher, more general symmetries naturally generated as the consequence of a more profound theory (say the generation of the symmetries out of some posited gauge field or other unified field account). The congruence of all of this with the positivistic notions of explanation is clear.

Now consider the alleged explanation of the symmetries of baryons and mesons given us by quark theory. Here we do seem to feel that a genuine explanatory account has been offered. Part of this institution is, I think, perfectly acceptable positivistically, quark theory being more general and profound than the symmetry accounts derivable from it. But part of the intuition rests on our viewing quarks "analogically" as tiny particles constituting larger particles in a manner understandable to us from the whole–part relationship and its explanatory value in the realm of observables. Once again we see that the invocation of terms whose meaning rests solely on the role they play in the appropriate theory suggests an account of explanation which is wholly positivistic, whereas the other kind of realism, the one which allows terms to have "excess meaning" attributable to them through semantic analogy seems to fit more harmoniously with the realistic notion that explanation by positing of theoretical structure is more than the subsumption of observational facts under broad generalities.

I do not, naturally, intend to pursue this problem of explanation from the realist point of view in any depth here. To repeat, my point is only this: A realist may take two attitudes toward the meaning of theoretical terms, either the holistic role–in–the–theory attitude or the alternative which grants to theoretical terms meaning accrual over and above the role the terms play in the theory. From the latter point of view it is fairly clear how we will be offered a realist account of explanation which takes explanation to be something over and above hierarchical subsumption of observational possibilities into generalities. But, from this point of view, it is hard to see what the realist account of equivalence of theories will amount to. Basically it must come down to the line that theories are equivalent only when they "say the same thing," and this notion will have to be explicated in terms of whatever theory of meaning (over and above role–in–theory) the realist offers us. From the alternative point of view regarding theoretical meaning a much simpler notion of theoretical equivalence can be constructed. This is the notion which simply takes the positivistic notion of equivalence as necessary and super–adds to it some notion of structural isomorphism between the theories at the theoretical level to obtain a sufficient condition for equivalence. But this notion of theoretical meaning, and of theoretical equivalence, besides borrowing many dubious presuppositions from the positivist position, suffers also from the difficulty of making it hard for us to see just what it is that is explanatory in a theory, or in any of its equivalents, which is not already there in the allegedly non–equivalent and non-explanatory Craigian reduction of the theory so beloved of positivists.

VI

To summarize: Positivism, for all its defects, offers us a theory of theoretical equivalence neatly integrated with its theory of confirmation and its theory of explanation. The realist is obliged to do the same. If the realist adopts a theory of the meaning of theoretical terms which attributes to these terms a meaning over and above that which accrues to them by means of the role they play in the theory, then his notion of theoretical equivalence will be complex and will depend in detail on just what theory of meaning for theoretical terms he offers us. If he adopts a theory of meaning for theoretical terms which takes their meaning to be fixed solely by the role they play in the theory, then his notion of theoretical equivalence is likely to be simpler. It will be the notion of two theories, first of all, sharing commonality of observational prediction and, secondly, having an appropriate structural isomorphism at the theoretical level.

One will first ask if the notion of equivalence offered corresponds with the realist's intuitions. Does it take as equivalent theories only those the realist wishes to count as equivalent and will it count as equivalent all those pairs pre–analytically thought equivalent?

Next, one will want to examine the interaction of the notion of equivalence offered with the "upward" notion of confirmation. Will equivalent theories always be equally confirmed by the same data or, if not, will the violation of the "equivalence condition" for confirmation receive an adequate explanation? Will the theory of confirmation allow for differential confirmation of theories designated non–equivalent, or instead, will it leave room for skepticism? Even if the theory of confirmation does always select from a group of non-equivalent alternatives a unique "best confirmed"member, will there be an appropriate *realist* rationale for adopting that notion of confirmation?

Finally will one want to look at the interaction of the notion of equivalence with the "downward" notion of explanation? Will equivalent theories always offer "the same explanation" of the phenomena and non–equivalent theories "different" explanations? More importantly what notion of explanation, over and above the positivist notion, does the realist have in mind? Will it be such as to allow us to understand why a theory is not, according to the realist, equivalent to its Craigian reduction and why the theory with its realistic posits is explanatory in a way in which the Craigian surrogate is not?

Far be it from me to claim that these questions cannot be answered by the realist. But until they are we should be reluctant to dismiss the positivist notions with a sneering reference to "outmoded verificationism." It is sometimes said (wrongly I believe) that scientists do not reject one theory until a better one is available to take its place. Just exactly what is the realist notion of equivalence, and which associated realist notions of confirmation and explanation are supposed to take the place of the positivist's integrated if implausible accounts?

NOTES

1. H. Field, "Conventionalism and Instrumentalism in Semantics," *Nous* 9 (1975), p. 392 and footnote 7, p. 404.

2. Reichenbach's allegation of the equivalence of curved spacetime with the flat spacetime plus universal forces theory can be found in W. Reichenbach, *The Philosophy of Space and Time* (New York: Dover, 1958), secs. 3–8. The contrasting "realist" attitude toward theories is in his *Experience and Prediction* (Chicago: University of Chicago, 1938). Specifically useful on the problem in Reichenbach, and generally useful as background to this paper throughout is C. Glymour, "Theoretical Realism and Theoretical Equivalence," *Boston Studies in the Philosophy of Science*, Vol. 8 (1971), 275–88. See also W. Quine, "On Empirically Equivalent Systems of the World," *Erkenntnis* 9 (1975), 313–28.

3. For an attempt at a theory of confirmation appropriate to this realist notion of equivalence see C. Glymour, *Theory and Evidence* (Princeton: Princeton University Press), esp. chaps. 5 and 9.

4. For a discussion of that notion of the meaning of theoretical terms which allows meaning to accrue to them over and above the role they play in the theory see the author's "Semantic Analogy," *Philosophical Studies* 38 (1980), 217–34.

5. On realistic theories as explaining by unifying see M. Friedman, "Explanation and Scientific Understanding," *Journal of Philosophy* 71 (1974), 5–19.

The Natural Ontological Attitude

> Let us fix our attention out of ourselves as much as possible; let us chace our imagination to the heavens, or to the utmost limits of the universe; we never really advance a step beyond ourselves, nor can conceive any kind of existence, but those perceptions, which have appear'd in that narrow compass. This is the universe of the imagination, nor have we any idea but what is there produced.
>
> Hume, *Treatise*, book 1, part II, section VI

Realism is dead. Its death was announced by the neopositivists who realized that they could accept all the results of science, including all the members of the scientific zoo, and still declare that the questions raised by the existence claims of realism were mere pseudo-questions. Its death was hastened by the debates over the interpretation of quantum theory, where Bohr's nonrealist philosophy was seen to win out over Einstein's passionate realism. Its death was certified, finally, as the last two generations of physical scientists turned their backs on realism and have managed, nevertheless, to do science successfully without it. To be sure, some recent philosophical literature has appeared to pump up the ghostly shell and to give it new life. I think these efforts will eventually be seen and understood as the first stage in the process of mourning, the stage of denial. But I think we shall pass through this first stage and into that of acceptance, for realism is well and truly dead, and we have work to get on with, in identifying a suitable successor. To aid that

My thanks to Charles Chastain, Gerald Dworkin, and Paul Teller for useful preliminary conversations about realism and its rivals, but especially to Charles—for only he, then, (mostly) agreed with me, and surely that deserves special mention. This paper was written by me, but cothought by Micky Forbes. I don't know any longer whose ideas are whose. That means that the responsibility for errors and confusions is at least half Micky's (and she is two-thirds responsible for "NOA"). Finally, I am grateful to the many people who offered comments and criticisms at the conference on realism sponsored by the Department of Philosophy, University of North Carolina at Greensboro in March 1982, where an earlier version of this chapter was first presented under the title "Pluralism and Scientific Progress." I am also grateful to the National Science Foundation for a grant in support of this research.

work I want to do three things in this essay. First, I want to show that the arguments in favor of realism are not sound, and that they provide no rational support for belief in realism. Then, I want to recount the essential role of nonrealist attitudes for the development of science in this century, and thereby (I hope) to loosen the grip of the idea that only realism provides a progressive philosophy of science. Finally, I want to sketch out what seems to me a viable nonrealist position, one that is slowly gathering support and that seems a decent philosophy for postrealist times.[1]

1. Arguments for Realism

Recent philosophical argument in support of realism tries to move from the success of the scientific enterprise to the necessity for a realist account of its practice. As I see it, the arguments here fall on two distinct levels. On the ground level, as it were, one attends to particular successes; such as novel, confirmed predictions, striking unifications of disparate-seeming phenomena (or fields), successful piggybacking from one theoretical model to another, and the like. Then, we are challenged to account for such success, and told that the best and, it is slyly suggested, perhaps, the *only* way of doing so is on a realist basis. I do not find the details of these ground-level arguments at all convincing. Neither does Larry Laudan (1984) and, fortunately, he has provided a forceful and detailed analysis which shows that not even with a lot of handwaving (to shield the gaps in the argument) and charity (to excuse them) can realism itself be used to explain the very successes to which it invites our attention. But there is a second level of realist argument, the methodological level, that derives from Popper's (1972) attack on instrumentalism, which he attacks as being inadequate to account for the details of his own, falsificationist methodology. Arguments on this methodological level have been skillfully developed by Richard Boyd (1981, 1984), and by one of the earlier Hilary Putnams (1975). These arguments focus on the methods embedded in scientific practice, methods teased out in ways that seem to me accurate and perceptive about ongoing science. We are then challenged to account for why these methods lead to scientific success and told

1. In the final section, I call this postrealism "NOA." Among recent views that relate to NOA, I would include Hilary Putnam's "internal realism," Richard Rorty's "epistemological behaviorism," the "semantic realism" espoused by Paul Horwich, parts of the "Mother Nature" story told by William Lycan, and the defense of common sense worked out by Joseph Pitt (as a way of reconciling W. Sellars's manifest and scientific images). For references, see Putnam (1981a), Rorty (1979), Horwich (1982), Lycan (1982), and Pitt (1981).

that the best and, (again) perhaps, the only truly adequate way of explaining the matter is on the basis of realism.

I want to examine some of these methodological arguments in detail to display the flaws that seem to be inherent in them. But first I want to point out a deep and, I think, insurmountable problem with this entire strategy of defending realism, as I have laid it out above. To set up the problem, let me review the debates in the early part of this century over the foundations of mathematics, the debates that followed Cantor's introduction of set theory. There were two central worries here, one over the meaningfulness of Cantor's hierarchy of sets insofar as it outstripped the number-theoretic content required by Kronecker (and others); the second worry, certainly deriving in good part from the first, was for the consistency (or not) of the whole business. In this context, Hilbert devised a quite brilliant program to try to show the consistency of a mathematical theory by using only the most stringent and secure means. In particular, if one were concerned over the consistency of set theory, then clearly a set-theoretic proof of consistency would be of no avail. For if set theory were inconsistent, then such a consistency proof would be both possible and of no significance. Thus, Hilbert suggested that finite constructivist means, satisfactory even to Kronecker (or Brouwer) ought to be employed in metamathematics. Of course, Hilbert's program was brought to an end in 1931, when Gödel showed the impossibility of such a stringent consistency proof. But Hilbert's idea was, I think, correct even though it proved to be unworkable. Metatheoretic arguments must satisfy more stringent requirements than those placed on the arguments used by the theory in question, for otherwise the significance of reasoning about the theory is simply moot. I think this maxim applies with particular force to the discussion of realism.

Those suspicious of realism, from Osiander to Poincaré and Duhem to the "constructive empiricism" of van Fraassen,[2] have been worried about the significance of the explanatory apparatus in scientific investigations. While they appreciate the systematization and coherence brought about by scientific explanation, they question whether acceptable explanations need to be true and, hence, whether the entities mentioned in explanatory principles need to exist.[3] Suppose they are right. Suppose, that is, that the usual explanation-inferring devices in scientific practice do not lead

2. Van Fraassen (1980). See especially pp. 97–101 for a discussion of the truth of explanatory theories. To see that the recent discussion of realism is joined right here, one should contrast van Fraassen with Newton-Smith (1981), especially chap. 8.

3. Cartwright (1983) includes some marvelous essays on these issues.

to principles that are reliably true (or nearly so), nor to entities whose existence (or near existence) is reliable. In that case, the usual abductive methods that lead us to good explanations (even to "the best explanation") cannot be counted on to yield results even approximately true. But the strategy that leads to realism, as I have indicated, is just such an ordinary sort of abductive inference. Hence, if the nonrealist were correct in his doubts, then such an inference to realism as the best explanation (or the like), while possible, would be of no significance—exactly as in the case of a consistency proof using the methods of an inconsistent system. It seems, then, that Hilbert's maxim applies to the debate over realism: to argue for realism one must employ methods more stringent than those in ordinary scientific practice. In particular, one must not beg the question as to the significance of explanatory hypotheses by assuming that they carry truth as well as explanatory efficacy.

There is a second way of seeing the same result. Notice that the issue over realism is precisely the issue of whether we should believe in the reality of those individuals, properties, relations, processes, and so forth, used in well-supported explanatory hypotheses. Now what *is* the hypothesis of realism, as it arises as an explanation of scientific practice? It is just the hypothesis that our accepted scientific theories are approximately true, where "being approximately true" is taken to denote an extratheoretical relation between theories and the world. Thus, to address doubts over the reality of relations posited by explanatory hypotheses, the realist proceeds to introduce a further explanatory hypothesis (realism), itself positing such a relation (approximate truth). Surely anyone serious about the issue of realism, and with an open mind about it, would have to behave inconsistently if he were to accept the realist move as satisfactory.

Thus, both at the ground level and at the level of methodology, no support accrues to realism by showing that realism is a good hypothesis for explaining scientific practice. If we are open-minded about realism to begin with, then such a demonstration (even if successful) merely begs the question that we have left open ("need we take good explanatory hypotheses as true?"). Thus, Hilbert's maxim applies, and we must employ patterns of argument more stringent than the usual abductive ones. What might they be? Well, the obvious candidates are patterns of induction leading to empirical generalizations. But, to frame empirical generalizations, we must first have some observable connections between observables. For realism, this must connect theories with the world by way of approximate truth. But no such connections are observable and, hence, suitable as the basis for an inductive inference. I do not want to labor the points at issue here. They amount to the well-known idea that realism commits one to an unverifiable correspondence with the world. So far as I am aware, no recent defender of realism has tried to make a case based on a Hilbert strategy of using suitably stringent grounds and, given the problems over correspondence, it is probably just as well.

The strategy of arguments for realism as a good explanatory hypothesis, then, *cannot* (logically speaking) be effective for an open-minded nonbeliever. But what of the believer? Might he not, at least, show a kind of internal coherence about realism as an overriding philosophy of science, and should that not be of some solace, at least for the realist?[4] Recall, however, the analogue with

4. Some realists may look for genuine support, and not just solace, in such a coherentist line. They may see in their realism a basis for general epistemology, philosophy of language, and so forth (as does Boyd 1981, 1984). If they find in all this a coherent and comprehensive worldview, then they might want to argue for their philosophy as Wilhelm Wien argued (in 1909) for special relativity, "What speaks for it most of all is the inner consistency which makes it possible to lay a foundation having no self-contradictions, one that applies to the totality of physical appearances." (Quoted by Gerald Holton, "Einstein's Scientific Program: Formative Years," in H. Woolf (1980), p. 58.) Insofar as the realist moves away from the abductive defense of realism to seek support, instead, from the merits of a comprehensive philosophical system with a realist core, he marks as a failure the bulk of recent defenses of realism. Even so, he will not avoid the critique pursued in the text. For although my argument above has been directed, in particular, against the abductive strategy, it is itself based on a more general maxim; namely, that the form of argument used to support realism must be more stringent than the form of argument embedded in the very scientific practice that realism itself is supposed to ground—on pain of begging the question. Just as the abductive strategy fails because it violates this maxim, so too would the coherentist strategy, should the realist turn from one to the other. For, as we see from the words of Wien, the same coherentist line that the realist would appropriate for his own support is part of ordinary scientific practice in framing judgments about competing theories. It is, therefore, not a line of defense available to the realist. Moreover, just as the truth-bearing status of abduction is an issue dividing realists from various nonrealists, so too is the status of coherence-based inference. Turning from abduction to coherence, therefore, still leaves the realist begging the question. Thus, when we bring out into the open the character of arguments *for* realism, we see quite plainly that they do not work. See Fine (1986) for a more detailed discussion.

In support of realism there seem to be only those "reasons of the heart" which, as Pascal says, reason does not know. Indeed, I have long felt that belief in realism involves a profound leap of faith, not at all dissimilar from the faith that animates deep religious convictions. I would welcome engagement with realists on this understanding, just as I enjoy conversation on a similar basis with my religious friends. The dialogue will proceed more fruitfully, I think, when the realists finally stop pretending to a rational support for their faith, which they do not have. Then we can all enjoy their intricate and sometimes beautiful philosophical constructions (of, e.g., knowledge, or reference, etc.), even though to us, as nonbelievers, they may seem only wonder-full castles in the air.

consistency proofs for inconsistent systems. That sort of harmony should be of no solace to anyone. But for realism, I fear, the verdict is even harsher. For, so far as I can see, the arguments in question just do not work, and the reason for that has to do with the same question-begging procedures that I have already identified. Let me look closely at some methodological arguments in order to display the problems.

A typical realist argument on the methodological level deals with what I shall call the problem of the "small handful." It goes like this. At any time, in a given scientific area, only a small handful of alternative theories (or hypotheses) are in the field. Only such a small handful are seriously considered as competitors, or as possible successors to some theory requiring revision. Moreover, in general, this handful displays a sort of family resemblance in that none of these live options will be too far from the previously accepted theories in the field, each preserving the well-confirmed features of the earlier theories and deviating only in those aspects less confirmed. Why? Why does this narrowing down of our choices to such a small handful of cousins of our previously accepted theories work to produce good successor theories?

The realist answers this as follows. Suppose that the already existing theories are themselves approximately true descriptions of the domain under consideration. Then surely it is reasonable to restrict one's search for successor theories to those whose ontologies and laws resemble what we already have, especially where what we already have is well confirmed. And if these earlier theories were approximately true, then so will be such conservative successors. Hence, such successors will be good predictive instruments; that is, they will be successful in their own right.

The small-handful problem raises three distinct questions: (1) why only a small handful out of the (theoretically) infinite number of possibilities? (2) why the conservative family resemblance between members of the handful? and (3) why does the strategy of narrowing the choices in this way work so well? The realist response does not seem to address the first issue at all, for even if we restrict ourselves just to successor theories resembling their progenitors, as suggested, there would still, theoretically, always be more than a small handful of these. To answer the second question, as to why conserve the well-confirmed features of ontology and laws, the realist must suppose that such confirmation is a mark of an approximately correct ontology and approximately true laws. But how could the realist possibly justify such an assumption? Surely, there is no valid inference of the form, "*T* is well-confirmed; therefore, there exist objects pretty much of the sort required by T and satisfying laws approximating those of T." Any of the dramatic shifts of ontology in science show the invalidity of this schema. For example, the loss of the ether from the turn-of-the-century electrodynamic theories demonstrates this at the level of ontology, and the dynamics of the Rutherford-Bohr atom vis-à-vis the classical energy principles for rotating systems demonstrates it at the level of laws. Of course, the realist might respond that there is no question of a strict inference between being well confirmed and being approximately true (in the relevant respects), but there is a probable inference of some sort. But of what sort? Certainly there is no probability relation that rests on inductive evidence here. For there is no independent evidence for the relation of approximate truth itself; at least, the realist has yet to produce any evidence that is independent of the argument under examination. But if the probabilities are not grounded inductively, then how else? Here, I think the realist may well try to fall back on his original strategy and suggest that being approximately true provides the best explanation for being well confirmed. This move throws us back to the ground-level realist argument, the argument from specific success to an approximately true description of reality, which Lauden (1984) has criticized. I should point out, before looking at the third question, that if this last move is the one the realist wants to make, then his success at the methodological level can be no better than his success at the ground level. If he fails there, he fails across the board.

The third question, and the one I think the realist puts most weight on, is why does the small-handful strategy work so well. The instrumentalist, for example, is thought to have no answer here. He must just note that it does work well and be content with that. The realist, however, can explain why it works by citing the transfer of approximate truth from predecessor theories to the successor theories. But what does this explain? At best, it explains why the successor theories cover the same ground as well as their predecessors, for the conservative strategy under consideration assures that. But note that here the instrumentalist can offer the same account: if we insist on preserving the well-confirmed components of earlier theories in later theories, then, of course, the later ones will do well over the well-confirmed ground. The difficulty, however, is not here at all but rather in how to account for the successes of the later theories in new ground, or with respect to novel predictions, or in overcoming the anomalies of the earlier theories. And what can the realist possibly say in this area except that the theorist, in proposing a new theory, has happened to make a good guess? For nothing in the approximate truth of the old theory can guarantee (or even make it likely) that modifying the theory in its less-confirmed parts will produce a progressive shift. The history of science shows well enough how such tinkering succeeds only now and again, and fails for the most part. This history of failures can scarcely be adduced to explain the occasional success. The idea that by extending what is approximately true one is likely to bring new approximate truth is a chimera. It finds support neither in the logic of approximate truth nor in the history of science. The problem for the realist is how to explain the *occasional success* of a strategy that *usually fails*.[5] I think he has no special resources with which to do this. In particular, his usual fallback onto approximate truth provides nothing more than a gentle pillow. He may rest on it comfortably, but it does not really help to move his cause forward.

The problem of the small handful raises three challenges: why small? why narrowly related? and why does it work? The realist has no answer for the first of these, begs the question as to the truth of explanatory hypotheses on the second, and has no resources for addressing the third. For comparison, it may be useful to see how well his archenemy, the instrumentalist, fares on the same turf. The instrumentalist, I think, has a substantial basis for addressing the questions of smallness and narrowness, for he can point out that it is extremely difficult to come up with alternative theories that satisfy the many empirical constraints posed by the instrumental success of theories already in the field. Often it is hard enough to come up with even one such alternative. Moreover, the common apprenticeship of scientists working in the same area certainly has the effect of narrowing down the range of options by channeling thought into the commonly accepted categories. If we add to this the instrumentally justified rule, "If it has worked well in the past, try it again," then we get a rather good account, I think, of why there is usually only a small and narrow handful. As to why this strategy works to produce instrumentally successful science,

5. I hope all readers of this essay will take this idea to heart. For in formulating the question as how to explain why the methods of science lead to instrumental success, the realist has seriously misstated the explanandum. Overwhelmingly, the results of the conscientious pursuit of scientific inquiry are failures: failed theories, failed hypotheses, failed conjectures, inaccurate measurements, incorrect estimations of parameters, fallacious causal inferences, and so forth. If explanations are appropriate here, then what requires explaining is why the very same methods produce an overwhelming background of failures and, occasionally, also a pattern of successes. The realist literature has not yet begun to address this question, much less to offer even a hint of how to answer it.

we have already noted that for the most part it does not. Most of what this strategy produces are failures. It is a quirk of scientific memory that this fact gets obscured, much as do the memories of bad times during a holiday vacation when we recount all our "wonderful" vacation adventures to a friend. Those instrumentalists who incline to a general account of knowledge as a social construction can go further at this juncture and lean on the sociology of science to explain how the scientific community "creates" its knowledge. I am content just to back off here and note that over the problem of the small handful, the instrumentalist scores at least two out of three, whereas the realist, left to his own devices, has struck out.[6]

I think the source of the realist's failure here is endemic to the methodological level, infecting all of his arguments in this domain. It resides, in the first instance, in his repeating the question-begging move from explanatory efficacy to the truth of the explanatory hypothesis. And in the second instance, it resides in his twofold mishandling of the concept of approximate truth: first, in his trying to project from some body of assumed approximate truths *to* some further and novel such truths, and second, in his needing genuine access to the relation of correspondence. There are no general connections of this first sort, however, sanctioned by the logic of approximate truth, nor secondly, any such warranted access. However, the realist must pretend that there are in order to claim explanatory power for his realism. We have seen those two agents infecting the realist way with the problem of the small handful. Let me show them at work in another methodological favorite of the realist, the "problem of conjunctions."

The problem of conjunctions is this. If T and T' are independently well-confirmed, explanatory theories, and if no shared term is ambiguous between the two, then we expect the conjunction of T and T' to be a reliable predictive instrument (provided, of course, that the theories are not mutually inconsistent). Why? challenges the realist, and he answers as follows. If we make the realist assumption that T and T', being well confirmed, are approximately true of the entities (etc.) to which they refer, and if the unambiguity requirement is taken realistically as requiring a domain of common reference, then the conjunction of the two theories will also be approximately true and, hence, it will produce reliable observational predictions. Q.E.D.

6. Of course, the realist can appropriate the devices and answers of the instrumentalist, but that would be cheating, and anyway, it would not provide the desired support of realism per se.

But notice our agents at work. First, the realist makes the question-begging move from explanations to their approximate truth, and then he mistreats approximate truth. For nothing in the logic of approximate truth sanctions the inference from "T is approximately true" and "T' is approximately true" to the conclusion that the conjunction "$T \cdot T'$" is approximately true. Rather, in general, the tightness of an approximation dissipates as we pile on further approximations. If T is within ϵ, in its estimation of some parameter, and T' is also within ϵ, then the only general thing we can say is that the conjunction will be within 2ϵ of the parameter. Thus, the logic of approximate truth should lead us to the opposite conclusion here; that is, that the conjunction of two theories is, in general, *less* reliable than either (over their common domain). But this is neither what we expect nor what we find. Thus, it seems quite implausible that our actual expectations about the reliability of conjunctions rest on the realist's stock of approximate truths.

Of course, the realist could try to retrench here and pose an additional requirement of some sort of uniformity on the character of the approximations, as between T and T'.[7] It is difficult to see how the realist could do this successfully without making reference to the distance between the approximations and "the truth." For what kind of internalist requirement could possibly insure the narrowing of this distance? But the realist is in no position to impose such requirements, since neither he nor anyone else has the requisite access to "the truth." Thus, whatever uniformity-of-approximation condition the realist might impose, we could still demand to be shown that this leads closer to the truth, not farther away. The realist will have no demonstration, except to point out to us that it all works (sometimes!). But that was the original puzzle.[8] Actually, I think the puzzle is not very difficult. For surely, if we do not entangle ourselves with issues over approximation, there is no deep mystery as to why two compatible and successful theories lead us to expect their conjunction to be successful. For in forming the conjunction, we just add the reliable predictions of one onto the reliable predictions of the other, having antecedently ruled out the possibility of conflict.

7. Paul Teller has made this suggestion to me in conversation.

8. Niiniluoto (1982) contains interesting formal constructions for "degree of truthlikeness," and related versimilia. As conjectured above, they rely on an unspecified correspondence relation to the truth and on measures of the "distance" from the truth. Moreover, they fail to sanction that projection, from some approximate truths to other, novel truths, which lies at the core of realist rationalizations.

There is more to be said about this topic. In particular, we need to address the question of why we expect the logical gears of the two theories to mesh. However, I think that a discussion of the realist position here would only bring up the same methodological and logical problems that we have already uncovered at the center of the realist argument.

Indeed, this schema of knots in the realist argument applies across the board and vitiates every single argument at the methodological level. Thus my conclusion here is harsh, indeed. The methodological arguments for realism fail, even though, were they successful, they would still not support the case. For the general strategy they are supposed to implement is just not stringent enough to provide rational support for realism. In the next two sections, I will try to show that this situation is just as well, for realism has not always been a progressive factor in the development of science and, anyway, there is a position other than realism that is more attractive.

2. *Realism and Progress*

If we examine the two twentieth-century giants among physical theories, relativity and the quantum theory, we find a living refutation of the realist's claim that only his view of science explains its progress, and we find some curious twists and contrasts over realism as well. The theories of relativity are almost single-handedly the work of Albert Einstein. Einstein's early positivism and his methodological debt to Mach (and Hume) leap right out of the pages of the 1905 paper on special relativity.[9] The same positivist strain is evident in the 1916 general relativity paper as well, where Einstein (in section 3 of that paper) tries to justify his requirement of general covariance by means of a suspicious-looking verificationist argument which, he says, "takes away from space and time the last remnants of physical objectivity" (Einstein et al. 1952, p. 117). A study of his tortured path to general relativity[10] shows the repeated use of this Machist line, always used to deny that some concept has a real referent. Whatever other, competing strains there were in Einstein's philosophical orientation (and there certainly were others), it would be hard to deny the importance of this instrumentalist/

9. See Gerald Holton, "Mach, Einstein, and the Search for Reality," in Holton (1973), pp. 219–59. I have tried to work out the precise role of this positivist methodology in chapter 2. See also Fine (1981a).

10. Earman and Glymour (1978). The tortuous path detailed by Earman is sketched by B. Hoffmann (1972), pp. 116–28. A nontechnical and illuminating account is given by John Stachel (1979).

positivist attitude in liberating Einstein from various realist commitments. Indeed, on another occasion, I would argue in detail that without the "freedom from reality" provided by his early reverence for Mach, a central tumbler necessary to unlock the secret of special relativity would never have fallen into place.[11] A few years after his work on general relativity, however, roughly around 1920, Einstein underwent a philosophical conversion, turning away from his positivist youth (he was forty-one in 1920) and becoming deeply committed to realism (see chapter 6). In particular, following his conversion, Einstein wanted to claim genuine reality for the central theoretical entities of the general theory, the four-dimensional space-time manifold and associated tensor fields. This is a serious business, for if we grant his claim, then not only do space and time cease to be real but so do virtually all of the usual dynamical quantities.[12] Thus motion, as we understand it, itself ceases to be real. The current generation of philosophers of space and time (led by Howard Stein and John Earman) have followed Einstein's lead here. But, interestingly, not only do these ideas boggle the mind of the average man in the street (like you and me), they boggle most contemporary scientific minds as well.[13] That is, I believe the majority opinion among working, knowledgeable scientists is that general relativity provides a magnificent organizing tool for treating certain gravitational problems in astrophysics and cosmology. But few, I believe, give credence to the kind of realist existence and nonexistence claims that I have been mentioning. For relativistic physics, then, it appears that a nonrealist attitude was important in its development, that the founder nevertheless espoused a realist attitude to the finished product, but that most who actually use it think of the theory as a powerful instrument, rather than as expressing a "big truth."

With quantum theory, this sequence gets a twist. Heisenberg's seminal paper of 1925 is prefaced by the following abstract, announcing, in effect, his philosophical stance: "In this paper an attempt will be made to obtain bases for a quantum-theoretical mechanics based exclusively on relations between quantities observable in principle" (Heisenberg 1925, p. 879). In the body of

11. I have in mind the role played by the analysis of simultaneity in Einstein's path to special relativity. Despite the important study by Arthur Miller (1981) and an imaginative pioneering work by John Earman et al. (1983), I think the role of positivist analysis in the 1905 paper has yet to be properly understood.

12. Roger Jones in "Realism about What?" (in draft) explains very nicely some of the difficulties here.

13. I think the ordinary, deflationist attitude of working scientists is much like that of Steven Weinberg (1972).

the paper, Heisenberg not only rejects any reference to unobservables, he also moves away from the very idea that one should try to form any picture of a reality underlying his mechanics. To be sure, Schrödinger, the second father of quantum theory, seems originally to have had a vague picture of an underlying wavelike reality for his own equation. But he was quick to see the difficulties here and, just as quickly, although reluctantly, abandoned the attempt to interpolate any reference to reality.[14] These instrumentalist moves away from a realist construal of the emerging quantum theory were given particular force by Bohr's so-called philosophy of complementarity. This nonrealist position was consolidated at the time of the famous Solvay Conference, in October 1927, and is firmly in place today. Such quantum nonrealism is part of what every graduate physicist learns and practices. It is the conceptual backdrop to all the brilliant successes in atomic, nuclear, and particle physics over the past fifty years. Physicists have learned to think about their theory in a highly nonrealist way, and doing just that has brought about the most marvelous predictive success in the history of science.

The war between Einstein, the realist, and Bohr, the nonrealist, over the interpretation of quantum theory was not, I believe, just a sideshow in physics, nor an idle intellectual exercise. It was an important endeavor undertaken by Bohr on behalf of the enterprise of physics as a progressive science. For Bohr believed (and this fear was shared by Heisenberg, Sommerfeld, Pauli, and Born—and all the big guys) that Einstein's realism, if taken seriously, would block the consolidation and articulation of the new physics and, thereby, stop the progress of science. They were afraid, in particular, that Einstein's realism would lead the next generation of the brightest and best students into scientific dead ends. Alfred Landé, for example, as a graduate student, was interested in spending some time in Berlin to sound out Einstein's ideas. His supervisor was Sommerfeld, and recalling this period, Landé (1974, p. 460) writes, "The more pragmatic Sommerfeld . . . warned his students, one of them this writer, not to spend too much time on the hopeless task of "explaining" the quantum but rather to accept it as fundamental and help work out its consequences."

The task of "explaining" the quantum, of course, is the realist program for identifying a reality underlying the formulas of the theory and thereby explaining the predictive success of the formulas as approximately true descriptions of this reality. It is this

14. See Wessels (1979), and chapter 5.

program that I have criticized in the first part of this chapter, and this same program that the builders of quantum theory saw as a scientific dead end. Einstein knew perfectly well that the issue was joined right here. In the summer of 1935, he wrote to Schrödinger, "The real problem is that physics is a kind of metaphysics; physics describes 'reality.' But we do not know what 'reality' is. We know it only through physical description. . . . But the Talmudic philosopher sniffs at 'reality,' as at a frightening creature of the naive mind."[15]

By avoiding the bogey of an underlying reality, the "Talmudic" originators of quantum theory seem to have set subsequent generations on precisely the right path. Those inspired by realist ambitions have produced no predictively successful physics. Neither Einstein's conception of a unified field, nor the ideas of the de Broglie group about pilot waves, nor the Bohm-inspired interest in hidden variables has made for scientific progress. To be sure, several philosophers of physics, including another Hilary Putnam and myself, have fought a battle over the last decade to show that the quantum theory is at least consistent with some kind of underlying reality. I believe that Hilary has abandoned the cause, perhaps in part on account of the recent Bell-inequality problem over correlation experiments, a problem that van Fraassen (1982) calls "the Charybdis of realism." My own recent work in the area suggests that we may still be able to keep realism afloat in this whirlpool.[16] But the possibility (as I still see it) for a realist account of the quantum domain should not lead us away from appreciating the historical facts of the matter.

One can hardly doubt the importance of a nonrealist attitude for the development and practically infinite success of the quantum theory. Historical counterfactuals are always tricky, but the sterility of actual realist programs in this area at least suggests that Bohr and company were right in believing that the road to scientific progress here would have been blocked by realism. The founders of quantum theory never turned on the nonrealist attitude that served them so well. Perhaps that is because the central underlying theoretical device of quantum theory, the densities of a complex-valued and infinite-dimensional wave function, are even harder to take seriously than is the four-dimensional manifold of relativity. But now there comes a most curious twist. For just as the practitioners of relativity, I have suggested, ignore the *realist* interpre-

15. Letter to Schrödinger, June 19, 1935.
16. See my (1982d) for part of the discussion and also Chapter Nine.

tation in favor of a more pragmatic attitude toward the space/time structure, the quantum physicists would appear to make a similar reversal and to forget their nonrealist history and allegiance when it comes time to talk about new discoveries.

Thus, anyone in the business will tell you about the exciting period, in the fall of 1974, when the particle group at Brookhaven, led by Samuel Ting, discovered the J particle, just as a Stanford team at the Stanford Linear Accelerator Center, under Burton Richter, independently found a new particle they called ψ. These turned out to be one and the same, the so-called ψ/J particle (Mass 3,098 MeV, Spin 1, Resonance 67 keV, Strangeness 0). To explain this new entity, the theoreticians were led to introduce a new kind of quark, the so-called charmed quark. The ψ/J particle is then thought to be made up out of a charmed quark and an anticharmed quark, with their respective spins aligned. But if this is correct, then there ought to be other such pairs antialigned, or with variable spin alignments, and these ought to make up quite new observable particles. Such predictions from the charmed-quark model have turned out to be confirmed in various experiments.

I have gone on a bit in this story in order to convey the realist feel to the way scientists speak in this area. For I want to ask whether this is a return to realism or whether, instead, it can somehow be reconciled with a fundamentally nonrealist attitude.[17] I believe that the nonrealist option is correct.

3. Nonrealism

Even if the realist happens to be a talented philosopher, I do not believe that, in his heart, he relies for his realism on the rather sophisticated form of abductive argument that I have examined and rejected in the first section of this chapter, and which the history of twentieth-century physics shows to be fallacious. Rather, if his heart is like mine, then I suggest that a more simple and homely sort of argument is what grips him. It is this, and I will put it in the first person. I certainly trust the evidence of my senses, on the whole, with regard to the existence and features of everyday objects. And I have similar confidence in the system of "check, double-check, check, triple-check" of scientific investiga-

17. The nonrealism that I attribute to students and practitioners of the quantum theory requires more discussion and distinguishing of cases and kinds than I have room for here. It is certainly not the all-or-nothing affair I make it appear in the text. I carry out some of the required discussion in chapter 9. My thanks to Paul Teller and James Cushing, each of whom saw the need for more discussion.

tion, as well as the other safeguards built into the institutions of science. So, if the scientists tell me that there really are molecules, and atoms, and ψ/J particles, and, who knows, maybe even quarks, then so be it. I trust them and, thus, must accept that there really are such things with their attendant properties and relations. Moreover, if the instrumentalist (or some other member of the species "nonrealistica") comes along to say that these entities and their attendants are just fictions (or the like), then I see no more reason to believe him than to believe that *he is* a fiction, made up (somehow) to do a job on me; which I do not believe. It seems, then, that I had better be a realist. One can summarize this homely and compelling line as follows: it is possible to accept the evidence of one's senses and to accept, *in the same way*, the confirmed results of science only for a realist; hence, I should be one (and so should you!).

What is it to accept the evidence of one's senses and, *in the same way*, to accept confirmed scientific theories? It is to take them into one's life as true, with all that implies concerning adjusting one's behavior, practical and theoretical, to accommodate these truths. Now, of course, there are truths, and truths. Some are more central to us and our lives, some less so. I might be mistaken about anything, but were I mistaken about where I am right now, that might affect me more than would my perhaps mistaken belief in charmed quarks. Thus, it is compatible with the homely line of argument that some of the scientific beliefs that I hold are less central than some, for example, perceptual beliefs. Of course, were I deeply in the charmed-quark business, giving up that belief might be more difficult than giving up some at the perceptual level. (Thus we get the phenomenon of "seeing what you believe," well known to all thoughtful people.) When the homely line asks us, then, to accept the scientific results "in the same way" in which we accept the evidence of our senses, I take it that we are to accept them both as true. I take it that we are being asked not to distinguish between kinds of truth or modes of existence or the like, but only among truths themselves in terms of centrality, degrees of belief, or such.

Let us suppose this understood. Now, do you think that Bohr, the archenemy of realism, could toe the homely line? Could Bohr, fighting for the sake of science (against Einstein's realism) have felt compelled either to give up the results of science, or else to assign its "truths" to some category different from the truths of everyday life? It seems unlikely. And thus, unless we uncharitably think Bohr inconsistent on this basic issue, we might well come to question whether there is any necessary connection moving us from accepting the results of science as true to being a realist.[18]

Let me use the term "antirealist" to refer to any of the many different specific enemies of realism: the idealist, the instrumentalist, the phenomenalist, the empiricist (constructive or not), the conventionalist, the constructivist, the pragmatist, and so forth. Then, it seems to me that both the realist and the antirealist must toe what I have been calling "the homely line." That is, they must both accept the certified results of science as on par with more homely and familiarly supported claims. That is not to say that one party (or the other) cannot distinguish more from less well-confirmed claims at home or in science; nor that one cannot single out some particular mode of inference (such as inference to the best explanation) and worry over its reliability, both at home and away. It is just that one must maintain parity. Let us say, then, that both realist and antirealist accept the results of scientific investigations as "true," on par with more homely truths. (I realize that some antirealists would rather use a different word, but no matter.) And call this acceptance of scientific truths the "core position."[19] What distinguishes realists from antirealists, then, is what they add onto this core position.

The antirealist may add onto the core position a particular analysis of the concept of truth, as in the pragmatic and instrumentalist

18. I should be a little more careful about the historical Bohr than I am in the text. For Bohr himself would seem to have wanted to truncate the homely line somewhere between the domain of chairs and tables and atoms, whose existence he plainly accepted, and that of electrons, where he seems to have thought the question of existence (and of realism, more generally) was no longer well defined. An illuminating and provocative discussion of Bohr's attitude toward realism is given by Paul Teller (1981). Thanks, again, to Paul for helping to keep me honest.

19. In this context, for example, van Fraassen's "constructive empiricism" would prefer the concept of empirical adequacy, reserving "truth" for an (unspecified) literal interpretation and believing in that truth only among observables. I might mention here that in this classification Putnam's internal realism comes out as antirealist. For Putnam accepts the core position, but he would add to it a Peircean construal of truth as ideal rational acceptance. This is a mistake, which I expect that Putnam will realize and correct in future writings. He is criticized for it by Horwich (1982) whose own "semantic realism" turns out, in my classification, to be neither realist nor antirealist. Indeed, Horwich's views are quite similar to what is called "NOA" below, and could easily be read as sketching a philosophy of language compatible with NOA. Finally, the "epistemological behaviorism" espoused by Rorty (1979) is a form of antirealism that seems to me very similar to Putnam's position, but achieving the core parity between science and common sense by means of an acceptance that is neither ideal nor especially rational, at least in the normative sense. (I beg the reader's indulgence over this summary treatment of complex and important positions. I have been responding to Nancy Cartwright's request to differentiate these recent views from NOA.) See chapter 8 for a discussion of these antirealisms.

and conventionalist conceptions of truth. Or the antirealist may add on a special analysis of concepts, as in idealism, constructivism, phenomenalism, and in some varieties of empiricism. These addenda will then issue in a special meaning, say, for existence statements. Or the antirealist may add on certain methodological strictures, pointing a wary finger at some particular inferential tool, or constructing his own account for some particular aspects of science (e.g., explanations or laws). Typically, the antirealist will make several such additions to the core.

What then of the realist, what does he add to his core acceptance of the results of science as really true? My colleague, Charles Chastain, suggested what I think is the most graphic way of stating the answer—namely, that what the realist adds on is a desk-thumping, foot-stamping shout of "Really!" So, when the realist and antirealist agree, say, that there really are electrons and that they really carry a unit negative charge and really do have a small mass (of about 9.1×10^{-28} grams), what the realist wants to add is the emphasis that all this is really so. "There really are electrons, really!" This typical realist emphasis serves both a negative and a positive function. Negatively, it is meant to deny the additions that the antirealist would make to that core acceptance which both parties share. The realist wants to deny, for example, the phenomenalistic reduction of concepts or the pragmatic conception of truth. The realist thinks that these addenda take away from the substantiality of the accepted claims to truth or existence. "No," says he, "they *really* exist, and not in just your diminished antirealist sense." Positively, the realist wants to explain the robust sense in which *he* takes these claims to truth or existence; namely, as claims about reality—what is really, really the case. The full-blown version of this involves the conception of truth as correspondence with the world, and the surrogate use of approximate truth as near-correspondence. We have already seen how these ideas of correspondence and approximate truth are supposed to explain what *makes* the truth *true* whereas, in fact, they function as mere trappings, that is, as superficial decorations that may well attract our attention but do not compel rational belief. Like the extra "really," they are an arresting foot thump and, logically speaking, of no more force.

It seems to me that when we contrast the realist and the antirealist in terms of what they each want to add to the core position, a third alternative emerges—and an attractive one at that. It is the core position itself, *and all by itself.* If I am correct in thinking that, at heart, the grip of realism only extends to the homely connection of everyday truths with scientific truths, and that good sense dictates our acceptance of the one on the same basis as our acceptance of the other, then the homely line makes the core position, all by itself, a compelling one, one that we ought to take to heart. Let us try to do so and see whether it constitutes a philosophy, and an attitude toward science, that we can live by.

The core position is neither realist nor antirealist; it mediates between the two. It would be nice to have a name for this position, but it would be a shame to appropriate another "ism" on its behalf, for then it would appear to be just one of the many contenders for ontological allegiance. I think it is not just one of that crowd but rather, as the homely line behind it suggests, it is for commonsense epistemology—the natural ontological attitude. Thus, let me introduce the acronym *NOA* (pronounced as in "Noah"), for *natural ontological attitude,* and, henceforth, refer to the core position under that designation.

To begin showing how NOA makes for an adequate philosophical stance toward science, let us see what it has to say about ontology. When NOA counsels us to accept the results of science as true, I take it that we are to treat truth in the usual referential way, so that a sentence (or statement) is true just in case the entities referred to stand in the referred-to relations. Thus, NOA sanctions ordinary referential semantics and commits us, via truth, to the existence of the individuals, properties, relations, processes, and so forth referred to by the scientific statements that we accept as true. Our belief in their existence will be just as strong (or weak) as our belief in the truth of the bit of science involved, and degrees of belief here, presumably, will be tutored by ordinary relations of confirmation and evidential support, subject to the usual scientific canons. In taking this referential stance, NOA is not committed to the progressivism that seems inherent in realism. For the realist, as an article of faith, sees scientific success, over the long run, as bringing us closer to the truth. His whole explanatory enterprise, using approximate truth, forces his hand in this way. But, a "NOAer" (pronounced as "knower") is not so committed. As a scientist, say, within the context of the tradition in which he works, the NOAer, of course, will believe in the existence of those entities to which his theories refer. But should the tradition change, say, in the manner of the conceptual revolutions that Kuhn dubs "paradigm shifts," then nothing in NOA dictates that the change be assimilated as being progressive, that is, as a change where we learn more accurately about *the same things*. NOA is perfectly consistent with the Kuhnian alternative, which counts such changes as wholesale changes of reference. Unlike the realist, adherents to NOA are free to examine the facts in cases of paradigm shift, and to see whether or not a convincing case for stability of reference across paradigms can be made without superimposing on these facts a realist-progressivist superstructure. I have argued elsewhere (Fine 1975) that if one makes oneself free, as NOA enables one to do, then the facts of the matter will not usually settle the case; and that this is a good reason for thinking that cases of so-called incommensurability are, in fact, genuine cases where the question of stability of reference is indeterminate. NOA, I think, is the right philosophical position for such conclusions. It sanctions reference and existence claims, but it does not force the history of science into prefit molds.

So far I have managed to avoid what, for the realist, is the essential point: what of the "external world"? How can I talk of reference and of existence claims unless I am talking about referring to things right out there in the world? And here, of course, the realist, again, wants to stamp his feet.[20] I think the problem that makes the realist want to stamp his feet, shouting "Really!" (and invoking the external world) has to do with the stance the realist tries to take vis-à-vis the game of science. The realist, as it were, tries to stand outside the arena watching the ongoing game and then tries to judge (from this external point of view) what the point is. It is, he says, *about* some area external to the game. The realist, I think, is fooling himself. For he cannot (really!) stand outside the arena, nor can he survey some area off the playing field and mark it out as what the game is about.

Let me try to address these two points: How are we to arrive at the judgment that, in addition to, say, having a rather small mass, electrons are objects "out there in the external world"? Certainly, we can stand off from the electron game and survey its claims, methods, predictive success, and so forth. But what stance could we take that would enable us to judge what the theory of electrons is *about*, other than agreeing that it is about electrons? It is not like matching a blueprint to a house being built, or a map route to a country road. For we are *in* the world, both physically and conceptually.[21] That is, *we* are among the objects of science, and the concepts and procedures that we use to make judgments of subject

20. In his remarks at the Greensboro conference, my commentator, John King, suggested a compelling reason to prefer NOA over realism; namely, because NOA is less percussive! My thanks to John for this nifty idea, as well as for other comments.

21. "There is, I think, no theory-independent way to reconstruct phrases like 'really there'; the notion of a match between the ontology of a theory and its 'real' counterpart in nature now seems to me illusive in principle." T. S. Kuhn (1970), p. 206. The same passage is cited for rebuttal by Newton-Smith (1981). But the "rebuttal" sketched there in chapter 8, sections 4 and 5, not only runs afoul of the

matter and correct application are themselves part of that same scientific world. Epistemologically, the situation is very much like the situation with regard to the justification of induction. For the problem of the external world (so-called) is how to satisfy the realist's demand that we justify the existence claims sanctioned by science (and, therefore, by NOA) as claims to the existence of entities "out there." In the case of induction, it is clear that only an inductive justification will do, and it is equally clear that no inductive justification will do at all. So too with the external world, for only ordinary scientific inferences to existence will do, and yet none of them satisfies the demand for showing that the existent is really "out there." I think we ought to follow Hume's prescription on induction with regard to the external world. There is no possibility for justifying the kind of externality that realism requires, yet it may well be that, in fact, we cannot help yearning for just such a comforting grip on reality.

If I am right, then the realist is chasing a phantom, and we cannot actually do more, with regard to existence claims, than follow scientific practice, just as NOA suggests. What then of the other challenges raised by realism? Can we find in NOA the resources for understanding scientific practice? In particular (since it was the topic of the first part of this chapter), does NOA help us to understand the scientific method, say, the problems of the small handful or of conjunctions? The sticking point with the small handful was to account for why the few and narrow alternatives that we can come up with, result in successful novel predictions, and the like. The background was to keep in mind that most such narrow alternatives are not successful. I think that NOA has only this to say. If you believe that guessing based on some truths is more likely to succeed than guessing pure and simple, then if our earlier theories were in large part true and if our refinements of them conserve the true parts, then guessing on this basis has some relative likelihood of success. I think this is a weak account, but then I think the phenomenon here does not allow for anything much stronger since, for the most part, such guesswork fails. In the same way, NOA can help with the problem of conjunctions (and, more generally, with problems of logical combinations). For if two consistent theories in fact have overlapping domains (a fact, as I have just

---

objections stated here in my first section, it also fails to provide for the required theory-independence. For Newton-Smith's explication of verisimilitude (p. 204) makes explicit reference to some unspecified background theory. (He offers either current science or the Peircean limit as candidates.) But this is not to rebut Kuhn's challenge (and mine); it is to concede its force.

suggested, that is not so often decidable), and if the theories also have true things to say about members in the overlap, then conjoining the theories just adds to the truths of each and, thus, *may*, in conjunction, yield new truths. Where one finds other successful methodological rules, I think we will find NOA's grip on the truth sufficient to account for the utility of the rules.

Unlike the realist, however, I would not tout NOA's success at making science fairly intelligible as an argument in its favor, vis-à-vis realism or various antirealisms. For NOA's accounts are available to the realist and the antirealist, too, provided what they add to NOA does not negate its appeal to the truth, as does a verificationist account of truth or the realist's longing for approximate truth. Moreover, as I made plain enough in the first section of this chapter, I am sensitive to the possibility that explanatory efficacy can be achieved without the explanatory hypothesis being true. NOA may well make science seem fairly intelligible and even rational, but NOA could be quite the wrong view of science for all that. If we posit as a constraint on philosophizing about science that the scientific enterprise should come out in our philosophy as not too unintelligible or irrational, then, perhaps, we can say that NOA passes a minimal standard for a philosophy of science.

Indeed, perhaps the greatest virtue of NOA is to call attention to just how minimal an adequate philosophy of science can be. (In this respect, NOA might be compared to the minimalist movement in art.) For example, NOA helps us to see that realism differs from various antirealisms in this way: realism adds an outer direction to NOA, that is, the external world and the correspondence relation of approximate truth; antirealisms (typically) add an inner direction, that is, human-oriented reductions of truth, or concepts, or explanations (as in my opening citation from Hume). NOA suggests that the legitimate features of these additions are already contained in the presumed equal status of everyday truths with scientific ones, and in our accepting them both as *truths*. No other additions are legitimate, and none are required.

It will be apparent by now that a distinctive feature of NOA, one that separates it from similar views currently in the air, is NOA's stubborn refusal to amplify the concept of truth by providing a theory or analysis (or even a metaphorical picture). Rather, NOA recognizes in "truth" a concept already in use and agrees to abide by the standard rules of usage. These rules involve a Davidsonian-Tarskian referential semantics, and they support a thoroughly classical logic of inference. Thus NOA respects the customary "grammar" of "truth" (and its cognates). Likewise, NOA respects the customary epistemology, which grounds judgments of truth in perceptual judgments and various confirmation relations. As with the use of other concepts, disagreements are bound to arise over what is true (for instance, as to whether inference to the best explanation is always truth-conferring). NOA pretends to no resources for settling these disputes, for NOA takes to heart the great lesson of twentieth-century analytic and continental philosophy, namely, that there *are* no general methodological or philosophical resources for deciding such things. The mistake common to realism and all the antirealisms alike is their commitment to the existence of such nonexistent resources. If pressed to answer the question of what, then, does it *mean* to say that something is true (or to what does the truth of so-and-so commit one), NOA will reply by pointing out the logical relations engendered by the specific claim and by focusing, then, on the concrete historical circumstances that ground that particular judgment of truth. For, after all, there *is* nothing more to say.[22]

Because of its parsimony, I think the minimalist stance represented by NOA marks a revolutionary approach to understanding science. It is, I would suggest, as profound in its own way as was the revolution in our conception of morality, when we came to see that founding morality on God and his order was *also* neither legitimate nor necessary. Just as the typical theological moralist of the eighteenth century would feel bereft to read, say, the pages of *Ethics*, so I think the realist must feel similarly when NOA removes that "correspondence to the external world" for which he so longs. I too have regret for that lost paradise, and too often slip into the realist fantasy. I use my understanding of twentieth-century physics to help me firm up my convictions about NOA, and I recall some words of Mach, which I offer as a comfort and as a closing. With reference to realism, Mach writes,

> It has arisen in the process of immeasurable time without the intentional assistance of man. It is a product of nature, and preserved by nature. Everything that philosophy has accomplished . . . is, as compared with it, but an insignificant and ephemeral product of art. The fact is,

22. No doubt I am optimistic, for one can always think of more to say. In particular, one could try to fashion a general, descriptive framework for codifying and classifying such answers. Perhaps there would be something to be learned from such a descriptive, semantical framework. But what I am afraid of is that this enterprise, once launched, would lead to a proliferation of frameworks not so carefully descriptive. These would take on a life of their own, each pretending to ways (better than its rivals) to settle disputes over truth claims, or their import. What we need, however, is less bad philosophy, not more. So here, I believe, silence is indeed golden.

every thinker, every philosopher, the moment he is forced to abandon his one-sided intellectual occupation . . . , immediately returns [to realism].

Nor is it the purpose of these "introductory remarks" to discredit the standpoint [of realism]. The task which we have set ourselves is simply to show why and for what purpose we hold that standpoint during most of our lives, and why and for what purpose we are . . . obliged to abandon it.

These lines are taken from Mach's *The Analysis of Sensations* (sec. 14). I recommend that book as effective realism-therapy, a therapy that works best (as Mach suggests) when accompanied by historicophysical investigations (real versions of the breakneck history of my second section). For a better philosophy, however, I recommend NOA.

The Reality of Causes in a World of Instrumental Laws

Nancy Cartwright

Stanford University

1. Introduction

Empiricists are notoriously suspicious of causes. They have not been equally wary of laws. Hume set the tradition when he replaced causal facts with facts about generalizations. Modern empiricists do the same. But nowadays Hume's generalizations are the laws and equations of high level scientific theories. On current accounts, there may be some question about where the laws of our fundamental theories get their necessity, but it is no question that these laws are the core of modern science. Bertrand Russell is well known for this view: "The law of gravitation will illustrate what occurs in any exact science ... . Certain differential equations can be found, which hold at every instant for every particle of the system ... . But there is nothing that could be properly called "cause" and nothing that could be properly called "effect" in such a system." For Russell, causes "though useful to daily life and in the infancy of a science, tend to be displaced by quite different laws as soon as a science is successful." (Russell 1913, p. 194).

It is convenient that Russell talks about physics, and that the laws he praises are its fundamental equations--Hamilton's equations or Schroedinger's, or the equations of general relativity. That is what I want to talk about too. But I hold just the reverse of Russell's view. I am in favor of causes and opposed to laws. I think that, given the way modern theories of mathematical physics work, it makes sense only to believe their causal claims and not their explanatory laws.

2. Explaining by Causes

From Bromberger (1965), Scriven (1962), and others we know there are various things one can be doing in explaining. Two are of importance here: in explaining a phenomenon one can cite the causes of that phenomenon; or one can set the phenomenon in a general theoretical

PSA 1980, Volume 2, pp. 38-48

framework. The framework of modern physics is mathematical, and good explanations will generally allow us to make quite precise calculations about the phenomena we explain. Rene Thom remarks the difference between these two kinds of explanation, though he thinks that only the causes really explain: "Descartes with his vortices, his hooked atoms, and the like explained everything and calculated nothing; Newton, with the inverse square law of gravitation calculated everything and explained nothing." (Thom 1972, p. 5).

Unlike Thom, I am happy to call both explanation, so long as we do not illicitly attribute to theoretical explanation features that apply only to causal explanation. There is a tradition, since the time of Aristotle, of deliberately conflating the two. But I shall argue that they function quite differently in modern physics. If we accept Descartes' causal story as adequate, we must count his claims about hooked atoms and vortices true. But we do not use Newton's inverse square law as if it were either true or false.

One powerful argument speaks against my claim and for the truth of explanatory laws--the <u>argument from coincidence</u>. Those who take laws seriously tend to subscribe to what Harman (1965) has called inference to the best explanation. They assume that the fact that a law <u>explains</u> provides evidence that the law is true. The more diverse the phenomena that it explains, the more likely it is to be true. For it would be an absurd coincidence if a wide variety of different kinds of phenomena were all explained by a particular law, and yet were not in reality consequent from the law. Thus the argument from coincidence supports a good many of the inferences we make to best explanations.

The method of inference to the best explanation is subject to an important constraint, however--the requirement of non-redundancy. We can infer the truth of an explanation only if there are no alternatives which account in an equally satisfactory way for the phenomena. In physics nowadays, I shall argue, an acceptable causal story is supposed to satisfy this requirement. But exactly the opposite is the case with the specific equations and models which make up our theoretical explanations. There is redundancy of theoretical treatment, but not of causal account.

There is, I think, a simple reason for this: causes make their effects happen. We begin with a phenomenon which, relative to our other general beliefs, we think would not occur unless something peculiar brought it about. In physics we often mark this belief by labelling the phenomena as effects--the Sorbet effect, the Zeeman effect, the Hall effect. An effect needs something to bring it about, and the peculiar features of the effect depend on the particular nature of the cause, so that--in so far as we think we've got it right--we are entitled to infer the character of the cause from the character of the effect.

But equations don't bring about the phenomenological laws we derive from them (even if the phenomenological laws are themselves equations). Nor are they used in physics as if they did. The specific equations we use to treat particular phenomena provide a way of casting the phenomena into the general framework of the theory. Thus we are able to treat a variety of disparate phenomena in a similar way, and to make use of the theory to make quite precise calculations. For both of these purposes, it is an advantage to multiply theoretical treatments.

Duhem (1906) used the redundancy requirement as an argument against scientific realism, and recently Putnam (1978) uses it as an argument against realism in general. Both propose that, in principle, for any explanation of any amount of data there will always be an equally satisfactory alternative. The history of science suggests that this claim may be right: we constantly construct better explanations to replace those of the past. Such arguments are irrelevant here, however, for they do not distinguish between causal claims and theoretical accounts. Both are likely to be replaced by better accounts in the future.

Here I am not concerned with alternatives which are at best available only in principle, but rather with the practical availability of alternatives within theories we actually have to hand. For this discussion, I want to take the point of view which Putnam calls "internal realism"; to consider actual physical theories which we are willing to account as acceptable, even if only for the time being, and to ask, "Relative to that theory, which of its explanatory claims are we to deem true?" My answer is that causal claims are to be deemed true (note: "deem true" is Paul Grice's nice expression); but to count the basic explanatory laws as true is to fail to take seriously how physics succeeds in giving explanations.

I will use two examples to show this. The first--quantum damping and its associated line broadening--is a phenomenon whose understanding is critical to the theory of lasers. Here we have a single causal story, but a fruitful multiplication of successful theoretical accounts. This contrasts with the unacceptable multiplication of causal stories in the second example.

But first I should make some remarks in answer to questions raised by Dan Hausmann and Ennis (1968) about how we are to distinguish the explanatory laws, which I argue are not to be taken literally, from the causal claims and more pedestrian statements of fact which are. The short answer is that there is no way. A typical way of treating a problem like this is find some independent criterion--ideally syntactical, but more realistically semantical--which will divide the claims of a theory into two parts. Then it is argued that claims of one kind are to be taken literally, whereas those of the other kind function in some different way.

This is not what I have in mind. I think of a physics theory as providing an explanatory scheme into which phenomena of interest can be fitted. I agree with Duhem here. The scheme simplifies and organizes the phenomena so that we can treat similarly, happenings that are phenomenologically different, and differently ones that are phenomenologically the same. It is part of the nature of this organizing activity that it cannot be done very well if we stick too closely to stating what is true. I give the argument for this in Cartwright (1980). Some claims of the theory must be literally descriptive (I think the claims about the mass and charge of the electron are a good example) if the theory is to be brought to bear on the phenomena, but I suspect that there is no general independent way of characterizing which these will be. What is important to realize is that if the theory is to have considerable explanatory power, most of its fundamental claims will not state truths, and that this will in general include the bulk of our most highly prized laws and equations.

3. Examples: Quantum Damping

In radiative damping, atoms de-excite, giving off photons whose frequencies depend on the energy levels of the atom. We know by experiment that the emission line observed in a spectroscope for a radiating atom is not indefinitely sharp, but rather has a finite line width--that is, there is a spread of frequencies in the light emitted. What causes this natural linewidth? Here is the standard answer which physicists give, quoted from a good textbook on quantum radiation theory: "There are many interactions which may broaden an atomic line, but the most fundamental one is the reaction of the radiation field on the atom. That is, when an atom decays spontaneously from an excited state radiatively, it emits a quantum of energy into the radiation field. This radiation may be reabsorbed by the atom. The reaction of the field on the atom gives the atom a linewidth and causes the original level to be shifted as we show. This is the source of the natural linewidth and the Lamb shift." (Louisell 1973, p. 285). Following his mathematical treatment of the radiative decay, Louisell continues: "We see that the atom is continually emitting and reabsorbing quanta of radiation. The energy level shift does not require energy to be conserved while the damping requires energy conservation. Thus damping is brought about by the emission and absorption of real photons while the photons emitted and absorbed which contribute to the energy shift are called virtual photons." (p.289). This account is universally agreed upon. Damping, and its associated line broadening, are brought about by the emission and absorption of real photons.

Here we have a causal story; but not a mathematical treatment. We have not yet set line broadening into the general mathematical framework of quantum mechanics. There are many ways to do this.

One of the Springer Tracts by G. S. Agarwal (1974) summarizes the basic treatments which are offered. He lists six different approaches in his table of contents: (1) Weisskopf-Wigner method; (2) Heitler-Ma method; (3) Goldberger-Watson method; (4) Quantum Statistical method: master equations; (5) Langevin equations corresponding to the master equation and a c-number representation; and (6) Neoclassical theory of spontaneous emission.

Before going on to discuss these six approaches, I will give one other example. The theory of damping forms the core of current quantum treatments of lasers. Figure 1 is a diagram from a summary article by

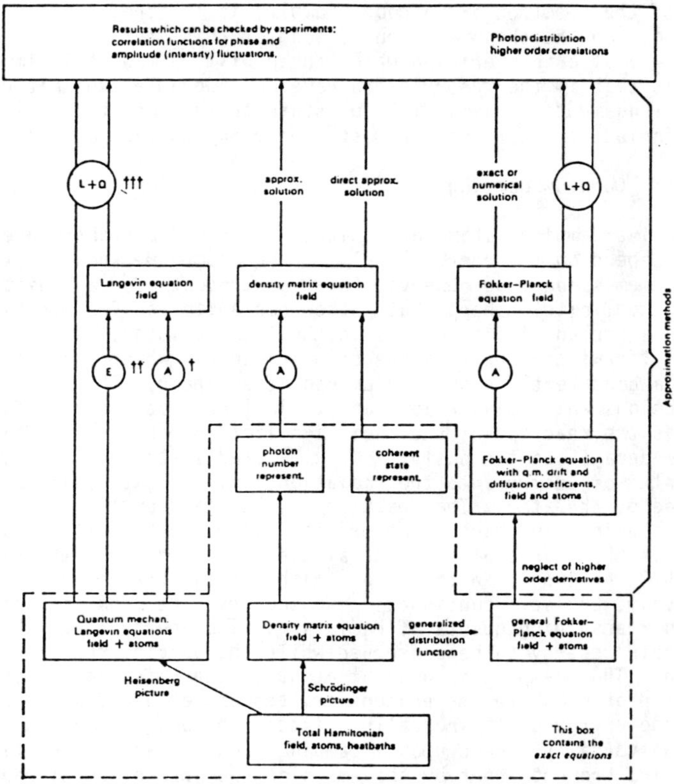

† A: adiabatic elimination of atomic variabies

†† E: exact elimination of atomic variables

††† L+Q: linearization and quantum mechanical quasilinearization

Family tree of the quantum theory of the laser

Fig. 1 from Haken (1970), page 244.

H. Haken on "the" quantum theory of the laser. We see that the situation I described for damping theory is even worse here. There are so many different treatments that Haken provides a "family tree" to set straight their connections. Looking at the situation Haken himself describes it as a case of "theory overkill". Laser theory is an extreme case, but I think there's no doubt that this kind of redundancy of treatment which Haken and Agarwal picture is common throughout physics.

Agarwal describes six treatments of line broadening. All six provide precise and accurate calculations for the shape and width of the broadened line. How do they differ? All of the approaches employ the basic format of quantum mechanics. Each writes down a Schroedinger equation; but it is a <u>different</u> equation in each <u>different</u> treatment. (Actually among the six treatments there are really just three different equations.) The view that I am attacking takes theoretical explanations to provide, as best they can, statements of objective laws. On this view, the six approaches that Agarwal lists compete with one another; they offer different laws for exactly the same phenomena.

But this is not Agarwal's attitude. Different approaches are useful for different purposes; they complement rather than compete. The Langevin and Master equations of 4) and 5), for instance, have forms borrowed from statistical mechanics. They were introduced in part because the development of lasers created an interest in photon correlation experiments. Clearly, if we have statistical questions, it is a good idea to start with the kind of equations from which we know how to get statistical answers.

Let us consider an objection to the point of view I have been urging. We all know that physicists write down the kinds of equations they know how to solve; if they can't use one approximation, they try another; and when they find a technique that works, they apply it in any place they can. These are commonplace observations that remind us of the pragmatic attitude of physicists. Perhaps contrary to my argument the multiplication of theoretical treatments says more about this pragamatic orientation than it does about how explanatory laws ought to be viewed. I disagree. I think that it does speak about laws, and in particular shows how laws differ from causes. For we do not have the same pragmatic tolerance of causal alternatives. We do not use first one causal story in explanation, then another, depending on the ease of calculation, or whatever.

The case of the radiometer illustrates. The radiometer was introduced by William Crookes in 1875, but it is still not clear what makes it work. There are three plausible theories. The first attributes the motion of the vanes to light pressure, but this explanation is now universally rejected. As M. Goldman remarks in "The Radiometer Revisited": "A simple calculation shows that on a typical British summer day, when the sky is a uniform grey (equally luminous all over) the torque from the black and silver faces exactly balance, so that for a perfect radiometer [i.e., a radiometer with a perfect vacuum] no motion would be possible." (1978, p. 428).

But two explanations still contend. The first is the more standard, textbook account, which is supported by Goldman's calculations. It supposes that the motion is caused by the perpendicular pressure of the gas in the perfect vacuum against the vanes. According to the calculations of James Clerk Maxwell, however, the perpendicular pressure on the vanes will be equal in both directions. The motion must instead be due to the tangential stress created by the gas slipping around the edge of the vanes.

There is a sense in which Maxwell and Goldman may both be right: the motion may be caused by a combination of tangential and perpendicular stress. But this is not what they claim. Each claims that the factor he cites is the single significant factor in bringing about the motion, and only one or the other of these claims can be accepted. This situation clearly contrasts with Agarwal's different theoretical treatments. Insofar as we are interested in giving a causal explanation of the motion, we must settle on one account or the other. We cannot use first one account, then the other, according to our convenience.

I know of this example through Francis Everitt, who thinks of building an experiment that would resolve the question. Everitt is an experimental physicist at Stanford, and the biographer of Maxwell for the Dictionary of Scientific Biography. I mention Everitt's experiment because it argues for the difference in objectivity which I urge between theoretical laws and causal claims. It reminds us that unlike theoretical accounts, which can be justified only by an inference to the best explanation, causal accounts have an independent test of their truth: we can perform controlled experiments to find out if our causal stories are right or wrong. Experiments of these kinds in fact play an important role in an example from which Salmon defends inferences to the best explanation.

4. The Argument from Coincidence

In a recent paper at the University of Southern California, Salmon (unpublished) considers Perrin's arguments for the existence of atoms and for the truth of Avagadro's hypothesis that there are a fixed number of molecules in any gram mole of a fluid. Perrin performed meticulous experiments on Brownian motion in colloids from which he was able to calculate Avagadro's number quite precisely. His 1913 tract, in which he summarizes his experiments and recounts the evidence for the existence of atoms, helped sway the community of physicists in favor of these hypotheses. Besides Brownian motion, Perrin lists 13 quite different physical situations which yield a determination of Avagadro's number. So much evidence of such a variety of kinds all pointing to the same value must surely convince us, urges Perrin, that atoms exist and that Avagadro's hypothesis is true.

For many, Perrin's exemplary reasoning is a paradigm of inference to the best explanation; and it shows the soundness of that method. I think this misdiagnoses the structure of the argument. Perrin does not make an inference to the best explanation, where explanation includes anything from theoretical laws to a detailed description of how the explanandum was brought about. He makes rather a more restricted inference--an inference to the most probable cause.

A well designed experiment is constructed to allow us to infer the character of the cause from the character of its more readily observable effects. Prior to Perrin, chemists focused their attention on the size and velocities of the suspended particles. But this study was unrewarding; the measurements were difficult and the results did not signify much. Perrin instead studied the height distribution of the Brownian granules at equilibrium. From his results, with a relatively simple model for the collision interactions, he was able to calculate Avagadro's number. Perrin was a brilliant experimenter. It was part of his genius that he was able to find quite specific effects which were peculiarly sensitive to the exact character of the causes he wanted to study. Given his model, the fact that the carrier fluids had just 6×10^{23} atoms for every mole made precise and calculable differences to the distribution he observed.

The role of the model is important. It brings out exactly what part coincidence plays in the structure of Perrin's argument. Our reasoning from the character of the effect to the character of the cause is always against a background of other knowledge. We aim to find out about a cause with a particular structure. What effects appear as a result of that structure will be highly sensitive to the exact nature of the causal processes which connect the two. If we are mistaken about the processes which link cause and effect in our experiment, what we observe may not result in the way we think from the cause under study. Our results may be a mere artifact of the experiment, and our conclusions will be worthless.

Perrin explicitly has this worry about the first of the 13 phenomena he cites: The viscosity of gases, which yields a value for Avagadro's number via Van der Waal's equation and the kinetic theory of gases. In his (1913, p. 82) he writes that "the probable error, for all these numbers is roughly 30 percent., owing to the approximations made in the calculations that lead to the Clausius-Maxwell and Van der Waal's equations." He continues: "The Kinetic Theory justly excites our admiration. [But] It fails to carry complete conviction, because of the many hypotheses it involves." (I take it he means "unsubstantiated hypotheses.") What sets Perrin's worries to rest? He tells us himself in the next sentence: "If by entirely independent routes we are led to the same values for the molecular magnitudes, we shall certainly find our faith in the theory considerably strengthened."

Here is where coincidence enters. We have 13 phenomena from which we can calculate Avagadro's number. Any one of these phenomena--if we were sure enough about the details of how the atomic behavior gives rise to it--would be good enough to convince us that Avagadro is right. Frequently we are not sure enough; we want further assurance that we are observing genuine results and not experimental artifacts. This is the case with Perrin. He lacks confidence in some of the models on which his calculations are based. But he can appeal to coincidence. Wouldn't it be a coincidence if each of the observations was an artifact, and yet all agreed so closely about Avagadro's number? The convergence of results provides reason for thinking that the various models used in Perrin's diverse calculations were each good enough. It thus reassures us that those models can legitimately be used to infer the nature of the cause from the character of the effects.

In each of Perrin's 13 cases we infer a concrete cause from a concrete effect. We are entitled to do so because we assume that causes make effects occur in just the way that they do, via specific, concrete causal processes of the kind that Salmon (1981) describes. The structure of the cause physically determines the structure of the effect. Coincidence enters Perrin's argument, but not in a way that supports inference to the best explanation in general. There is no connection analogous to causal propagation between theoretical laws and the phenomenological generalizations which they bring together and explain. Explanatory laws summarize phenomenological laws; they don't make them true. Coincidence won't help with laws. We've no ground for inferring from any phenomenological law that an explanatory law must be just so; multiplying cases can't help. [I discuss this point also in "When Can We Infer the Best Explanation?" read at the 1980 Western Washington Annual Philosophy Conference, unpublished.]

I mentioned that Harman (1965) introduced the expression "inference to the best explanation." Harman uses two examples in his original paper. The first is the example which Salmon presented at U.S.C.--coming to believe in atoms. The second is a common and important kind of example from everyday life--inferring that the butler did it. Notice that these are both cases in which we infer facts about concrete causes; they are not inferences to the laws of some general explanatory scheme. Like Perrin's argument, these do not vindicate a general method for inferring the truth of explanatory laws. What they illustrate is a far more restrictive kind of inference--inference to the best cause.

5. Conclusion

Perrin didn't make an inference to the best explanation, only an inference to the most probable cause. This is typical of modern physics. "Competing" theoretical treatments--treatments which write down different laws for the same phenomena--are encouraged in physics but only a single causal story is allowed. Although philosophers generally believe in laws and deny causes, explanatory practice in physics is just the reverse.

References

Agarwal, G.S. (1974). Quantum-Statistical Theories of Spontaneous Emission and their Relation to Other Approaches. Berlin: Springer.

Bromberger, Sylvain. (1965). "An Approach to Explanation." In Analytical Philosophy, 2nd Ser. Edited by R.J. Butler. Oxford: Basil Blackwell. Pages 72-105.

Cartwright, Nancy. (1980). "Do the Laws of Physics State the Facts?" Pacific Philosophical Quarterly 61: 75-84.

-----------------. (unpublished). "When Can We Infer The Best Explanation?" Paper read at the 1980 Western Washington Annual Philosophy Conference.

Duhem, Pierre. (1906). La théorie physique: son objet et sa structure. Paris: Chevalier & Riviere (As translated by P. Wiener as The Aim and Structure of Scientific Theory. Princeton: Princeton University Press, 1975.)

Ennis, Robert H. (1968). "Enumerative Induction and Best Explanation." Journal of Philosophy 65: 523-529.

Goldman, M. (1978). "The Radiometer Revisited." Physics Education 13: 427-429.

Haken, H. (1970). "The Semiclassical and Quantum Theory of the Laser." In Quantum Optics. Edited by S.M. Kay and A. Maitland. London: Academic Press. Pages 201-321.

Harman, G. (1965). "The Inference to the Best Explanation." Philosophical Review LXXIV: 88-95.

Louisell, William H. (1973). Quantum Statistical Properties of Radiation. New York: Wiley Interscience.

Perrin, Jean. (1913). Les Atomes. Paris: Alcan. (As translated by D. Hammick as Atoms. (1916). London: Constable.)

Putnam, H. (1978). Meaning and the Moral Sciences. London: Routledge & Kegan Paul.

Russell, Bertrand. (1913). "On the Notion of Cause." Proceedings of the Aristotelian Society. New Series Vol 13: 1-26. (As Reprinted in Mysticism and Logic. London: Allen and Unwin, 1917. Pages 180-208.)

Salmon, Wesley. (unpublished). "Realism and Scientific Explanation." Paper read at University of Southern California 1980.

--------------. (1981). "Causality: Production and Propagation." In

PSA 1980. Volume 2. Edited by P.D. Asquith and R.N. Giere. East Lansing: Philosophy of Science Association. Pages 49-69.

Scriven, Michael. (1962). "Explanations, Predictions, and Laws." In Scientific Explanation, Space, and Time. (Minnesota Studies in the Philosophy of Science, Vol. III). Edited by Herbert Feigl and Grover Maxwell. Minneapolis: University of Minnesota Press. Pages 170-230.

Thom, Rene. (1972). Stabilite Structurelle et Morphogénèse. Reading, Mass.: Benjamin. (As translated by D.H. Fowler as Structural Stability and Morphogenesis. Reading, Mass.: Benjamin, 1975.)

THE CONVENTIONALIST THESIS AND ITS FIRST CRITICS

In this section I will expound a famous epistemological critique of geometry by Henri Poincaré. His position is usually labeled a doctrine of geometric conventionalism, but as we shall see, understanding just what a conventionalist claim amounts to is far from trivial. Part 1 will outline Poincaré's position, Part 2 will offer a reply to it originally formulated by Eddington and developed at great length by Reichenbach. (The conflict between the views of these two men and that of Poincaré forms the bulk of the investigations of [II,H].) Part 3 of this section will be a preliminary explanation of the meaning of conventionalist doctrine when it is extended beyond the conventionality of the metric features of spacetime into a doctrine of the conventionality of topological features. The treatment here will be quite brief, as I shall have to return to this question in some detail at a later point (IV, D,3).

1. *Poincaré's Thesis of Conventionalism*

Poincaré's epistemological critique of geometry is best expressed in three famous articles on geometry collected in *Science and Hypothesis*. In outline, his argument goes something like this: (1) It might be thought that non-Euclidean geometries can be rejected out of hand because they are infected with internal inconsistency. (2) This line is clearly refuted by the relative consistency proofs that demonstrate that the non-Euclidean geometries are at least as consistent as the older Euclidean theory. (3) It might then be contended that, consistent as they are, non-Euclidean geometries can have no applicability as empirically establishable theories about the actual world because the world of our sense experience obeys the Euclidean laws. (4) A proper understanding of how the geometric theory of the nature of the physical world is founded upon our sensory experience shows that this argument deeply misconstrues the relation between the immediate data of experience and the postulated geometric theory. The relationship shows conclusively that sensory data compatible with a non-Euclidean physical world are perfectly possible and, in fact, they can be easily described. (5) The above two arguments seem to show clearly that an empiricist epistemology for geometry is to be preferred. If the consistent geometries are all possibly compatible with some sensory experiences and incompatible with others, the means for establishing which of the geometries truly describes the actual world seems clear—perform a large range of experiments and on the basis of the observations made choose the geometry that is compatible with these actual results. In this process, of course, one may have to make "inductive leaps" from the data to the geometric hypothesis, but this is clearly a feature of the empirical establishment of theories in general and in no way shows an inadequacy in the empiricist account in this case. (6) Nevertheless, the empiricist account is wrong. For, given any collection of empirical observations, a multitude of geometries, all incompatible with one another, will all be *equally compatible with the experimental results.* (7) The inevitable consequence of such an "under-determination" of hypothesis by facts, no matter how rich a collection of observational facts we accumulate, suggests a *conventionalist* doctrine. One must simply *choose* the geometry one uses to describe the world, *by convention.* (8) Since Euclidean geometry is "simpler" than the non-Euclidean geometries we will always, in fact, conventionally choose to use Euclidean geometry to describe the actual physical world. Let us examine each stage of this somewhat complex argument in turn.

Steps (1) and (2). This requires simply the application of the well-known relative consistency proofs of the axiomatic non-Euclidean geometries to Euclidean geometry. If one is considering as possible candidates for the "real" geometry of the world those Riemannian geometries of nonconstant curvature which are not representable by the earlier axiomatic geometric theories, one needs additional assurance of the consistency of the more general analytic geometries; but the models for two-dimensional Riemannian geometry in terms of curved two-surfaces embedded in a Euclidean three-space should convince us that they too are consistent. Poincaré himself, incidentally, offers in these papers his own relative consistency proof of Lobachevskian geometry in terms of a nonisometric model in a fragment of Euclidean three-space.

Steps (3) and (4). Can we reject non-Euclidean geometries as descriptive of the actual physical space of the world because our sensations of the world are prima facie Euclidean in nature? This would be, I suppose, a sort of neo-Kantian apriorism. Poincaré answers with a resolute, "No!"

It is grossly misleading, he argues, even if one believes the physical space of the world to be the traditional Euclidean three-space, to take Euclidean three-space as the "structural form" of the perceptions that inform us about the nature of the space in which we live. For the time being, think of our sensory access to space as being visual and tactile sensations. Do these have "intrinsically" a three-dimensional Euclidean structure? This seems implausible in the visual case, for if we can describe a "geometry" of the "subjective visual field" at all, shouldn't it be more a bounded two-dimensional order than the unbounded three-dimensional structure of the hypothesized physical space? It is a difficult problem to characterize the psychological notion of "visual space" at all. But whatever we decide to say about this, the constraints imposed by the physical geometry of our eyes' retinas should persuade us that the geometry of the visual field can hardly be identified, if it is well defined at all, with the geometry of the physical space of the world. What I have just said about visual space holds even more clearly for "tactile space." Poincaré argues that if tactile space could be said to have any dimensionality at all, it would be the number of muscles whose motion generates the kinesthetic sensations of touch, rather than the three dimensions of the space of the visual world.

But if we do not develop the hypothesis about the three-dimensional Euclidean structure of space from direct apprehension of this structure in the space of sensation, how do we frame the hypothesis on the basis of our experience? The answer, according to Poincaré, is this: The sensations that we experience are not random or incoherent. They have an order and coherence that we can describe by saying that there is a lawlike structure to them. For example, consider some sensations due to an object placed in front of our eyes. We change the object and the sensations change. We note that some changes of sensation due to causal manipulation of the object are such that we can restore the original sensation pattern by leaving the object alone and moving our bodies. This is true for rigid motions. By experimentation we can discover the possible rigid motions of objects and their structure, as would be described, for example, by the mathematical group theory of rigid motions.

The hypothesis about the actual geometrical structure of the physical world is an *inference* from the lawlike structure of our sensational experience. We do not perceive space directly as three-dimensional Euclidean; rather, we *infer* that this is its structure from the lawlike systematization that we find we can impose upon the true, immediate sensory data.

Steps (5) *and* (6). A single imaginative parable offered by Poincaré is designed to serve two distinct purposes. First, we are provided with a picture of a world where the sensory data of the inhabitants would naturally lead them to infer that the physical space of their world was not Euclidean, but Lobachevskian. This shows that the kind of sensational experience compatible with a non-Euclidean geometry is perfectly possible, and that we can easily describe it. But, second, we will then see that this "inference" to a non-Euclidean structure for physical space is not the only one that can be made from the data, and that, in fact, one could equally well explain the totality of sensory experience by postulating instead that the space of the world is Euclidean. This feature of Poincaré's "parable" shows us that the naïve empiricist account of geometry, plausible as it may seem in the light of the failure of the aprioristic accounts, is itself inadequate.

With powers actually beyond us we construct in the laboratory the "imaginary world" of Poincaré (Fig. 22). It is a closed Euclidean two-dimensional disk, heated to a constant temperature at the center while the circumference is uniformly cooled to 0° absolute. The temperature gradient along a radius is given by $R^2 - r^2$, where R is the radius of the disk and r is the distance of a point from the center of the disk. We populate the disk with two-dimensional creatures who are interested in determining the intrinsic geometry of their world. We equip them with rigid rods, except that the rods all contract uniformly with diminishing temperatures and in proportion to the drop in temperature, and all the rods have length 0 when their temperature is 0° absolute. We assume ideal thermal contact, so that the temperature of a point on a rod is equal to its location on the disk.

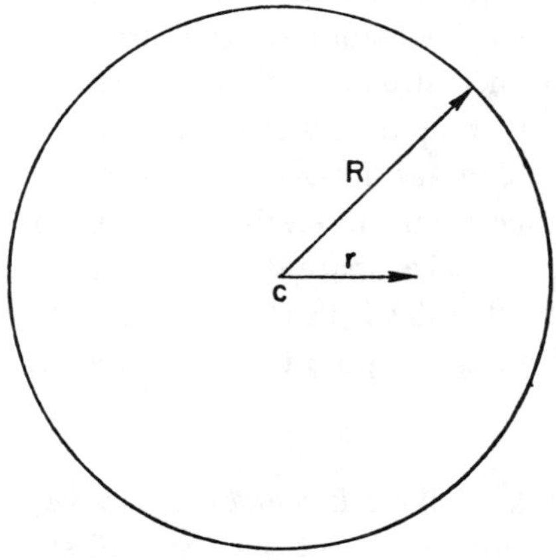

Fig. 22. The world of Poincaré's parable. The center of the disk is kept at temperature TR^2. At a distance r from the center, c, the temperature is $T(R^2 - r^2)$, where R is the radius of the disk. Consequently the temperature at the edge of the disk is zero. The index of refraction of the medium of transmission of light on the disk at any point r from the center is $1/(R^2 - r^2)$.

The inhabitants are equipped only with light beams and rigid rods whose length at a point is directly proportional to the temperature at the point. Unaware of the temperature distortion of their rods and the bending of their light rays, the inhabitants calculate their world to be a Lobachevskian plane of infinite extent.

The inhabitants of this world proceed to measure distances between points on it so as to determine its intrinsic geometry. They assume, falsely, that a rod remains invariant in its length upon transport. After a short period they come to the following conclusion: Their world is a Lobachevskian plane of infinite extent. For example, geodesics are not straight lines but the geodesics of Lobachevski, and the ratio of circumference to radius of a circle is always greater than 2π.

Suppose we expand the experimental repertoire of the inhabitants by providing them with light rays, but at the same time giving their world a medium of transmission for light whose index of refraction varies as $1/(R^2 - r^2)$. If they triangulate their world with the light rays, assuming that the light travels along geodesics and that the index of refraction of the "vacuum" is constant, they will once again falsely infer that their plane is Lobachevskian, unbounded and infinite.

But now, suppose some clever denizen of the disk realizes that all the experimental data that seemed to lead to the Lobachevskian hypothesis can instead be taken to indicate that the plane is a finite Euclidean disk, equipped with "shrinking" and "light-bending" fields of the kind we have described. How could the issue between his hypothesis and that held by the majority of scientists be resolved? *We* know, utilizing our "superior" standpoint out in Euclidean three-space, that the innovator is correct and the orthodoxy wrong. But how could *they* decide? According to Poincaré, they cannot. They must simply, by *convention, choose* one of the alternative hypotheses.

Now imagine ourselves in three-space performing a similar mapping of the intrinsic geometry of our space. Can't we see that an exactly similar situation could arise? And in this new case, there simply is no "right" theory. Both alternatives are equally correct alernative hypotheses, with only a "conventional" choice to make between them. The parable has shown us that there are describable experimental results that would lead us, making our ordinary assumptions about the behavior of rigid rods upon transporation and about the straight-line passage of light in a vacuum, to infer that our world was non-Euclidean, and that we can conventionally avoid discarding our Euclidean geometry in favor of an alternative, if only we are willing to make sufficient changes elsewhere in our hypotheses about the world.

Step (8). Step (8) is usually considered the weakest point in Poincaré's argument. What he has not seen is that whereas Euclidean geometry may be in some sense simpler than the non-Euclidean alternatives, in just the same sense of "simpler" the total non-Euclidean-geometry-plus-traditional-remaining-theory hypothesis may be as a whole, simpler than the Euclidean-geometry-plus-revised-remaining-hypotheses theory. So let us ignore step (8).

If we make the necessary changes to carry Poincaré's parable from one about space to one about spacetime, we will see that it contains the core elements of all the basic questions we will have to face about the proper analysis of the epistemic foundations of geometry. In addition, it will throw important light upon empiricist accounts of theory-acceptance in general and will lead ultimately to difficult questions of ontology and metaphysics, as well.

I have been describing Poincaré as a "conventionalist," but I have not made it very clear just what a commitment to a conventionalist doctrine amounts to. I think one can find evidence that Poincaré held at different points doctrines that we will want to maintain as distinct. Since I will later be looking at alternative reactions to the Poincaréan argument in some detail (II,H), I will not yet pursue this line of inquiry. In the next section, when we examine the important replies to Poincaré of Eddington and Reichenbach, we will begin to see what some of these alternative philosophical accounts of the Poincaréan situation are.

2. *The Empiricist Reply to Poincaré*

The physicist Arthur Eddington and the philosophers Moritz Schlick and Hans Reichenbach provide epistemological critiques of geometry that are largely directed to refuting Poincaré's allegation that the selection of a geometric hypothesis can only be a matter of convention. Eddington's views are contained in the "Prologue: What is Geometry?" to his *Space, Time, and Gravitation,* an elementary treatise on the general theory of relativity. The section is quite brief, but the fundamental argument is clearly and precisely laid out. Reichenbach's quite similar views are developed at much greater length, especially in Chapter 1 of his *The Philosophy of Space and Time.* In this section I will summarize their position and go into Reichenbach's development in some detail. Part 3 and Section II,G, to follow will be in the nature of digressions from the main argument, but I shall return to the conflict between Poincaré's position and those expressed in this section shortly in (II,H).

Poincaré's conventionalist doctrine applies to the geometry of space. Eddington and Reichenbach are writing in the context of an already-existing general theory of relativity; consequently, the problem they are concerned with is whether the adoption of a geometry for *spacetime* is a matter of "framing a hypothesis from observational data," as the empiricists would have us believe, or "merely making a conventional choice," as Poincaré alleges. Although Poincaré never contemplates the possibility of non-Euclidean spacetimes, as opposed to non-Euclidean spaces, the application of his doctrine to this new "manifold" of the world is clear.

Eddington's prologue is in the form of a three-way discussion between an experimental physicist, a mathematician, and an exponent of the new theory of general relativity. The experimental physicist gets the role of a rather naïve traditionalist, maintaining the "self-evidence" of Euclidean geometry and the unintelligibility of the alternative world pictures. Naturally enough, the "relativist" has little difficulty in disposing of his arguments. The non-Euclidean picture is consistent. If geometry is meant to characterize the empirically observable interrelationships among the material objects (rods, clocks, freely-moving particles, light rays) that we would intuitively use to "map" the structure of spacetime, then a set of such observed relationships supportive of a non-Euclidean spacetime structure is just as possible, and conceivable, as the sets that would reaffirm traditional belief in the Euclidean (or, since we are dealing with spacetime, more properly Minkowskian) nature of the manifold.

The debate between the relativist and the mathematical exponent of Poincaréan conventionalism is handled more briefly, but it is sufficiently important to quote at some length. In italicized whole clauses and sentences, the italics are mine:

> *Mathematician:* The view has been widely held that space is neither physical nor metaphysical, but conventional. Here is a passage from Poincaré's *Science and Hypothesis,* which describes this alternative idea of space:
>
> "If Lobatchewsky's geometry is true, the parallax of a very distant star will be finite. If Riemann's is true it will be negative. These are the results which seem within the reach of experiment, and it is hoped that astronomical observations will enable us to decide between the two geometries. But what we call a straight line in astronomy is simply the path of a ray of light. If, therefore, we were to discover negative parallaxes, or to prove that all parallaxes are higher than a certain limit, we should have a choice between two conclusions: we could give up Euclidean geometry, or modify the laws of optics, and suppose that light is not rigorously propagated in a straight line. It is needless to add that everyone would look upon this solution as the more advantageous. Euclidean geometry, therefore, has nothing to fear from fresh experiments."
>
> *Relativist:* Poincaré's brilliant exposition is a great help in understanding the problem now confronting us. He brings out the interdependence of geometrical laws and physical laws which we have to bear in mind continually. We can add on to one set of laws that which we subtract from the other set. I admit that space is conventional—for that matter, *the meaning of every word in the language is conventional.* More-

> over, we have actually arrived at the parting of the ways imagined by Poincaré, though the crucial experiment is not precisely the one he mentions. But I deliberately adopt the alternative, which, he takes for granted, everyone would consider less advantageous. I call the space thus chosen *physical space,* and its geometry *natural geometry,* thus *admitting that other conventional meanings of space and geometry are possible.* If it were only a question of the meaning of space—a rather vague term—these other possibilities might have some advantages. But the meaning assigned to length and distance has to go along with the meaning assigned to space. Now these are quantities which the physicist has been accustomed to measure with great accuracy; and they enter fundamentally into the whole of our experimental knowledge of the world. We have a knowledge of the so-called extent of the stellar universe, which, whatever it may amount to in terms of ultimate reality, is not a mere description of location in a conventional and arbitrary mathematical space. *Are we to be robbed of the terms in which we are accustomed to describe that knowledge?*
>
> The law of Boyle states that the pressure of a gas is proportional to its density. It is found by experiment that this law is only approximately true. A certain mathematical simplicity would be gained by conventionally redefining *pressure* in such a way that Boyle's law would be rigorously obeyed. But it would be high-handed to appropriate the word pressure in this way, unless it had been ascertained that the physicist had no further use for it in its original meaning.
>
> A. S. Eddington, *Space, Time and Gravitation* (Cambridge, 1921), pp. 9, 10.

The doctrine espoused by Eddington in the voice of the relativist is only briefly characterized here, but its fundamental themes are clear and important. Basically the argument is running something like this: Poincaré has *apparently* shown us that we can construct alternative, incompatible geometries of space (or spacetime) which are equally compatible with the same observational data, no matter how much data we accumulate. The two theories he constructs each have a geometric part and an additional "physical" part; so, by making suitable modifications in the physical part of the total theory, we can "save" any geometric hypothesis we like from refutation by any amount of experimental data. Now Poincaré thinks that this forces us to the conclusion that we must simply choose one alternative theory or the other by making a "conventional" choice. In a sense this is correct, but in a trivial and not particularly interesting sense. The conventional choice we must make is neither more nor less than a conventional choice of using a certain verbal sound or written symbol with a certain *meaning*. The appearance of two "incompatible" total theories is misleading. Properly speaking there is only one theory, although it is written in two different ways. The confusion comes about because, in the *two expressions of the same theory,* words are being used with different meanings.

What the physicist means by "straightest line between two points" is "path of a light ray *in vacuuo* between those points." When we "save" Euclidean geometry by changing the physical theory we are not really saving Euclidean geometry at all. We are changing the usage of such terms as 'geodesic,' 'distance between *P* and *Q* along *C*,' etc., in such a way that given these new meanings the old sentences of Euclidean geometry remain true. But they no longer assert the same *propositions,* since the meanings of the terms in the sentences have been changed by the revision in the nongeometric portion of the theory. Poincaré no more establishes the conventional truth of Euclidean geometry than I could establish the truth of the assertion, 'lions have stripes,' by simply changing *my* meaning of the word 'lion' to mean what everybody else means by 'tiger' in English.

Reichenbach presents us with a similar reply to the conventionalist thesis. It does not, I believe, differ significantly in fundamentals from Eddington's briefly-related account; but since Reichenbach has presented us with what has become a convenient (if sometimes misleading) vocabulary for discussing these issues, it will be helpful to offer his version of this reply to Poincaré in a little detail.

If we wish to use a term in science, says Reichenbach, we must first *define* it, or at least, we should realize when we are using it that our assertions have full meaning only if the proper definition for the terms therein is forthcoming upon reflection and analysis. Now definitions can be of quite varied kinds. For example, we can introduce a novel term into a theory by offering a stipulative definition of it in terms of previously accepted meaningful terms of the theory; say, by defining 'bachelor' as 'an unmarried adult male.' But the kind of definition we are concerned with is not quite like this:

> Physical knowledge is characterized by the facts that concepts are not only defined by other concepts, but are also coordinated to real objects. This coordination cannot be replaced by an explanation of meanings, it simply states that *this concept* is coordinated to this *particular thing*. In general this coordination is not arbitrary. Since the concepts are interconnected by testable relations, the coordination may be verified as true or false, if the requirement of uniqueness is added, i.e., the rule

> that the same concept must always denote the same object. The method of physics consists in establishing the uniqueness of this coordination, as Schlick has clearly shown. But certain preliminary coordinations must be determined before the method of coordination can be carried through any further; these first coordinations are therefore definitions which we shall call *coordinative definitions.* They are *arbitrary,* like all definitions; on their choice depends the conceptual system which develops with the progress of science.
>
> Wherever metrical relations are to be established, the use of coordinative definitions is conspicuous. If a distance is to be measured, the unit of length has to be determined beforehand by definition. This definition is a coordinative definition.
>
> H. Reichenbach, *The Philosophy of Space and Time* (New York: Dover, 1958), p. 14.

So picking as the meaning of 'one meter' the length of a particular rigid rod in a vault in Paris is a coordinative definition. But, he goes on to argue, choosing to call two objects separated from one another 'the same length,' because one can transport a rigid rod congruent when in coincidence with the first object to the second object and find it is congruent with the second object, is again supplying a coordinative definition. So is taking 'geodesic curve between two points' to be 'path of a light ray *in vacuuo* connecting those points,' and so on.

Given the notion of coordinative definition, let us look at Poincaré's "conventional alternative" theories once again. We save Euclidean geometry in the face of evidence that apparently establishes the non-Euclidean nature of space, or Minkowski spacetime in the face of evidence that apparently establishes that spacetime is Riemannian and not flat, by making certain changes in the remainder of the total physical theory. Like Eddington, Reichenbach would interpret these changes as simply adopting new coordinative definitions for certain important terms of the theory. To refuse to accept the principle that light rays travel null-geodesics in a vacuum is simply to change the coordinative definition for, and hence the meaning of, 'null-geodesic.' To maintain that all "rigid rods" expand or shrink uniformly when transported from place to place, so that they remain congruent to equally transported rods but cease to be congruent with previously congruent rods that were left behind, is simply to adopt a new coordinative definition for the concept 'same length for objects at a spatial separation from one another.' We thought we had two incompatible theories that were equally compatible with the class of possible observational results. We discover that we really have, properly speaking, one theory written in two misleadingly different ways. Since the coordinative definitions for terms in the "two" theories are different, these terms simply have different meanings in the two contexts. The fact that one theory says '*S*' and the other 'Not *S*' no more indicates their true incompatibility than does our disagreeing with a German who thinks ill of "Gift," because to him 'Gift' means a poison and to us 'gift' means a present.

One additional concept of Reichenbach's is that of *universal forces*. The terminology is misleading, but the concept is fruitful. Suppose we wish to construct two "conventionally alternative" geometries of spacetime. They differ in what they apparently say about the structure of spacetime, but how in addition must they differ if they are to remain equally compatible with exactly the same experimental observations?

To discover this we need only reflect back on the ways we can experimentally map out the structure of spacetime, as discussed above (II,D,3). Four kinds of observations were discussed: (1) We can measure spatial separations between events, relative to a particular "state-of-motion" coordinate frame, by using transported rigid rods that are at rest in the designated coordinate frame. (2) We can similarly determine time intervals between events "at a single place," relative to a state-of-motion frame, by using a clock at that place, i.e., at rest in the coordinate frame. (3) We can determine the timelike-geodesics of the spacetime by observing the paths of point masses whose motion is affected only by gravity and inertia. (4) We can determine the null-geodesics of the spacetime by observing the paths of light rays in a vacuum.

So to make a total theory "come out right experimentally" while changing its geometric portion, we would have to introduce a change in the remaining portion of the theory which had the following features: (1) It postulated a "stretching and shrinking" field that preserved congruences between transported rods, but destroyed the congruence of one rod to another "left behind" when the one rod was moved about in space. This is just like Poincaré's temperature field in the two-dimensional parable of (II,F,1). (2) The field would also have to affect congruence in the temporal sense of clocks. Clocks congruent at a place-time would remain congruent when equally moved about to new place-times, but they would no longer "tick at the same rate" with a clock to which they were initially synchronized but which stayed behind. (3) The field would affect the motion of point masses, deflecting them from their "inertial geodesic motion," but this deflection would be independent of the particular mass or internal constitution of the massive particle. (4) The field must deflect light rays in vacuums from their otherwise null-geodesic path, and the deflection must be independent of the amplitude or frequency of the light.

With regard to clause (3), the classical gravitational force is a field of this kind. Some confusion has arisen because Reichenbach refers to gravity as being just such a field as I am describing, and it is important to remember that, in relativity, gravity is *not* just a *force,* but something that affects light rays and the metrical features of spacetime as well. That is, the field I am attempting to describe must satisfy clauses (1), (2), and (4), as well as clause (3).

Such a field Reichenbach calls a *universal force.* To save a particular geometry by postulating a universal force is not to save one part of physics by making "real changes" in the remainder. It is simply to change one's coordinative definitions of such notions as 'same length for objects at a distance,' 'null-geodesic,' etc. The "natural geometry" of Eddington, the geometry of the usual expression of general relativity, is the one we obtain by setting the universal forces to zero; i.e., by "postulating" that there are no such universal forces. To postulate a nonzero universal force would simply be to change one's coordinative definitions. The modified theory would not really hypothesize a new geometry for spacetime at all, it would simply rewrite the old geometric theory in a potentially misleading way.

One gains clearer understanding of the relationship of one "alternative theory" to the other, says Reichenbach, if he makes the simple distinction between *genuine inductive simplicity* and *mere descriptive simplicity.* In choosing a scientific hypothesis in the light of evidential data, we must frequently add to the data some methodological principle of simplicity in order to select out as "preferred" one of the many different possible hypotheses, all compatible with the specified data.

For example, in attempting to fit a curve to a number of data points, the scientist will first try a straight line in preference to curves with more complex algebraic descriptions. This is genuine inductive simplicity at work. It is characterized by the fact that additional data can force us to reject our simplest hypothesis in favor of one more complex. One additional datum point, for example, may force us to reject the straight line that we drew in favor of some more complicated curve, which then becomes the simplest compatible with the extended data. But no data whatever, no matter how novel or how extensive, can ever force a decision between one or another of a pair of Poincaréan alternatives. What this shows us is that, properly speaking, the two alternatives are not really alternatives at all. They are the same theory differently written. They differ from one another only in the form that they use to express a common theory of the world, or in Reichenbach's terms, merely in their *descriptive simplicity*.

This by no means concludes the discussion of these issues. All of (II,H), in fact, will be devoted to a reconsideration of the claims of Poincaré, Eddington, and Reichenbach, and to a discussion of the fundamental epistemological, metaphysical, and semantic issues underlying their disagreement. But in the next section, and in (II,G), that follows, I will digress.

3. *The Conventionality of Nonmetric Features of Spacetime*

In (II,B,6) above I noted some important connections between the metrical features of a non-Euclidean space and its overall topological structure. Questions of intrinsic geometry, therefore, are not totally disjoint from questions about the compactness or noncompactness, simple-connectedness versus multiple-connectedness, etc., of the manifold as a whole. As soon as one contemplates the physical possibility of the space or spacetime of the world being non-Euclidean in its metric structure, one immediately begins to speculate about the possibility of its deviating from the Euclidean model in topological features as well. And as soon as one has seen the Poincaréan allegation of conventionality for the metric features, one begins to wonder if a conventionalist account of topology will also go through. To what extent can we claim the attribution of a particular topological form to spacetime, on the basis of the empirical evidence, to be a matter of convention?

The subtle nature of spacetime, as opposed to space-through-time, when combined with these topological questions gives rise to a rather intricate set of questions. These are quite fascinating to the philosopher because anomalous models of the topology of spacetime introduce rather disturbing pictures of the structure of *time*. Variant spatial topologies seem to require at most a little flexibility of the imagination. Variant structures for temporal order, however, introduce impressive stresses into our naïve conceptual scheme. For this reason, I intend to postpone detailed discussion of possible alternative topological structures for spacetime until later (IV,D). At this point I will offer only one concrete instance of "alternative topologies" and do this for a spatial, as opposed to spacetime, structure. The simple instance, however, will give us access to some general issues that arise whenever alternative topological hypotheses are considered, and whenever the question of the conventionality of the alternatives is brought forth.

Consider the two-dimensional space that is the surface of a cylinder. As we have seen above, the intrinsic geometry on this surface, for a sufficiently small region about a point, is indistinguishable to an inhabitant of this surface from a small region about a point on the Euclidean plane (II,B,4). But, in the large, the surfaces certainly differ in their topological structure and this difference is determinable by a two-dimensional inhabitant with no access to the three-dimensional embedding space. Any simple closed curve on the plane can be shrunk continuously to a point, i.e., the plane is simply-connected. But this is certainly not true on the cylinder, as even the geodesic curves that are circles girdling the cylinder cannot be so contracted while kept confined to the cylindrical surface. So we could, being such two-dimensional creatures, determine on the basis of empirical experiment whether the surface to which we were confined was that of a cylinder or that of a plane in its topology, the local equivalence of the two surfaces with regard to metric features notwithstanding.

But is the situation so simple? Considerations clearly stated by Reichenbach show us that it is not. Consider the surface that is a plane divided up into regions by a set of parallel straight lines drawn at constant perpendicular distances from one another. Imagine such a space that has the feature that each such strip on the surface is, at any time, qualitatively identical to any other strip. How could an inhabitant of a surface distinguish between living (1) on a cylindrical surface or (2) on such a repetitious plane? Would not all his observations be exactly the same?

Of course, in order to accommodate the evidence that would naturally lead one to postulate a cylindrical universe for this new picture, one must allow for the changed topology of the world by making suitable modifications elsewhere in one's theory. On the cylindrical model, we can easily imagine an event happening which is like no other event elsewhere in the space at the same time. In the alternative picture, each and every event occurring in the space at a time is accompanied by an infinite number of qualitatively exactly similar events; these events differ from one another only in their *numerical* diversity. In addition, the necessity for a given change in one strip to be accompanied by exactly similar changes in all of the other strips forces us to assume a sort of "causal interdependence for events at a distance from one another." Reichenbach calls this accepting a theory that tolerates causal anomaly. So it is in the principles of individuation of events and in the accepted principles of causal interaction that the plane "repeated-strip" model exacts its price for being made compatible with data that from the other point of view suggest that the structure of the space is a cylinder.

In mathematical terms, the plane of repeated strips is called a *covering space* for the cylinder, and the cylinder is a *quotient space* for the repeated-strip plane. They are related to one another by the many-one mapping, which associates with each member of the infinite set of "qualitatively identical" points on the plane a single point on the cylinder.

Other alternative topologies require different modifications in the remaining physical theory to be made equally compatible with the data. Reichenbach shows, for example, that to try to give an account of data that naturally suggest that one inhabits a two-dimensional spherical surface to the alternative plane topology, we must accept a world in which points at infinity can be reached in a finite time by transporting a finite rigid rod, and in which "events at infinity" can causally affect events at a finite distance. As we have remarked, the possibilities for topological structures and the varieties of conventional alternatives multiply when we consider not only spatially interesting topological structures, but the topology of spacetime itself. I will, indeed, pursue this in some detail in (IV,D); but the material here is enough to make us keep in mind, in the discussion of conventionality which is to follow, that conventionality can be attributed not merely to the metrical features of spacetime but to its topological features as well.

THE FRAMEWORK OF PHILOSOPHICAL PERPLEXITY

1. *Poincaré's Parable Revisited*

For the remainder of this Chapter I will be concerned with the most fundamental questions in the epistemology of geometry: (1) On what basic features do the arguments for a conventionalist position rest? (2) How are we to evaluate the two positions in the debate between conventionalists and the defenders of the empiricist position against the conventionalist critique? (3) What implications for the philosophy of science and epistemology in general can we draw from the specific debate about the status of geometric hypotheses?

We can set the stage for the discussion by paying a return visit to Poincaré's two-dimensional world, which he used to establish the conventionalist claim in the first place. Previously we had considered two groups of physical scientists inhabiting this world. Both groups were determined to map the intrinsic geometry of the world, and both had the same evidential basis from which to extrapolate their geometric hypotheses. One group took the evidence to support the thesis that the world was Lobachevskian and of infinite extent. The other claimed instead that the world was the interior of a finite Euclidean disk, and that the appearance of a non-Euclidean structure was due to the presence of distorting fields whose effects on rigid rods, upon transportation, and upon the propagation of light gave rise to the misleading appearance of a non-Euclidean world.

But instead of concentrating on this debate among the physicists, let us instead focus on the parasitical controversy inflicted on the *philosophers* of this world by their reflections on the physicist's quandary. So far I have delineated two schools of philosophy; we shall soon see that there are more than two contending positions that may be defended. One school, the conventionalist, has argued that one can easily construct two alternative total theories of the world which, although they are incompatible with one another, are so constructed that any evidence supporting one theory necessarily supports the other equally well, and any evidence tending to refute the one also tends necessarily to lead to the rejection of the other to an equal degree. But the only rational grounds for accepting or rejecting a physical theory are its conformities and nonconformities with the observational data, so we could not possibly perform any experiment leading us to accept one of these theories as preferable to the other. In such a situation our only recourse is simply to *choose,* by making a decision or adopting a convention, one theory or the other as the "true" picture of the world. But such truth is merely conventional truth, and a careful epistemological critique will keep us from misleadingly confusing such truth-by-convention with the kind of truth that receives its warrant from "genuine" confrontation with the observational facts.

The other school, the defenders of empiricism against the conventionalist critique, replied that the *appearance* of incompatible alternative theories equally compatible or incompatible with any possible experimental results is merely an appearance. Theories are hypotheses adopted to explain the observation facts. A correct doctrine of *meaning,* including a correct understanding of the way in which the terms of a theory obtain their meaning by being associated with experience, shows us that in the case we have been considering there are not, properly speaking, two alternative theories at all. Insofar as the two theories are demonstrably equivalent with respect to predicted experimental outcomes, they are the same theory. The appearance of a multiplicity of theories is due to a simple neglect of the possibility of ambiguity in the use of terms. Once considerations of the *meaning* of terms is brought in, we see that the situation is much simpler than the conventionalist claims. It is, of course, possible to have two incompatible theories that we cannot at a given time decide on between the basis of the observational data yet accumulated. But this situation, common to the frontiers of scientific investigation, shouldn't disturb the empiricist in the least. What is impossible is the existence of two incompatible theories immune to empirical discrimination and test in the face of all possible observations and experiments. Such a "pair" of theories is only a pair of expressions of a single theory, and there is no need to make a choice between them, conventional or otherwise, since there is no choice to be made.

The remainder of Chapter II will be devoted to investigating in a little more detail the features of the scientific situation which give rise to this *philosophical* controversy, the delineation in a little more depth of what the alternative philosophical groups intend to assert along with the presentation of some alternative stances different from these two, and an examination of the way this issue plays a crucial role in the philosophy of science and epistemology in areas outrunning the specific problem of the status of geometry. I won't pretend, however, to offer, a final "resolution" of the philosophical dispute, although I will try to offer what I take to be the best arguments for the contending positions and the most difficult objections they must face.

2. *Observational Basis and Theoretical Superstructure*

The claim is frequently made by philosophers of science that for any physical theory a distinguished subclass of the class of propositions entailed by the theory can be selected, which encompasses all of the propositions that follow from the theory which can be said to report results of possible observational and experimental tests. Geometric theories differ from theories in general only in that, within say the context of a presentation of a general-relativistic theory of curved spacetime, we can not only speculate about the existence of such a set of "observational" consequences, but we can also more fully characterize it.

Given the usual presentations of relativistic spacetime theories, what are the features of the world open to observational test? Our "idealized test apparatus" consists of idealized rigid rods capable of transportation throughout the spacetime, idealized clocks, light rays traveling through a vacuum, and point masses traveling unconstrained by any but gravitational "forces." With regard to the rods, we can experimentally determine whether two rods brought into coincidence are the same length. For the clocks, we can determine whether two clocks at the same place are tick-ing at a uniform rate with regard to each other. And for the light rays and freely-moving point masses, we can observe what classes of events constitute their trajectories.

But to fully characterize the geometry of spacetime we need "facts" that outrun those determinable by direct observation. We need to know, for example, of four separated points, *a*, *b*, *c*, *d*, whether *a* is as far from *b* as *c* is from *d*. This requires knowing whether rods at a distance from one another are congruent (i.e., the same length), not merely knowing the congruence or noncongruence of coincident rods. Similarly, we need to know time separations for events at a spatial separation from one another, and, of successive events, *a*, *b*, *c*, *d*, at a place, whether the temporal interval from *a* to *b* is the same as that from *c* to *d*. This means that we need to be able to establish synchrony for clocks at a spatial separation from one another, and synchrony of a clock through one period with itself at a later time. If you wish, our observational data using rods and clocks is all of a purely "local" kind; but it is "global" information that we need to fully determine the metric.

Switching to the use of light rays and free particles, a similar "leap beyond the observable" is needed to establish the geometry on the basis of the observations. We need to know what the null-geodesics and timelike-geodesics of the spacetime are. We can observe paths of light rays and of free particles. Only an additional "assumption"—that light rays travel null-geodesics when in a vacuum and acted upon only by gravity, and that free particles in a similar situation follow timelike-geodesics—can lead us to a knowledge of the geodesics themselves. The observation is of the behavior of certain material entities, the theory is about the structure of spacetime itself. So two leaps beyond observation are necessary to go from "all possible observational data" to "the nature of spacetime itself." The observational basis is the local (coincidence) behavior of material objects; but the nature of spacetime includes features of a global nature and features not of the behavior of material objects but of the structure of the spacetime itself.

As I have pointed out, it is frequently claimed that this outrunning of the data by the theory, which we have noted in the case of geometry, is a characteristic of theories in general. So why has conventionalism become an issue with regard to geometry, but not in the case of physical theory in general? I can only offer a psychological explanation: First, whereas the assumption of a designated subclass of the consequences of the theory in question which exhausts all the theory's observational consequences is frequently made, in the case of geometric theories the specific characterization of this class of consequences is most easily made. It is, in fact, available to the philosopher offering an epistemological critique of the theory by inspection of the *physical* literature. It is as though in the case of geometry the observational-nonobservational disinction among consequences of the theory is worn on the sleeve of the physicists' theoretical speculation. Second, the allegation is frequently made by philosophers that for any full-blown physical theory an alternative theory can be constructed which saves all the same observational consequences. In the case of geometry, however, we can not only speculate about such alternatives but, as the Poincaré parable shows, actually construct them. Once again, geometry differs from theories in general in the forthright, explicit way the conventionalist thesis can be defended.

The step from a theory, which by inductive generalization upon actual observations characterizes the lawlike structure of the observables, to the theory that gives the full specification of the spacetime structure can, in Reichenbach's terms, be made by picking a hypothesis about the universal forces. But we have still not made it clear what such a hypothesis about universal forces amounts to. In particular, is it making a physical assumption about the world, or rather, is it simply adopting a set of coordinative definitions? The next part is in the nature of a temporary diversion from my main concern, but I shall return to this major issue immediately thereafter.

3. *Duhemian and Super-Duhemian Alternatives*

There is a well-known argument, of Pierre Duhem, concerning the possibility of crucial experiments in physics. It has sometimes been thought to be crucial to the sort of issue we are now facing. At this point it would be wise to rehearse his thesis and see just how related to, and how distinct from, the argument we have been attending to it is.

Duhem offers an important critique of the notion of a *crucial experiment* in science. Suppose we have a hypothesis, *H*, of general form, "All. . . .", and we wish to confront the hypothesis with an experimental test. Suppose, on the basis of the hypothesis, that we predict an outcome, *E*, for the test, but upon performing the experiment not-*E* is found to be the result. Can we then affirm that the hypothesis *H* has been conclusively found to be false?

It might seem, says Duhem, that the situation is just like that in mathematics where general propositions can be refuted by the construction of a single counterexample. For example, the assertion that all continuous functions are nondifferentiable at, at most, a finite number of points is rejected once and for all by Weirstrass's construction of a continuous function on the reals which is nowhere differentiable. But, continues Duhem, the assimilation of the method of crucial experiments in physics to that of counterexamples in mathematics is spurious. For in order to infer an observational result from the hypothesis *H* we need a vast amount of additional theory, *A*, as well. This theory involves, among other things, the laws governing the behavior of the measuring instruments, etc. The observation of not-*E* upon performing the experiment informs us only that something is wrong with our total theory, *H* and *A*; but this failure of the theory to correctly predict the observational facts can be attributed either to the falsity of *H*, or instead, to the falsity of *A*. In other words, if we are motivated strongly enough to retain our belief in *H* in the face of apparently confuting observational evidence, not-*E*, we can do so at the cost of making sufficiently broad changes in our "auxiliary hypotheses," *A*.

We can broaden Duhem's claim somewhat. In bringing observations to bear on the test of a hypothesis we need not only a set of auxiliary generalizations, *A*, but also a set of beliefs about the particular facts governing the experimental situation as well. For example, in "testing" the Newtonian theory of gravitation by observing the orbits of the planets, we need to assume not only the correctness of some additional laws, say Newton's mechanical laws of general dynamics, but some particular facts about the planets as well. We must assume, for instance, that the planets carry no net electric charge sufficient to introduce electrostatic forces among them which would make our observations unreliable for testing the form of the gravitational force law; because if there were such charges the planets would not, as we assumed, be moving solely under the influence of gravitational forces.

Notice that Duhem is certainly not asserting the skeptical claim that no experiment ever gives us any reason for doubting the truth of any hypothesis. We may very well have good reason, in some specific case, to blame the failure of a total theory to correctly predict the outcome of an experiment on a particular hypothesis *H*, rather than on the auxiliary theory or upon the specific factual assumptions made. His only claim is that such a "refutation by crucial experiment" cannot be assimilated to the mathematical method of counterexample where the existence of the counterexample is taken as logically sufficient to guarantee the falsity of the hypothesis.

Are Poincaréan conventionality arguments just a subspecies of Duhemian claims about refuting experiments? We must see clearly that they are not. Suppose we have been testing a hypothesis, *H*, and have found apparently-refuting experimental data. We recognize some questionable elements in our auxiliary hypotheses, *A*, and are puzzled as to whether we should attribute the unexpected results to the falsity of *H* or to the mistaken assumption of a false auxiliary premise. Nothing in the Duhemian argument makes it impossible for us to perform a new experiment, designed to discriminate between whether the earlier results were due to the failure of *H* or to the failure, instead, of the auxiliary proposition. Of course, in this new experiment we will perhaps need new auxiliary premises, and the situation will never be one where the data logically forces a decision upon us, as do counterexamples in mathematics. But that is not the point. Although the alternatives, not-*H* and *A*, or *H* and not-*A*, may be equally compatible with some particular experimental outcome, *E*, there is no reason to believe that they will be equally compatible or incompatible with *all possible experimental results.*

It is just this last feature that gives the Poincaréan situation its particular importance. The alternatives, non-Euclidean geometry and no universal forces or Euclidean geometry plus non-null universal forces, are supposed to be equally compatible with *all possible experimental results.* We can say that what Poincaré has given us is a "super-Duhemian" argument. It is this feature that leads Poincaré to his conventionalist claim, and just this feature that leads Eddington and Reichenbach to deny that one really is presented here with genuine scientific alternatives at all.

4. *The Taxonomy of Controversy*

The perplexity we have been led to by the considerations of the epistemology of geometry will be familiar to those acquainted with the history of philosophy since Descartes. Deciding what to say in the face of such examples as the Poincaréan parable is just one instance of the general problem: How should one react, philosophically, in the face of an apparent demonstration of the existence of incompatible hypotheses equally capable of explaining all possible observational results?

Consider, for example, Descartes's well-known invocation of the "malevolent deceiver." Imagine a world governed by an omnipotent being who possesses the power of control over our minds. With malevolent, or perhaps prankish, motivation he fills our previously blank minds with just the sort of sensory data that we ordinarily take to indicate the independently-existing material world causally responsible for our sensations.

Nonetheless, no such material world exists. Were we to live in such a world, and who is to say that we do not, our sensory experience as a whole would be indiscriminable from the sensory experience we actually have. On what rational grounds, then, do we base our disbelief in the existence of such a materially empty world containing a grand deceiver and our belief in the existence of "the external material world"?

Suppose we try to characterize very briefly some possible responses to this familiar argument. Having done so, we can carry over these responses to the case of the epistemology of geometry, to see how they look applied to this particular instance of the general problem. My examination of each possible position in turn will be cursory at best. To try to encompass in a short space the details of all the major philosophical positions that have been proposed, to outline carefully the detailed arguments brought forth to support them, and to rehearse precisely all of the major objections against them, would be to attempt the obviously impossible. All I hope to do is to put some perspective on the issues involved in the epistemology of geometry by offering what amounts to little more than a *classification* of the alternative philosophical options. To a certain extent, some of the deficiencies will be made up later in the book, because the fundamental epistemological issues outlined here, and their metaphysical and semantic consequences, will provide the overall theme of the discussion in later sections. Each of these sections will treat a distinct issue in the philosophy of space and time, and the pursuit of each specific problem will be found, sometimes by different routes, to lead back to some of these same basic concerns.

So in arriving over and over at the same basic philosophical perplexities, some additional details about "ways out" will come to the fore. Let me once again warn, however, that the reader who hopes for a resolution of the perplexities, and not merely some clarification of them and some indications of the particular form they take when the issues are those suggested by theories of space and time, will be disappointed.

Let us now look at some possible "resolutions" of the problem of Descartes's Demon. The treatment will be extremely cursory, but I will go at least a little more into detail when we carry over the schematism into the particular issues of geometric conventionality.

First I must make a preliminary distinction, of overriding importance, between "schools" of philosophy. The Cartesian argument, like that of geometric conventionalists, initially assumes that we can make coherent sense of the notion of "all possible observational consequences" of a theory. So we must first distinguish between those who accept this claim, and those who reject it maintaining that the very concept of "all possible observational consequences of a theory," as contrasted with "all consequences of a theory of whatever kind," is simply not well formulated or even intelligible. Let us call these philosophers believers and disbelievers in an observation basis. Within the class of those who believe in the intelligibility of the notion of an observation basis, we next draw the distinction between those who believe that the total theory is not in any sense "reducible" to its set of observational consequences, and those who believe that, given the distinction of observational consequences from consequences in general, the total theory must be said, in some important sense, to be reducible to its set of observational consequences. Let us call these schools the antireductionist and the reductionist, respectively.

The antireductionists again split into contending schools. First there are those we can call skeptics: The total theories do not reduce to their observational consequences. Yet we can formulate alternative incompatible theories with exactly the same observational consequences. But the only rational grounds for choosing one theory over another to be believed are functions of its observational consequences. We must believe, therefore, that we will frequently be in the position of facing alternative incompatible theories between which we simply have no rational grounds for choice whatsoever.

Next there are the conventionalists. Faced with the same situation as the skeptics, they argue somewhat as follows: We must choose one theory or another, even where we cannot discriminate among the theories in terms of differentiating observational consequences. The only option left for us is to simply choose, by a convention or decision, one of the contending theories from the class equally compatible with the data. We accept one of these theories as true by decision. If this truth is merely "truth by convention," so be it. There is no reason for lamenting this, since it is the best we can do anyway.

Finally there are those we might call apriorists, and they argue like this: Granted that there can be alternative, incompatible theories between which we cannot rationally choose solely on the basis of the conformity or nonconformity of the theory with the experimental data. But choosing which theory to believe in is never a simple matter of examining the conformity of the theory with the data anyway. Considerations of simplicity, systematic power, elegance, etc., are a fundamental feature of any scientific inference from data to hypothesis. In the cases we are considering, such methodological principles must once again be brought into play. Once we see their role clearly we will see that we do have rational grounds for preferring, i.e., believing to be true, one of the many contending theories.

The reductionist position, on the other hand, looks like this: Presented with the set of alternative, *apparently* incompatible theories, all with the same observational consequences, it seems as though we are in the position of having to make a choice, but having to make it on no grounds of rationality whatsoever. But this is merely an appearance. A proper understanding of what theories are, and what the theoretical assertions contained in them *mean,* will show us that all of the "alternative" theories are not proper alternatives at all—they are simply the same theory expressed in different ways. The appearance of multiplicity is simply the confused consequence of failure to note ambiguity of linguistic usage. Insofar as the theories predict the same observational results, they are one and the same theory, and there is no choice to be made.

Exploring any of these positions in depth is, once again, beyond my capabilities here. Let me give a cursory examination to their application in the Cartesian case, however, and then rehearse them one more time, in a little fuller detail, in the case of the debate about the epistemology of geometry.

THE CARTESIAN DILEMMA

I. Believers in an observation basis: In the context of the Cartesian problem the observation basis is taken to be the immediate content of perception. By means of all too familiar arguments (illusion and hallucination, dependence of perceptual content on the state of the perceiver, etc.), it is maintained that one must distinguish between the objective features of material objects and the immediate contents of sense perception. Only the latter are open to "immediate inspection" by an observer, and assertions about them constitute the body of propositions knowable to him by perception alone.

ANTIREDUCTIONISTS: Proponents of the following three positions, while disagreeing with each other, all agree to the claim that whatever propositions about the existence and nature of the material world are, they are claims that cannot in any sense be "reduced to" or "translated without loss of content" into claims about the content of sensation. The objects and properties of the material world, whatever it is really like, are distinct from the contents and features of any observer's perceptual experience. To go from knowledge about the content of sensation, no matter how broad and how general, to knowledge about the material world and its features is to make an inferential leap.

Skeptics: Our empirical knowledge is constrained to knowledge about the immediate contents of perception. Even if we allow inductive generalization from our perceptual data, we obtain at best general knowledge of the lawlike structure of the world of sensory contents. But we can never obtain knowledge of the nature of the material world "beyond the veil of perception." We have no knowledge of this world by immediate perception, and to obtain inferential knowledge of it would require at some point "direct access" into the material world to establish the correct correlation between physical reality and our sensory contents. But again, such direct access is impossible. We must, then, remain in a perennial skeptical withholding of judgment about the nature of the material world, if our body of accepted belief is to be constricted to those beliefs for which we can offer rational justification.

Conventionalists: Let us agree that we can offer a variety of alternative hypotheses about the material world, all equally compatible with the possible totalities of sensory data (that the data are caused by an ordinary material world, that they are the result of the psychic manipulations of a malevolent demon, that they are the contents of an ongoing dream, etc.). To accept one of these hypotheses is as rational as to accept another. They are all "equally true." What we need is a correct understanding of *truth,* realizing that it is a matter of convention; but since conventional truth in these matters is the best obtainable, there is no reason to disparage it because of its conventionality. (It isn't clear whether anyone really held this position, but some idealist exponents of "coherence theories of truth" seem to come quite close.)

Apriorists: Let us grant that we can have a rational foundation for beliefs about the material world neither on the basis of direct perception nor on the basis of ordinary inductive inference from the perceptual basis. There are a priori arguments, however, that allow us to bridge the gap between the content of perception and material reality. Our belief in these principles of inference is itself founded in no way upon perceptual experience or induction from it. Descartes himself maintained this position, arguing that one could have good (indeed, indubitable) grounds for believing in the existence of an omnipotent, benevolent deity, and that the existence of such a deity guaranteed on a priori grounds the veracity of our empirical data as evidence of a material reality in general. Not, of course, that the evidence of the senses might not sometimes mislead us;

but only that we had as good a reason as possible for believing that such alternative general hypotheses about the origins of our sensory experience as the malevolent demon hypothesis and the eternal dream hypothesis were excludable on the grounds of this a priori argument.

REDUCTIONISTS: The disturbing conclusions one is led to by adopting any of the three preceding positions are unnecessary. All of these positions rest upon a fundamental error, for they all assume that some kind of leap is necessary to go from general knowledge about the contents of perception to knowledge about the material world. But consider any proposition about the material world and its features. What could serve as evidence for the truth of the proposition? Well, nothing but some complex of propositions about the content of the perceivable world of sensory contents. And what use could such a proposition have? Once again, it could serve no useful purpose other than to lead us about in the perceivable world, say to allow us to predict one complex of sensory contents as the likely successor to a complex of a specified kind.

Reflections such as these lead us to assert that what the proposition about the material world *says* is fully analyzable in terms of propositions about the world of sensory contents. Whatever propositions about the material world and its features may seem to say, a clear understanding of what it is for a proposition to have meaning, and what the meaning of a proposition is, will soon assure us that no mysterious inferential leaps are necessary. Once we understand what material object language-assertions say, we see that they are nothing but shorthand for complex assertions about sensations.

Now the body of assertions about sensory contents which even a simple material object proposition abbreviates may be quite complex. Even such a simple claim as, 'There is a table in this room,' may have to be "unpacked" or "translated" into a rather elaborate set of propositions about sensory contents. And some of the sentences of the translation may lead us to realize that not just *actual* sensory contents but *possible* ones count as well. To say that there is a table is to say what sensations an observer *would* have *were* he in a particular sensory position. To invoke such counterfactuals is to enter the realm of possibilia. But the complexity of the translation is not to be confused with the necessity of an inferential leap requiring some sort of justification, less we fall into skepticism or conventionalism; for translations, no matter how complex, simply unpack meanings and require no "justification" for the assertion of their truth.

II. Disbelievers in an observation basis: All of the above positions are the untenable results of an initial epistemological error. We need not be forced to choose between equally implausible lines of argument, that there is an inferential leap from knowledge of sensory contents to knowledge of the material world—with its almost equally intolerable sub-options of skepticism, conventionalism, or apriorism—or the grossly implausible claim that perfectly ordinary simple assertions about the features of the material world are intelligible only as incredibly prolix and complex propositions about the world of sensory contents.

The initial arguments designed to prove to us the "confinement" of our sensory knowledge to that of some mysterious realm of sensory contents are mistaken. Without these arguments, however, we would never have been prompted to distinguish "propositions about the immediate content of sensation" from "propositions about the material world." And without this distinction, the very possibility of "alternative hypotheses all equally capable of saving the empirical data" would never have arisen. The solution to the dilemma of choosing among any of the equally distasteful options explicated under (I) is to realize the implausibility and untenability of some coherent distinction between "consequences of a theory describing possible observations" and "consequences outrunning the realm of the observable." The detailed objections to each of the positions taken under (I) are of little importance compared to this general objection, which, if correct, makes the problem of a choice between the options under (I) appear as what it is: a pseudo-problem introduced on the basis of a fundamental epistemological error at the most basic level.

This sketch gives merely the barest bones of an argument that has been fleshed out in overwhelming detail in the last 400 years of philosophy. Worse, while stating the positions I have done nothing to provide the arguments used to support them. But rather than pursue this most fundamental issue of epistemology in detail, let me carry the schematism over to the case of the epistemology of geometry. Let us try to see how the arguments about the status of geometric beliefs resemble, and differ from, those about "our knowledge of the external world," and in the process look at the most crucial features of a few of the arguments that have been brought for and against each possible position.

THE GEOMETRIC DILEMMA

I. Believers in an observation basis: The alleged observation basis in the case of geometric theory differs in some important ways from the "immediate content of perception" basis of the Cartesian dilemma. It consists in the kinds of knowledge expressible in propositions that are, first of all, about material objects and their behavior—rods, clocks, light rays, and paths of material particles—and secondly, the expressions of local congruence relations. We are supposed to be able to determine, observationally, whether or not a class of events constitutes the path of a free particle or light ray, and whether or not two rods (clocks) in coincidence are congruent to one another. All other relations among material objects, for example distant congruence among the rigid rods, and all the features of spacetime itself are in the "inferred" realm of theoretical features of the world.

One difference between this observation basis and that discussed earlier is its *publicity.* Whether or not two coincident rods are congruent, etc., are supposed to be questions answerable by observation in an intersubjective way. We do not, at least at this point, have the additional Cartesian perplexity that is the result of making the observation basis the *private* awarenesses of some particular observer, with the resulting tendency to lead to a solipsistic skepticism. Later, in discussing the reductionist approach, we shall see that the geometric case is not as immune to this difficulty as it first appears.

An additional feature distinguishing the present observation basis from that in the Cartesian context is the way it appears to be "extricable" from the theory as normally presented. What I mean is that in the more general philosophical context, the postulation and description of the observation basis is a philosophical construction imposed upon the naïvely accepted realistic theory of the material world. In the case of geometry, however, the consciousness of the problem of the distinction between features of the observation basis and those of the "theoretical superstructure" is again the result of philosophical reflection, but the reflection is much closer to the postulation of the physical theory itself. Perhaps this is only a psychological fact, due to the fact that, whereas the material object theory is a prescientific theory embedded in our ordinary language, the theories of non-Euclidean geometric structure for the world are self-conscious creations of science. Still, there is a sense in which the geometric theories "naturally" characterize their appropriate observation bases in a more direct way than that in which, by philosophical reflection, the "sensory content" observation basis is extracted from the prescientific conception of the material world.

ANTIREDUCTIONIST APPROACHES: All of these positions agree that it makes sense to ask about such features of the world as the nonlocal congruences of rods and clocks and the structural features of spacetime itself. And, all agree, to ask these questions is not simply to inquire into the general or lawlike features of such things as the local congruence relations among rods and clocks, the paths of particles and light rays, etc. The propositions of the full theory and those of the evidential basis for it are no more reducible the one to the other than the proposition that someone is a murderer is translatable into the set of propositions about the evidence for his commission of the crime. Let me postpone the presentation of arguments for this and for the opposing claim to the discussion of the reductionist position, and move on to an exposition of some features of the various subpositions of the antireductionist position.

Skepticism: The skeptic has the virtue of epistemic humility on his side. Why should we assume, after all, that the world must be such that the resources available to human beings are sufficient to fully allow them to determine its nature, at least in principle? If the actual structure of spacetime is beyond reach of human knowledge, why should this surprise us? What a priori reason could we have, after all, to believe that all of the features that constitute the real nature of the world should be accessible to empirical investigation? Why shouldn't there be features of the world which are simply unknowable, in principle, to us?

At this point we should be careful to make a distinction of some importance. The opponents of the skeptic, in particular the reductionist opponents, are not denying the possibility of features of the world being immune to empirical investigation. What they are denying, however, is that we can meaningfully characterize just what these features are. The reductionist hardly denies the claim that there could be some features, "I know not what," of the world of which humans will, in principle, remain totally unaware; but, he claims, the resources for giving meaning to propositions and those for determining the truth or falsity of propositions are fundamentally the same. The objection to skepticism of the kind I am describing is not that it postulates unknowable features of the world, but that it attempts to say just what these features are. It is this that is objectionable, according to the reductionist. If the truth or falsity of a proposition, or at any rate its rational believability or disbelievability, are totally immune to human investigation, then it is misleading to speak of the proposition as being meaningful at all. If there are things we can't know, we can't meaningfully say what they are. I will return to this objection to the skeptic when I discuss the reductionist position.

There is one important difference between skepticism with regard to geometric theory and skepticism with regard to the existence of the external world that I might remark on. Hume has an interesting approach to skepticism. Frequently, he says, we are in the following situation: We believe a certain body of propositions. On philosophical reflection we realize that we have no rational basis for our beliefs, and indeed, that no rational justification for them could be given. Nonetheless, the *psychological* hold of the beliefs on us is sufficiently strong that, think as skeptically with the learned as we will, we cannot do otherwise in our unreflective moments, which constitute the bulk of our lives, than "act with the vulgar"; that is, to unreflectively believe the unjustifiable propositions. We can't help believing in the reality of the external world, as skeptical as we may be in our philosophical study reflecting on a possible Cartesian demon.

In the case of geometry, however, no such psychological necessity seems to hold. Poincaré thought that we would stick to the Euclidean theory, come what empirical data may. But this wasn't because of an ineluctable psychological hold the theory had on us, but because of a self-conscious choice of the descriptively simpler alternative. As a matter of fact, as we know in the light of general relativity, our preference might be for the curved spacetime alternative. But in any case, the choice is ours to make. If we are theoretical skeptics, we need not be psychologically-determined nonskeptical believers in one of the particular alternatives. Perhaps it is this openness of our minds to the various alternatives that makes the tenability of a genuinely skeptical position seem so much more plausible in the case of geometry than it does in the case of belief in the external world in general.

Conventionalism: The conventionalist maintains that the difference between the alternative theories is real, that no choice between them can be given a rational basis, either empirical or a priori, and that, since a choice must be made, we might as well make one "by decision." The basic problem with this view is the difficulty of characterizing it in a way that truly distinguishes it from skepticism on the one hand, or reductionism on the other. I think that Poincaré, although claiming that he espouses a conventionalist position, can usually on examination be found to be holding something much closer to what I shall describe as a reductionist view.

Take the problem of characterizing *truth* from a conventionalist point of view, and the accompanying problem of characterizing *reality*. Suppose our choice of a Euclidean as opposed to non-Euclidean world really is merely a decision to be made on our part. If our decision really is arbitrary, what useful purpose is served by calling the theory of our choice true and saying that reality has the features attributed to it by our theory? Shouldn't we either face up to the inevitability of skepticism (we simply can't know which theory is true or what the reality of the world is like) if we really hold to the position that the theories are genuinely incompatible alternatives, or to reductionism (the theories are equally true and attribute the same features to reality) if we slide instead into the reductionist position of taking the theories to be, properly speaking, equivalent? The conventionalist position seems to be too ambiguous to maintain its integrity as an independent position, and on reflection seems to turn rapidly into either skepticism or reductionism.

Apriorism: The difficulty with conventionalism that I have just brought out does not infect the aprioristic position. According to the apriorist the alternatives are really incompatible. To be sure, empirical evidence cannot decide for us which of the theories to believe, but this shouldn't disturb us because any theoretical or general belief always requires the invocation of "methodological principles of rational belief," which do not rest upon empirical experience for their acceptability as do principles of rational belief in science.

Before I continue I should carefully delineate what calling this position apriorist means. In particular, it does not mean that the holder of such a position is maintaining that we can a priori determine the nature of the world, nor that the theory of the world we rationally hold is immune to possible refutation by empirical experience. The apriorists I am discussing reserve a crucial role for observation and experiment in directing us to the theory we should believe; unlike the "pure rationalist" who, thinking that the correct description of the world can be inferred by deductive reasoning from self-evident first principles, allows no role for observational data whatsoever.

Some light on this kind of apriorism can be obtained, perhaps, if we reflect upon the role of so-called a priori probabilities in the Bayesian school of decision making. Suppose we must choose one of a number of hypotheses, $H_1, \ldots, H_n$, that exhaust all possible choices and are exclusive choices, i.e., we must adopt some H_i and we cannot adopt two distinct H_i's. According to a standard version of rational decision making, our procedure works like this: First, we know some utilities, i.e., gains and losses that accrue upon adopting H_i when H_j is actually the case. Next, we can perform experiments whose possible outcomes have various probabilities *given* the truth of the various H_i's. Finally, let us suppose that prior to any experiment we can attribute various *probabilities* to the H_i's, our "degree of partial belief" in their truth which we hold independently of any observation or experiment.

With these resources we act as follows: We perform the experiment and observe its outcome. Then, utilizing a theorem of the probability calculus, Baye's Theorem, we compute the a posteriori probabilities of all our hypotheses. We begin with $p(H_i)$'s, the a priori probablities of the H_i's, and $p(O_j/H_i)$'s, the probabilities of the various outcomes relative to the various hypotheses being true. We compute $p(H_i/O_j)$'s, the new probabilities of the hypotheses relative to the experiment, as a matter of fact, resulting in some particular outcome O_j. Finally, using the new a posteriori probabilities, we compute an "expected utility" for choosing each H_i. This is the sum of the new probability of each H_k times the utility of picking H_i when H_k is the case summed over all the k's. Then we choose the H_i for which this "act of choosing H_i" has the maximum expected utility.

Here is a rule for adopting a hypothesis. It takes into account the results of experimental observations in that the choice recommended depends upon the a posteriori probability of each hypothesis, and this depends upon the actual outcome of the performed experiment. But the rule also requires for its applicability an assignment of "probability of truth" to each hypothesis independent of any experimental outcome, the initial a priori probability assigned to the hypothesis.

So the apriorist argues like this: Granted that we cannot adopt a correct geometric theory of the world on the basis of a priori considerations alone, and granted that the empirical data cannot by itself determine for us which geometric hypothesis to adopt. But empirical data can never by itself force the adoption of a hypothesis. In any kind of reasoning from data to theory, a priori considerations of "initial plausibility" of the theories in question must be invoked. So they must be invoked in this case as well. All the alternative theories are equally compatible with the possible observational consequences, but they are not all equally a priori plausible. For a given set of observational consequences, the rational man will believe as true that one of the total theories compatible with these observational consequences whose initial a priori plausibility is highest. This way the hopelessness of skepticism, the unintelligibility of conventionalism, and the antirealism of reductionism are all avoided.

How reasonable is the apriorist's case? Not, perhaps, as convincing as the above makes it appear. To begin with, let us consider a particular case. Suppose a scientific investigator finds himself faced with the kind of observational data that could lead him to postulate a Euclidean world with imposed universal forces, or instead, a non-Euclidean world with no such forces. Which of these alternative theoretical possibilities seems to have greater "a priori probability"? It is hard to predict what the individual reader's response to this is likely to be, but one guesses that the overwhelming majority of readers would find the question unanswerable.

Now those who believe in the importance of a priori probabilities for scientific inference might have two quite distinct views about them. "Subjectivists" are likely to maintain that each investigator has some such a priori probabilities that he assigns to theories antecedent to experiment and observation. There is no criterion of rationality involved here, though, according to them. One has such a priori probabilities, and one believes or disbelieves partly in accordance with them, but there is no standard or rightness or wrongness for one's a priori plausibility assignments, over and above the constraint that the totality of one's such assignments should obey certain constraints of compatibility. From this point of view, if two investigators differ markedly in their a priori probability assignments, there is no question of one investigator being right and the other wrong. They simply differ.

"Objectivists," on the other hand, are likely to maintain that certain standards of rationality govern not only our responses to the results of observation and experiment, but also our initial assumptions of plausibility for theories as well. Since we are ultimately concerned with the question of whether the selection of one of the possible alternative theories is governed by standards of rationality, it is this position that would seem more interesting to us. The subjectivist position comes quite close to being simply conventionalism restated, with some details as to how the conventional choice is made superadded.

But what standards of rationality could govern choices of a priori probabilities when the alternative hypotheses are the kind we have been considering? It is quite clear that the kind of apriorism with which we are here concerned is rather different in kind from the Cartesian "apriorism through theology" we looked at earlier. What one could expect to find is something like this: We must have the alternative theories differing in their a priori plausibility. But, ignoring the comparison of theory with data, the only thing left to work with is the "internal structure" of the theory itself. This suggests that we look for some structural feature of theories, a feature that characterizes them independently of their compatibility with the experimental data and governs their degree of a priori plausibility.

Immediately several features come to mind. The *simplicity* of a theory is frequently alleged to be an important consideration in its scientific acceptability. So is its *generality* or its *explanatory power*. Everything else being equal, i.e., the alternative theories proving equally compatible with the observational data, simpler theories are preferred in scientific practice to more complex theories and more powerful theories to those less great in range and generality.

But now we can see why the apriorist position runs into difficulty in the cases we are considering. The argument is familiar, for it is just Reichenbach's distinction of inductive from descriptive simplicity, rehearsed (II,F,2). Consider the scientist presented with the task of drawing a curve relating two parameters with his data restricted to a finite number of observational "points." An infinity of curves pass through the determined points, but he must select one as appropriate. Here is where simplicity considerations are usually invoked. He should, it is argued—for want of any additional grounds for different action—choose the simplest curve through the points. Now already a number of different methodological problems appear. First there is the problem of characterizing just what simplicity is. A fully general "measure" of the simplicity of a theory seems a methodological pipe dream. Worse, the simplicity of a theory does not seem to be something preserved under transformations of the expression of the theory into a different but logically equivalent form. For example, a curve that appears simple given Cartesian orthogonal coordinates might have quite a complex algebraic description in terms of polar coordinates, but it is still the same curve. If simplicity is invoked as a rule of acceptability, the acceptability of a theory seems to depend on which of the logically equivalent means of expression we choose for it.

Again, simplicity is objected to since the connection between *simplicity* and *truth* seems so dubious. What the scientist wants, the argument goes, is the *correct* curve. But what kind of reasoning can we offer to convince someone that the simplest hypothesis compatible with the data is most likely to be the true hypothesis? And if no argument can establish such a connection, what reasons do we have in the first place for invoking simplicity in our mechanism for choosing hypotheses?

What is important for us, however, is that even if we accept the rationality of utilizing simplicity considerations in inductive inference, we still do not have the kind of a priori principle of selectivity we need for our case. The reason is this: Consider two curves of differing simplicity which the scientist is considering as candidates for his hypothesis. Both curves do go through all the presently-available data points, otherwise we would not still be considering them as candidates for the "true" theory. But the possibility of rejecting at least one of these curves on the basis of further experimentation remains open, for the curves do not make the same predictions for all possible experimental tests. It is this feature that makes the rule, "Adopt the simplest hypothesis compatible with the present data," seem somewhat more innocuous than might first appear. For even if we do make this choice, we are not stuck with it, in the sense that ongoing experimentation can "test" our choice and, conceivably, reject it in favor of some more complex hypothesis.

But it is just this possibility of further experimentation, discriminating between the alternative hypotheses, which is lacking in the case we have been considering; for the alternative total theories we have in mind are supposed to be identical in their totality of observational consequences. So if they differ in simplicity at all, it is not the same kind of difference that discriminates, for example, the straight line through two points from any of the parabolas that also pass through the points. For any straight line and any parabola there are at least some points on one but not on the other, but for our alternative hypotheses all the "points" confirming one confirm the other as well. The two hypotheses, as Reichenbach says, differ only in their *descriptive* simplicity, not in their *inductive* simplicity. And this difference appears more like that between the two *descriptions* of a straight line, one in Cartesian and the other in polar coordinates, than like that between a straight line and a parabola.

If the apriorist is to make out his case, it will have to be in a manner that does not consist of the observation that simplicity considerations are always necessary for inductive inference, and so there is no reason not to invoke them in the question of deciding between the alternative hypotheses here. For the use of simplicity as an a priori rule for selecting among hypotheses that are merely equally compatible with *present* observational data is quite a different matter from invoking it as a standard for selecting among hypotheses that have *all possible* observational consequences in common.

REDUCTIONISTS: I think that reflection upon the difficulties of the various antireductionist positions we have just considered will make considerably clearer and more intense the "pressures" toward adopting a reductionist position. If, as the reductionist claims, the various alternative theories under consideration can all be shown to be, properly speaking, a single theory expressed in differing ways, all the problems of motivating and rationalizing a choice vanish at once.

One can view the reductionist position as the result of expanding upon the following theme: The difficulties faced by the skeptic, conventionalist, and apriorist are the result of their belief that one can formulate alternative hypotheses that have identical observational consequences. But this is an illusion. To frame a hypothesis is to formulate a set of propositions. Propositions are not "marks on paper" or "sounds in the air," but meaningful assertions. We must reflect not only upon how we *test* a meaningful proposition for truth or falsity once we have it in mind, but how we formulate genuinely meaningful propositions in the first place. Once we have done this, we realize that the source of meaning is the same as the source of experimental adjudication—observation and experience.

A proper theory of meaning will have as one of its most fundamental consequences the awareness that the meaning of a scientific proposition is exhausted by the experimental or observational consequences derivable from it. If this is so, two theories with identical observational consequences can be seen to be identical in their meaning, whatever variation in their overt linguistic appearance there may be. To speak of adjudicating *between* these theories is as misleading as if one were to try to decide whether Newtonian mechanics expressed in English or Newtonian mechanics expressed in German was the correct dynamical theory. One can, to be sure, experimentally test the correctness of Newtonian mechanics, but one tests this theory, not a particular linguistic expression of it. Appearances to the contrary, the same thing holds in our case of "alternative" geometric theories. Euclidean geometry combined with a physics that postulates certain universal forces is the same theory as the corresponding observationally equivalent non-Euclidean theory with no universal forces. There is no deciding between them to be done.

This doctrine about meaning, and the accompanying thesis of the reducibility of theories to their observational consequences, is at least as old as Berkeley's rejection of "substance" as a meaningless metaphysical term; the detailed construction of reductionist programs has provided a great bulk of such various philosophical "programs" as phenomenalism, behaviorism, and logical positivism. Obviously I will not be able to do remote justice to all of the claims that have been made. Instead I will attempt to characterize just a few of the most fundamental features of the reductionist approach, emphasizing, as usual, not the general philosophical approach, but rather the way in which the controversy over geometry seems to serve as an illuminating paradigm of the more general philosophical issues. I will make some remarks about three issues: (1) Some comments about the doctrine of meaning fundamental to the reductionist claim. (2) Some consequences of the reductionist claim with regard to metaphysical and ontological issues. (3) Some observations about the most fundamental objections to the reductionist claim that one might raise.

(1) The reductionist requires a doctrine of meaning which has the consequence that two theories with exactly the same observational consequences are said by the doctrine to have the same meaning. Actually, this is a rather weak constraint upon a theory of meaning. It requires only that the notion of 'same meaning' be made clear, and this only when it is *theories* whose meaning is in question. Most reductionists go much further than this, however. They usually elaborate a doctrine that, first of all, attributes meaning to *terms,* not merely to propositions or theories, and secondly, attempts to say what the meaning of a term is, not merely to tell us when two terms have the same meaning. This more elaborate theory has a number of consequences that people have found objectionable. Let me very briefly indicate what such a fuller theory of meaning usually looks like, expand on the difficulties it is frequently alleged to have, and then make some comments upon the possibility of a "weaker" theory sufficient to ground the reductionist claims about "same meaning" for theories, but to which the standard objections are perhaps not applicable.

The standard theory usually begins with a distinction between observational and theoretical *terms*. This is a bifurcation of the vocabulary of the theory or, perhaps, its set of concepts. Observational terms are said to get their meaning by being associated directly with some element of sensory experience, say in the manner that 'red' is claimed to receive its meaning in our language by being associated with reddish sense-data. Given the meanings of the basic terms of the observational vocabulary, meanings are assigned to other terms by means of definitions. The defined term appears alone on the left-hand side of the definition, and on the right-hand side appears a meaning-assigning phrase containing only terms of the observational vocabulary or previously defined terms. These are Reichenbach's coordinative definitions—for example: 'null-geodesic' = (def) 'path of a light ray in a vacuum.' The theory is hierarchical in the sense that there is a natural order by which terms over and above the observational level are introduced.

One standard objection to this theory is the difficulty in practice of reconstructing these definitions by examining already existing theories with their already meaningful theoretical vocabulary. Frequently the following claim is made: For the "theoretical terms" of actual theories, no "definition" of each term in terms of the observational vocabulary alone can be given. But if the hierarchical theory of meaning were correct, we should be able to reach a definition of each theroetical term in terms of the observational vocabulary alone by following a chain of definitions whose ultimate result has the single theoretical term on the left-hand side and a meaning-assigning phrase, using only the pure observational vocabulary, on the right-hand side. The reason we cannot find such definitions, it is claimed, is that the network of theoretical terms is frequently introduced in a holistic way. To define one such term requires reference to the other theoretical terms of the theory, and no term-by-term elimination to the observational vocabulary is in fact possible.

The plausibility of this objection varies from case to case. In our geometric case, such "explicit" definitions of the theoretical terms seem fairly easy to obtain, witness the definition of null-geodesic above. In broader reductions, however, like that of "material object language" to "pure sensory-content language" the possibility of such term-by-term translations of the theoretical vocabulary does seem more doubtful. Try, for example, to fully "unpack" the meaning of 'There is a desk in this room' in terms of what "pure sensations" one does or would have in situations themselves characterized entirely in "pure sensory-content" language. What we should note here, though, is the extent to which this particular "term-by-term definition" theory of the meaning of theoretical terms outruns the initial claim, which was simply that, taken as a whole, theories have the same meaning if they have the same total sets of observational consequences.

Another alleged difficulty of the "term-by-term explicit definition" theory is this: Consider any of the propositions that give the explicit definition of a theoretical term in terms of the observational vocabulary. Since such propositions are definitive of the meaning of the term, the assertion of the propositions can only amount to asserting an analytic truth. Thus if 'null-geodesic' means 'path of a light ray in vacuum,' then to assert that all light rays in vacuums follow null-geodesics is to assert a proposition whose truth depends at most upon logic and the meanings of terms alone. Such propositions then are not revisable. At least they are as irrevocably true as the truths of pure logic. To deny such a proposition is simply to change the meaning of the theoretical term involved.

Now many of the philosophers we have been concerned with would have no objection to this consequence at all. For example, Reichenbach would welcome it as an explicit realization of the role played in science by coordinative definitions. According to him, some of the propositions of any theory must be analytic, for otherwise the theoretical terms would have no meaning attributed to them, and the possibility of meaningful synthetic propositions using these same terms would be blocked. Some propositions containing theoretical terms must be analytic, according to him, in order that any meaningful synthetic propositions using these terms exist at all.

Others, W. V. Quine for example, whose philosophical views in general are not totally out of sympathy with at least portions of the reductionist claim, would be strongly averse to the way the particular theory we have been discussing "localizes" the meaning-assigning features of a theory to some particular subset of its propositions, taking the remaining propositions as unabashedly "purely synthetic." Their doctrine is a more holistic one, and they argue that the best one could do is to look at a theory as a whole and examine its totality of observational consequences. To go further and distinguish the propositions of the theory into the two classes, analytic and synthetic, is to try to impose on our knowledge of theories an unjustifiable theory of meaning. The theory is there, and so are its observational consequences, but there is no need to try to sort the propositions of the theory into "meaning-attributing" and "fact-stating" ones; for realistically, all the propositions of the theory are on a par in that they all function as strands in a network that functions to bring together the observational consequences that "tie the network down at its periphery."

Some light can be thrown on these questions by asking how various philosophers would respond to changes in accepted scientific theory. Suppose we move from one theory to another with different observational consequences. All would agree that this constituted a genuine theoretical shift. Suppose we switch from one theory to another whose total set of observational consequences is identical to the former theory. The minimum condition we must impose to call a philosopher a reductionist is that he take the new theory as being, properly speaking, merely a novel formulation of, or perhaps a novel expression of, the same theory as was the older set of propositions. Philosophers like Reichenbach would insist that this "shift" be explicable as the change in meaning of some specific terms of the theories, and that one could localize the shift in meanings by locating which analytic coordinative definitions of the older theory are now denied truth in the new theory. The more holistically-inclined philosophers, on the other hand, would deny the possibility of localizing the meaning changes to specific terms, just as they deny the possibility of locating meaning-attribution in the theory to some specific propositions. What we must be clear about here, though, is that one could quite easily accept the most crucial reductionist claim, that same observational consequences imply the same theory, without accepting the particular theory of meaning-attribution familiar from Reichenbach and similar philosophers' works.

(2) I think it is important at this point to realize two truths: (*a*) Adopting a theory of meaning which takes the meaning of theoretical terms to accrue to these terms by means of the role they play in a theory that has observational consequences that can be expressed in an observational vocabulary whose meanings are, in some sense, "primitive," does not *in general* commit one to a doctrine that the theoretical entities and properties referred to or expressed by the terms in the theoretical vocabulary are somehow "unreal" or "convenient fictions." (*b*) On the other hand, the reductionist doctrine, as it is usually formulated, whatever particular theory about the meaning of theoretical terms accompanies it, does seem to have some such "instrumentalistic" or "fictionalistic" consequences with regard to the theory about the reality of theoretical entities or properties.

To see that a doctrine about the accrual of meaning to theoretical terms does not in itself commit one to the unreality of theoretical entities and features, consider a theory that looks something like this: Suppose we consider a theory with a number of observational terms and one theoretical term, say 'is electrically charged.' The theory will tell us about connections between being electrically charged and having various observational characteristics. Take the propositions of the theory and conjoin them into a single sentence. Consider the new sentence we get by replacing the term 'is electrically charged' wherever it appears by a property variable, 'is *F*,' and putting the expression 'There is a property *F* such that . . .' in front of the now single-sentence theory. This is a modified version of replacing a theory by its so-called Ramsey sentence. The claim is, among other things, that what 'is electrically charged' *means* can be determined by looking at the Ramsey sentence. It means 'having a property *F* such that, etc. . . .' So, in this sense, meaning accrues to the predicate 'is electrically charged' by the total role it plays in a theory whose other terms, being observational, have their meaning "primitively given."

But does such a theory deny that there *is* a theoretical property, that of being electrically charged, which has just the features attributed to it by the theory? Of course not; for the existence of such a theoretical property with these features is just what the modified Ramsey sentence asserts—"There is a property *F* such that, etc. . . ." So a doctrine that claims merely that meaning accrues to theoretical terms through their role in a theory that contains observational terms does not in any way commit one to the claim that the entities and properties referred to or expressed by the theoretical terms are unreal, convenient fictions, or anything of the sort.

How would an exponent of the Ramsey sentence type of reductionism view our present dilemma? He would say this: We have two incompatible theories with the same observational consequences. What each of these theories really says can be seen by replacing each theory by its appropriate Ramsey sentence. You get these by (1) conjoining all the axioms of the theory; (2) replacing each theoretical term in the theory by a variable; and (3) putting an existential quantifier for each new variable in front of the new one-sentence theory.

Now each Ramsey sentence will have the same observational consequences as the theory it replaces. But, while the older theories were incompatible with one another, the new Ramsey sentence theories will not be. Of course, they may not be identical either. Each of the new theories will be devoid of theoretical terms, so the reduction in *meaning* of the theoretical to the observational terms will have been accomplished. The new theories will still postulate the existence of theoretical entities and properties as explanatory of the observational facts. But the new explanations will no longer contradict one another.

The reasonableness of replacing theories by their Ramsey sentences, the consequences of doing so, and the ability of this kind of reductionism to solve the dilemma of apparently incompatible theories all "saving the phenomena" are all questions still requiring much exploration. The reader should remember, though, while thinking about the more radical kind of reductionism which we shall next explore, that there is this more modest kind of reductionism which, while claiming that theories reduce in their meanings to theories stated purely in observational terms, allows the theories the legitimacy of their ontological claims about the reality of theoretical entities and properties.

But what about the more usual reductionist doctrine? I think that we can see that it does seem to have deep ontological consequences. Consider the two alternative theories with identical observational consequences which have constituted our paradigm pair: Euclidean geometry plus universal forces or non-Euclidean geometry with the universal forces set to zero. Take the propositions from these two theories respectively which assert (1) geodesics are straight lines and (2) geodesics are not straight lines. According to the first alternative (1) is the case, according to the second alternative (2). But according to the reductionist theory, the two alternative theories are not really alternatives at all, but merely alternative expressions of the same theory. To believe that there "really are" geodesics and that they "really are straight lines" or "really are not straight lines" seems to many philosophers grossly incompatible with the reductionist view about the equivalence of the two alternative theories, since one asserts some feature as holding of geodesics and the other denies it.

The way out of this dilemma for this kind of reductionist seems obvious. It is to reject the "realist" interpretation for the theoretical entities and properties referred to by the theories. There *really* are, according to this view, the entities and properties talked about in the body of propositions which constitutes the *common* set of observational consequences of the two theories. Since all of the alternative theories say the same things about these observational entities and properties, we can adopt a "realistic" attitude toward them without fear of being put in the apparently absurd position of maintaining that certain entities exist, maintaining that they have certain features, and yet maintaining that two theories are equally true which assert quite incompatible things about these entities.

I think we can see that whatever detailed theory of meaning we adopt as explicative of the reductionist position, this version of the reductionist doctrine cannot remain simply an epistemological and semantic theory. It has inevitable metaphysical and ontological consequences as well. Since the bulk of Chapter III and significant portions of Chapters IV and V are devoted to just these basic issues of ontology, I will forgo pursuing this issue any further here.

(3) The reductionist position as we have just seen it explicated has many obvious virtues. The dilemmas of the antireductionist are simply dissolved. At some cost, to be sure, since we must now abandon a naïvely realistic view of theoretical entities and properties; but this seems a small price to pay. Since the theory does not seem to require the particular term-by-term theory of meaning which usually accompanies it, many of the usual objections to reductionist positions, in terms of the difficulty of finding explicit definitions for particular theoretical terms and in terms of the illegitimate splitting of propositions into analytic and synthetic, fail to be convincing objections to the general reductionist position. They are interesting objections only to some of its particular forms.

Before proceeding it is important to once again emphasize the relation between reductionism and skepticism. The reductionist does not maintain that he has refuted "problematic skepticism" in general. Nothing that he says is incompatible with the claim that there may be features of the world totally beyond the possibility of human knowledge. What is impossible however, according to the reductionist, is the formulation of *meaningful* propositions that cannot be confirmed or disconfirmed empirically (unless, of course, they are merely analytic). The reductionist does not claim that the limits of our knowledge necessarily constitute the limits of the world, but only that the limits of our knowledge constitute the limits of what we can meaningfully assert or deny to be the case.

The most problematic feature of reductionist accounts, I believe, is one that is independent of the detailed formulation of the position; for it rests upon a questioning of the nature and status of the so-called observation basis. In the geometric case we have been considering, the specification of the observation basis is fairly clear, since it is "read off," as it were, from the presentation of the geometric theory. We obviate the problems faced by all of the antireductionist positions by the twofold process of (1) adopting a theory of meaning which attributes direct meaning only to terms descriptive of things and features in the observation basis, and allowing meaning to the theoretical structure only insofar as it ties together observational propositions and (2) differentiating between the "reality" of the observational world and the "mere convenient fictionality" of the theoretical entities and features.

But should we stop with an observation basis of, say, coincident congruence among rods and clocks and observable trajectories of light rays and free particles? Do these features of the world, in other words, constitute the genuinely irreducible observational basis? Once one has started on the job of reducing the meaning of theories, ultimately, to the meanings of their observational terms, and the reality of the world to the reality of the observational entities and their observational features, it seems very important to carry out this reduction as fully as possible. To carry out "one stage" of the reduction, reducing a portion of a theory to some subsection of it but not continuing until all reductions have been made to the "ultimate" observational vocabulary and "ultimate" observational class of things and properties, seems like trying to set up housekeeping halfway down philosophy's notoriously slippery slope. The only true equilibrium, one feels, is to be found in a reduction to the lowest and irreducible level.

But once one has acceded to this point, the difficulties are manifold and all too familiar. The only truly observable features of the world soon come to be identified with "pure sensory contents." If one takes these as "ideas in the mind," following Berkeley, one soon is faced with all the traditional difficulties of solipsism: The only world I can truly be sure of the existence of is the world of my internal consciousness, and the only meaningful assertions I can frame are those about its features; if their is a genuine external world I can never have good reason for believing that this is so, nor can I even intelligibly assert propositions about it, irrespective of the limitations on my knowability of their truth value.

If, instead, the sensory contents are the atoms of a "neutral monist" or "phenomenalist" picture, where mind as well as matter is taken to be a logical construction out of sensory contents, although we are not afflicted with the solipsism of the idealists we are certainly afflicted with all of the rather well-known difficulties of coherently characterizing the phenomenalist position.

In any case, it seems that the price to be paid for avoiding the difficulties of the antireductionist position is rather higher than we anticipated. It is not only the reality of spacetime features themselves which go, eliminated in terms of the real local relations among the idealized material objects that we use to map the spacetime, it is the whole of the "external world" in its usual sense which must be sacrificed, leaving us with a world whose only genuine realities are immediate sensory contents and their observable interrelations. This is not the place to argue the virtues and vices of traditional philosophical idealism and phenomenalism. I can only show how quickly these "deep" questions come to the surface once an initial reductionist stance is tolerated.

Reductionism requires, if not the "reduction" of the whole world to the world of the immediately apprehended, at least such a reduction for all in the world whose nature can be meaningfully discussed by us. A program that starts by denying real existence to a small fragment of the naïve realist's external world seems innocuous enough. But it soon leads us to conclude that nearly all our ontological commitments to the reality of the material world must be taken as *façon de parler,* properly intelligible only in terms of a commitment to the somewhat spare reality of the world of immediate contents of sensation. At this point the ontological prejudice in favor of those features of the world which, from the naïve realist's point of view, merely happen to be apprehensible directly by the human perceptual apparatus becomes, for many, unbearable.

II. Denying the observation basis: Having seen the uncomfortable positions we are led to both by antireductionist and reductionist fundamental premises, it should no longer come as a surprise that many philosophers have attempted to undercut the whole controversy by arguing that antireductionists and reductionists suffer from the common mistake of assuming the intelligibility of the notion of observation basis in the first place. If the distinction between propositions verifiable by direct observation and those establishable only by theoretical connection to the observational level is denied to begin with, the very questions about how to treat the "nonobservational" propositions epistemologically and metaphysically will never arise. Given a dichotomy resulting in perplexity, a quite natural response is to deny the intelligibility and coherence of the dichotomy in the first place.

The details of various attempts to cast doubt upon the observation-basis/theoretical-superstructure dichotomy will be beyond our scope. However, there are two distinct approaches, each designed to attack different versions of the observation basis thesis. If the observation basis is taken to be some set of entities and features in the "external" physical world, the attack usually rests on the gradualness of the transition in the world from observable to unobservable features, arguing that this actually slow transition is grossly misconstrued by the philosophical theories that make the distinction one that is hard and fast enough to constitute a distinction "in principle." If the distinction is instead between immediate sensory contents as observables and all physical features as unobservables, the claim is more likely to be one that attempts to show that the postulation of a world of immediately observable sensory contents is simply unintelligible given the assumption that all physical things and features are forever immune to direct observation.

What I would like to do, instead of presenting these critiques of the observational/theoretical distinction in detail, and offering the best arguments I can find for and against their soundness, is to offer just one argument designed to show that there must be something coherent in the postulation of the dichotomy, no matter how persuasive the critical objections may be to detailed unpackings of the observational/theoretical distinction. I am arguing, in other words, that as painful as the adoption of either the reductionist or antireductionist positions may be, the pressures toward their common assumption, that one can coherently characterize the notion of the observational consequences of a theory as a proper subset of its total consequences, are very hard to resist.

My argument is fairly simple. In the case of "alternative geometric theories" we feel philosophically perplexed. Faced with all the alternative philosophical attitudes possible to the results of Poincaréan parables, we feel strong tensions in trying to decide just what the parable would be taken to show. But there are other cases of alternative theories saving the data in physics where the response of the scientific community is clear and unequivocal. It is, in fact, to adopt a straightforward reductionist position, maintaining that what we have before us is an unquestionable case of two alternative presentations of the same theory and not two genuinely alternative theories at all. But if the reductionist interpretation of the situation is clearly the correct one, the assumptions of the reductionist position, in particular the assumption that one can clearly differentiate the observational from the nonobservational consequences of the theory, must also be justified.

So the structure of my argument is really like this: If the denial of the possibility of characterizing the observation basis of a theory is utilized to undercut both the reductionist and antireductionist positions we have been examining, it can only be at the price of maintaining that a scientific position universally adopted by the scientific community actually rests upon an incoherent notion. Naturally, this will only convince one of the intelligibility of the assumption to which the rejector of the notion of an observation basis denies intelligibility, to the extent that one is convinced that the scientists are correct in maintaining the position they unquestioningly and universally hold.

The scientific example I have in mind comes from the development of quantum mechanics, in particular from the universally accepted position that the Heisenberg and Schrödinger versions of the theory constitute not two alternative theories between which we must choose, but rather two alternative *representations* of one and the same theory.

In the quantum-mechanical framework one "prepares" systems by some physical operation. One leaves the system alone for a while and later makes observations upon it. The outcomes of various experiments performed at later times are attributed certain probabilities by the theory, which probabilities depend upon the preparation of the system, its internal constitution, and the experiment performed. Into the mathematical apparatus of the theory go two symbols characterizing the state of a system at a time and the observable to be measured by a particular experiment. The probabilities that an experiment at a time on a specific system will result in the various possible outcomes of the experiment are a mathematical function of the state of the system at the time and the "state" of the observable at that time. In the Schrödinger representation, the state of a system is initially fixed by the preparation of the system. But as time elapses the state changes according to a dynamical law and the internal constitution of the system. The mathematical representatives of the observables are operators, constant in time. So the changing probabilities of outcomes as the system dynamically evolves are attributed to its time-dependent state. In the Heisenberg picture the state of a system is characterized by its preparation, and is represented by a temporally unchanging mathematical form, the unchanging state-vector for the prepared system. The operators for each observable, on the other hand, are time-dependent mathematical structures, and the changing probabilities of outcomes are now captured in the dynamical evolution of the operators.

Since the evolution of the probabilities predicted by both the Heisenberg and Schrödinger representations are identical, depending in both cases upon the preparation of the system, its internal constitution, and the experiment to be performed, it is universally accepted that the Schrödinger and Heisenberg theories constitute merely alternative representations of the same theory. The moral is clear. "States" and "observables" are not themselves directly inspectable quantities. They are, rather, constituents of the theoretical superstructure of the theory. So the fact that states are time-dependent in the one picture and time-independent in the other is irrelevant. The correlations between preparations and probabilities of experimental outcomes is the same in both accounts, mediated in the same way by the dynamics determined by the internal constitution of the system. The unquestioning adoption of the view that the Schrödinger and Heisenberg representations are merely representations of the same theory is a clear indication that, at least in this case, the intelligibility of the observation-basis/theoretical-superstructure distinction is assumed by all. And if it makes sense here, why should it be denied intelligibility for physical theories in general?

Of course the observation basis of quantum theory is not the philosopher's ordinary basis of "directly perceivable entities and qualities," for the correlations that are preserved by the alternative representations may be among such "theoretical" properties as the magnetic moment of an electron. Yet the analogy is sufficiently strong, I think, and the phenomena "saved" in the switch from one representation to another sufficiently like the observational phenomena saved in the switch from one Poincaré alternative to another, to see the strength of this argument for making an at least plausible case for the ultimate characterizability of an observation basis.

If the above argument is correct, the attempt to outflank the dilemmas faced by reductionists and antireductionists alike by denying their common assumption of the intelligibility of the observational/theoretical distinction is far too facile. It appears that we shall have to take the questions raised by the Poincaré parable and the philosophical responses to it seriously. We shall have to worry about what constitutes the observation basis of theories, how we are to interpret the relation of theoretical superstructure to observational basis, how we should respond to the questions raised by the possibility of apparently incompatible theories sharing a common observation basis, and what consequences for our ontology and our semantics we are to draw from this epistemological puzzle.

So far our response to these issues has been inconclusive. And, alas, so it shall remain. But the remainder of the book will perhaps enlighten us a bit further, for we shall investigate in some detail additional ways these general philosophical issues impinge upon and are impinged upon by specific questions of physical theory.

REALISM AND REASON

In one way of conceiving it, realism is an empirical theory.[1] One of the facts that this theory explains is the fact that scientific theories tend to 'converge' in the sense that earlier theories are, very often, limiting cases of later theories (which is why it is possible to regard theoretical terms as preserving their reference across most changes of theory). Another of the facts it explains is the more mundane fact that language-using contributes to getting our goals, achieving satisfaction, or what have you.

The realist explanation, in a nutshell, is not that language mirrors the world but that *speakers* mirror the world – i.e. their environment – in the sense of *constructing a symbolic representation of that environment*. In 'Reference and Understanding'[2] I argued that a 'correspondence' between words and sets of things (formally, a *satisfaction relation*, in the sense of Tarski) can be viewed as part of an *explanatory model* of the speakers' collective behaviour.

I'm not going to review this in this Address; but let me refer to realism in this sense – acceptance of this sort of scientific picture of the relation of speakers to their environment, and of the role of language – as *internal* realism.

Metaphysical realism, on the other hand, is less an empirical theory than a model – in the 'colliding billiard balls' sense of 'model'. It is, or purports to be, a model of the relation of *any* correct theory to all or part of THE WORLD. I have come to the

Presidential Address to the Eastern Division of the American Philosophical Association, Boston, Mass., 29 December 1976.

conclusion that this model is incoherent. This is what I want to share with you.

Let us set out the model in its basic form.

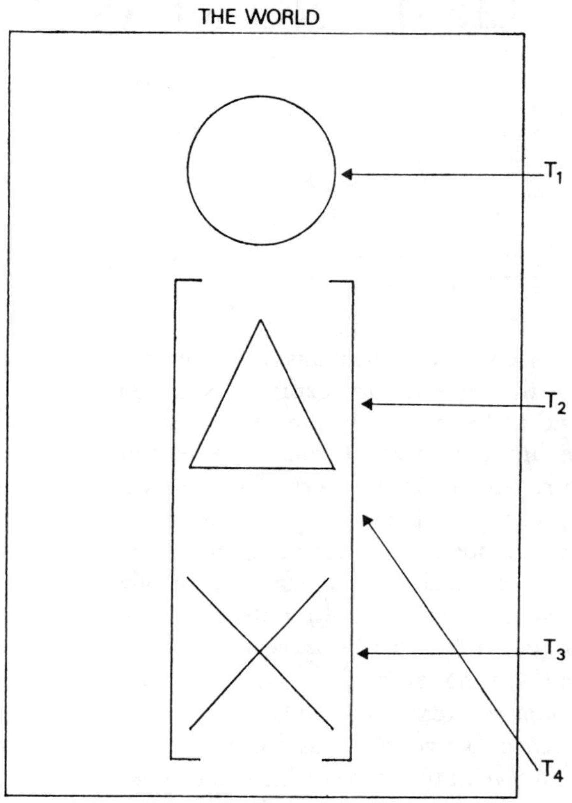

In its primitive form, there is a relation between each term in the language and a piece of THE WORLD (or a *kind* of piece, if the term is a general term).

This relation, the relation of reference, is given *by the truth – conditional semantics* for the language, in the canonical versions of the theory – i.e. *understanding* a term, say, T_1, consists in knowing what piece of THE WORLD it refers to (or in knowing a necessary and sufficient condition for it to refer to a piece of THE WORLD, in some versions). I shall not assume this account of understanding to be part of the picture in what follows, although it certainly was assumed by metaphysical realists in the past.

Minimally, however, there has to *be* a determinate relation of *reference* between terms in L and pieces (or sets of pieces) of THE WORLD, on the metaphysical realist model, whether *understanding* L is taken to consist in 'knowing' that relation or not. What makes this picture different from *internal* realism (which employs a similar picture *within* a theory) is that (1) the picture is supposed to apply to *all* correct theories at once (so that it can only be stated with 'typical ambiguity' – i.e. it transcends complete formalization in any one theory); and (2) THE WORLD is supposed to be *independent* of any particular representation we have of it – indeed, it is held that we might be *unable* to represent THE WORLD correctly at all (e.g. we might all be 'brains in a vat', the metaphysical realist tells us).

The most important consequence of metaphysical realism is that *truth* is supposed to be *radically non-epistemic* – we might be 'brains in a vat' and so the theory that is 'ideal' from the point of view of operational utility, inner beauty and elegance, 'plausibility', simplicity, 'conservatism', etc., *might be false*. 'Verified' (in any operational sense) does not imply 'true', on the metaphysical realist picture, even in the ideal limit.

It is this feature that distinguishes metaphysical realism, as I am using the term, from the mere belief that there *is* an ideal theory (Peircean realism), or, more weakly, that an ideal theory is a regulative ideal presupposed by the notions 'true' and 'objective' as they have classically been understood. And it is this feature that I shall attack!

So let T_1 be an ideal theory, by our lights. Lifting restrictions on our actual all-too-finite powers, we can imagine T_1 to have every property *except objective truth* – which is left open – that we like. E.g. T_1 can be imagined complete, consistent, to predict correctly all observation sentences (as far as we can tell), to meet whatever 'operational constraints' there are (if these are 'fuzzy', let T_1 seem *clearly* to meet them), to be 'beautiful', 'simple', 'plausible', etc. The supposition under consideration is that T_1 might be all this *and still be* (in reality) *false*.

I assume THE WORLD has (or can be broken into) infinitely many pieces. I also assume T_1 *says* there are infinitely many things (so in *this* respect T_1 *is* 'objectively right' about THE WORLD). Now T_1 is *consistent* (by hypothesis) and has (only) infinite models. So by the completeness theorem (in its model theoretic form),

T_1 has a model of every infinite cardinality. Pick a model M of the same cardinality as THE WORLD.[3] Map the individuals of M one-to-one into the pieces of THE WORLD, and use the mapping to define relations of M directly in THE WORLD. The result is a satisfaction relation SAT – a 'correspondence' between the terms of L and sets of pieces of THE WORLD – such that the theory T_1 comes out *true* – true of THE WORLD – provided we just interpret 'true' as TRUE(SAT).[4] So what becomes of the claim that even the *ideal* theory T_1 might *really* be false?

Well, it might be claimed that SAT is not the *intended* correspondence between L and THE WORLD. What does 'intended' come to here?

T_1 has the property of meeting all *operational* constraints. So, if 'there is a cow in front of me at such-and-such a time' belongs to T_1, then 'there is a cow in front of me at such-and-such a time' will certainly *seem* to be true – it will be 'exactly as if' there were a cow in front of me at that time. But SAT is a *true* interpretation of T_1. T_1 is TRUE(SAT). So 'there is a cow in front of me at such-and-such a time' is 'True' in this sense – TRUE(SAT).

On the other hand, if 'there is a cow in front of me at such-and-such a time' is *operationally* 'false' (falsified) then 'there is a cow in front of me at such-and-such a time' is FALSE(SAT). So, the interpretation of 'reference' in L as SAT certainly meets all *operational* constraints on reference. But the interpretation of 'reference' as SAT certainly meets all *theoretical* constraints on reference – it makes the *ideal* theory, T_1, come out *true*.

So what *further* constraints on reference are there that could single out some other interpretation as (uniquely) 'intended', and SAT as an 'unintended' interpretation (in the model-theoretic sense of 'interpretation')? The supposition that even an 'ideal' theory (from a pragmatic point of view) might *really* be false appears to collapse into *unintelligibility.*

Notice that a 'causal' theory of reference is not (would not be) of any help here: for how 'causes' can uniquely refer is as much of a puzzle as how 'cat' can, on the metaphysical realist picture.

The problem, in a way, is traceable back to Occam. Occam introduced the idea that concepts are (mental) *particulars.* If concepts are particulars ('signs'), then any concept we may have of the *relation* between a sign and its object is *another sign*. But it is unintelligible, from my point of view, how the sort of relation the metaphysical realist envisages as holding between a sign and its object can be singled out either by holding up the sign itself, thus

COW

or by holding up yet another sign, thus

REFERS

or perhaps

CAUSES

If concepts are not particulars, on the other hand, the obvious possibility is that (in so far as they are 'in the head') they are *ways of using* signs. But a 'use' theory, while intelligible (and, I believe, correct) as an account of what *understanding* the signs consists in, *does not single out a unique relation* between the terms of T_1 and the 'real objects'. If we don't think concepts are *either* particulars (signs) *or* ways of using signs, then, I think we are going to be led back to direct (and mysterious) grasp of Forms.

Suppose we (and all other sentient beings) are and always were 'brains in a vat'. Then how does it come about that *our* word 'vat' refers to *noumenal* vats and not to vats in the image?

If the foregoing is not to be just a new antinomy, then one has to show that there is at least one intelligible position for which it does *not* arise. And there is. It does not arise for the position Michael Dummett has been defending. Let me explain:

Dummett's idea[5] is to do the *theory of understanding* in terms of the notions of *verification* and *falsification*. This is what he calls 'non-realist semantics'.

What makes this different from the old phenomenalism is that there is no 'basis' of hard facts (e.g. sense data) with respect to which one ultimately uses the truth conditional semantics, classical logic, and the *realist* notions of truth and falsity. The analogy is with mathematical intuitionism: the intuitionist uses *his* notion of 'truth' – constructive provability – *even when talking about constructive proof itself*. Understanding a sentence, in this semantics, is knowing what constitutes a proof (verification) of it. And this is true *even of the sentences that describe verifications*. Thus, I might take 'I have a red sense datum' as a primitive sentence, or I might take 'I see a cow', or, if I do the semantics from the point of view of the brain rather than the person, I might take 'such and such neurons fired'.

Whatever language I use, a primitive sentence – say, 'I see a cow' – will be assertible if and only if *verified*. And we say it is verified *by saying the sentence itself*, 'I see a cow'. To use a term of Roderick Firth's, 'I see a cow' is 'self-warranting' in this kind of epistemology – not in the sense of being *incorrigible*, not even necessarily in the sense of being fully determinate (i.e. obeying strong bivalence – being determinately true or false). (Facts are 'soft all the way down' on this picture, Dummett says.) The important point is that the realist concepts of truth and falsity are not used in this semantics at all.

Now the puzzle about what singles out one correspondence as *the* relation of reference does not arise. The notion of 'reference' is not used in the semantics. We can introduce 'refers' into the language *à la* Tarski, but then

(1) 'Cow' refers to cows

will simply be a tautology – and the *understanding* of (1) makes no reference to the metaphysical realist picture at all.

One important point. It is no good to do the nonrealist semantics (I would rather call it *verificationist semantics* – because it is not incompatible with *internal* realism) in terms of any level of 'hard facts', even sense data. For if sense data are treated as 'hard data' – if the verificationist semantics is given in a meta-language *for* which itself we give the *truth-conditional* account of understanding – then we can repeat the whole argument against the intelligibility of metaphysical realism (as an argument against the intelligibility of the *meta*-language) – just think of the *past* sense data (or the *future* ones) as the 'external' part of THE WORLD. (This is a reconstruction of one aspect of Wittgenstein's private language argument.) This is why Dummett's move depends upon using the verificationist semantics all the way up (or down) – in the meta-language, the meta-meta-language, etc.

The reason I got involved in this problem is this: in 'Reference and Understanding' I argued that one could give a *model of a speaker* of the language in terms of the notion of 'degree of confirmation' (which might better be called 'degree of verification' when it has this understanding-theoretic role). And I contended that the realist notions of truth and reference come in not in explaining what goes on 'in the heads' of speakers, but in *explaining the success* of language-using. Thus I urged that we accept a species of 'verificationist' semantics. (Though not in the sense of verificationist theory of *meaning* – for, as I have argued elsewhere,[6] 'meaning' is not just a function of what goes on 'in our heads', but also of *reference*, and reference is determined by *social* practices and by actual physical paradigms, and not just by what goes on inside any individual speaker.) But, I claimed, one can still be a *realist* even though one accepts this 'verificationist' model. For the realist claim that there is a correspondence between words and things is not *incompatible* with a 'verificationist' or 'use' account of understanding. Such a correspondence, in my view, is part of an *explanatory theory* of the speakers' interaction with their environment.

The point is that Dummett and I *agree* that you can't treat understanding a sentence (in general) as knowing its truth conditions; because it then becomes unintelligible what *that* knowledge *in turn* consists in. We both *agree* that the theory of understanding has to be done in a verificationist way. (Although I don't think that theory of understanding is all of theory of *meaning* that is of no help *here* – theory of meaning, on my view, presupposes theory of understanding *and reference* – and reference is what the problem is all about!) But now it looks as if in conceding that *some* sort of verificationist semantics must be given as our account of understanding (or 'linguistic competence', in Chomsky's sense), I have given Dummett all he needs to demolish metaphysical realism – a picture I was wedded to!

So what? At this point, *I* think that a natural response would be the following: So metaphysical realism collapses. But internal realism – the empirical theory of 'Reference and Understanding' – doesn't collapse (I claim). Metaphysical realism was only a *picture* anyway. If the picture is, indeed, incoherent, then the moral is surely *not* that something is wrong with realism *per se*, but simply that realism *equals* internal realism. *Internal realism is all the realism we want or need.*

Indeed, I believe that this is true. But it isn't *all* the moral. Metaphysical realism collapsed *at a particular point.* (I am going to argue that it also collapses at other points.) And the point at which it collapsed tells us something. Metaphysical realism collapses just at the point at which it claims to be distinguishable from Peircean realism – i.e. from the claim that there is an ideal theory (I don't mean that even *that* claim isn't problematical, but it is problematical in a different way). Since Peirce himself (and the verificationists) always *said* metaphysical realism collapses into incoherence at *just* that point, and realists like myself thought they were *wrong*, there is no avoiding the unpleasant admission that 'they were right and we were wrong' on at least one substantive issue.

I now want to talk about other points at which the metaphysical realist picture is incoherent. Consider the following simple universe: let 'the world' be a *straight line*, thus

(If you want, there can be one-dimensional people – with apologies to Marcuse – on the line. How you tell the boys from the girls, I don't know.)

Consider the following two stories about THE WORLD:[7]

Story 1. *There are points* – i.e. the line has parts which are line *segments*, and also infinitely small parts called 'points'. The *same relation* – 'part of' – holds between points and line segments which contain them, and between line segments and bigger line segments (and between any piece of the line and the whole line).

Story 2. *There are no points* – the line and its parts all have *extension*. 'Of course,' the teller of this story says, 'I'm not saying Story 1 is *false*. You just have to understand that *points* are logical constructions out of line segments. Point talk is highly derived talk about convergent sets of line segments.'

A 'hard-core' realist might claim that there is a 'fact of the matter' as to which is true – Story 1 or Story 2. But 'sophisticated realists', as I have called them, concede that Story 1 and Story 2 are 'equivalent descriptions'. In effect, this concedes that line segments are a suitable set of 'invariants' – a description of

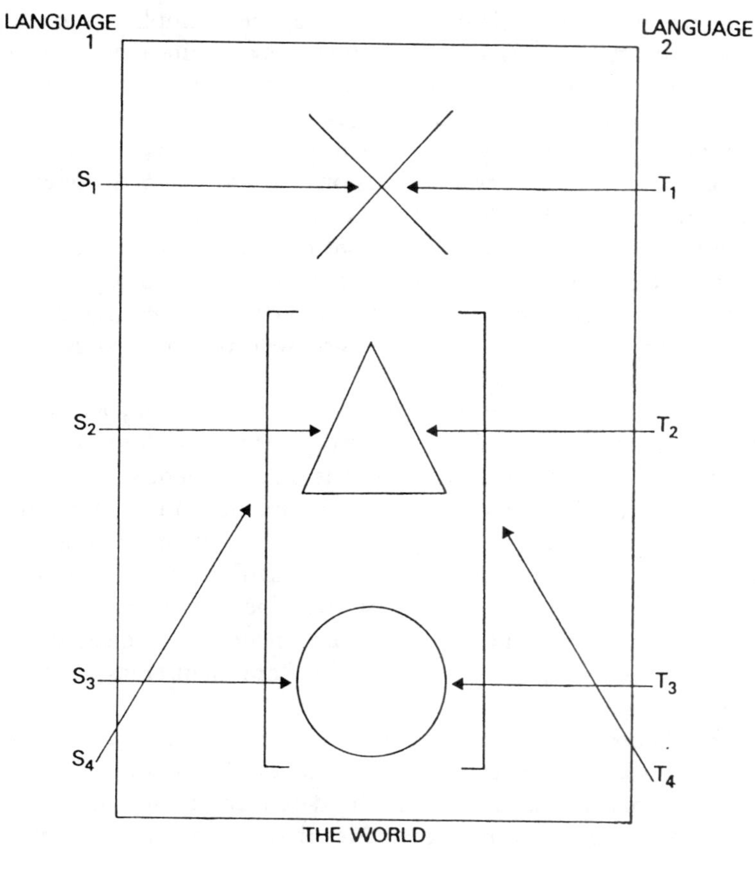

THE WORLD which says what is going on in every line segment is a *complete* description. In the past, I argued that this is no problem for the realist – it's just like the fact that the earth can be mapped by different 'projections', I said (Mercator, Polar, etc.). The metaphysical realist picture now looks as on p. 131.

In particular, I believed, it can happen that what we picture as 'incompatible' terms can be mapped onto the same real object – though not, of course, within the same theory. Thus the real object that is labelled 'point' in one theory might be labelled 'set of convergent line segments' in another theory. And the *same* term might be mapped onto one real object in one theory and onto a different real object in another theory. It is a property of the world itself, I claimed – i.e. a property of THE WORLD itself – that it 'admits of these different mappings'.

The problem – as Nelson Goodman has been emphasizing for many, many years – is that this story may retain THE WORLD but at the price of giving up any intelligible notion of *how* THE WORLD is. Any sentence that changes truth-value upon passing from one correct theory to another correct theory – e.g. an equivalent description – will express only a *theory-relative* property of THE WORLD. And the more such sentences there are, the more properties of THE WORLD will turn out to be theory-relative.

For example, if we concede that Story 1 and Story 2 are equivalent descriptions, then the property *being an object* (as opposed to a class or set of things) will be theory-relative. Consider now a third story:

Story 3. *There are only line segments with rational end points* (i.e. since there aren't 'points', in this story, except as logical constructions, (1) every line segment has rational length; (2) the piece of the line[8] between any two line segments is a line segment, and so has rational length; (3) every line segment is divisible into n equal pieces, for every integer n; (4) there is at least one line segment; and (5) the union of two line segments is a line segment). Irrational line segments are treated as logical constructions – sets of 'points', where 'points' are themselves Cauchy convergent sets of *rational* line segments.

A 'hard-core' realist might again object, this time because this story makes an irrational line segment of a different logical type than a rational line segment. But the defender of this story can reply: Isn't it common in mathematics that objects are identified with sets of other objects which are *pre-analytically* of the same logical type? Thus, negative integers and positive integers, whole numbers and rationals, rationals and reals, reals and imaginaries are pre-analytically all 'numbers', but in *formalizing* mathematics we are used to treating negative numbers (or more generally, 'signed numbers') as *ordered pairs* of 'natural numbers', *rational* numbers as *ordered pairs* of 'signed numbers', irrationals as *sets* of rationals, etc. So what is wrong with treating irrational line segments as sets of sets of rational line segments? After all, the rational line segments are a basis for the topology; if you know what is going on in every rational line segment, you have a complete description of all events, etc.

If we accept Story 3 as yet another equivalent description of 'the world', however, then even the *cardinality* of the world becomes theory-relative! For there are only denumberably many objects in Story 3, and non-denumerably many in Stories 1 and 2! (We might try to avoid this by treating *sets* as 'objects', too – but, as I've shown elsewhere, 'set' talk can be translated away into *possibility* talk.)

All this isn't an artifact of my simple example: actual physical theory is rife with similar examples. One can construe space-time points as objects, for example, or as properties. One can construe fields as objects, or do everything with particles acting at a distance (using retarded potentials). The fact is, *so many* properties of THE WORLD – starting with *just* the *categorial* ones, such as cardinality, particulars, or universals, etc. – turn out to be 'theory-relative' that THE WORLD ends up as a Kantian 'noumenal' world, a *mere* 'thing in itself'. If one cannot say *how* THE WORLD is theory-independently, then talk of all these theories as descriptions of 'the world' is empty.

ANOTHER POINT AT WHICH THE METAPHYSICAL REALIST PICTURE RUNS INTO TROUBLE

This has to do with what Quine calls 'ontological relativity'. Suppose we confine attention, for the moment, to *complete* theories. If T is a complete theory, we can define an equivalence relation on its terms – *provable co-extensiveness* – with the property that if two terms belong to different equivalence classes, then in *no* model of the theory do they refer to the same referent, whereas if they belong to the same equivalence class, then they have the same referent in *every* model of the theory. So, for our purposes, we may count terms as the same if they lie in the same equivalence class – i.e. if they are 'co-extensive taking the theory at face value'. With this preliminary identification made, we notice that if our picture is correct – I repeat the picture

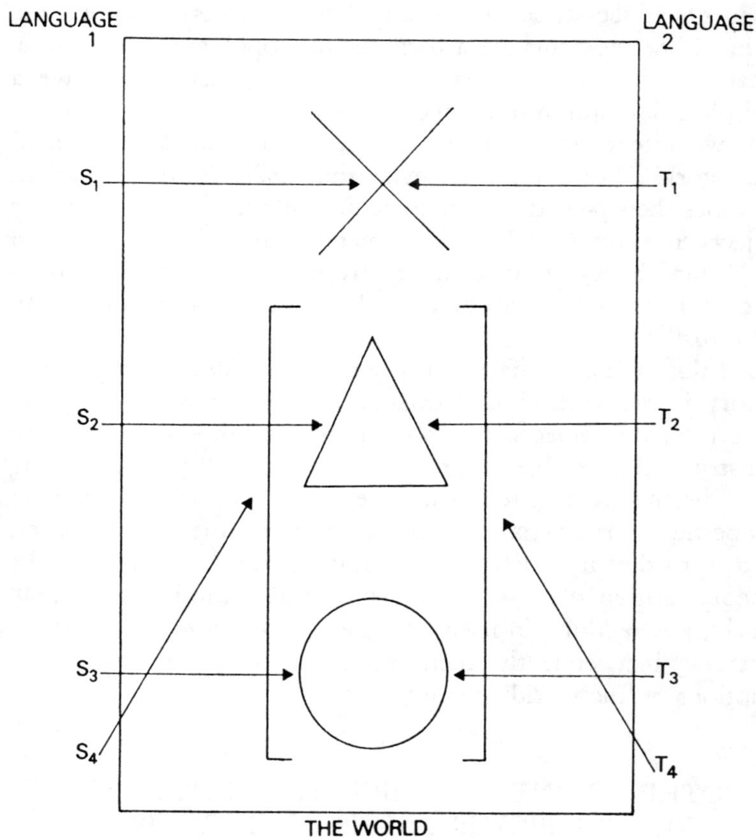

– then there is a *unique* reference-preserving 'translation' connecting the Languages.

But it is notorious that there are often *inequivalent* relative interpretations of one theory in another. Story 1 can be interpreted in Story 2 (in the case of our example) in *many* different ways.

'Points' can be sets of line segments whose lengths are negative powers of 2, for example, or sets of line segments whose lengths are negative powers of 3.

If the picture as I drew it were correct, there would have to be a 'fact of the matter' as to *which* translation *really* preserves reference in every such case!

Just as we complicated the picture by allowing the same term to be mapped onto different real objects when it occurs in different theories to meet the previous objection, so we could complicate the picture *again* to meet the second objection: we could say that the language has *more than one* correct way of being mapped onto THE WORLD (it must, since it has more than one way of being correctly mapped onto a language which is *itself* correctly mapped onto the world). But now *all* grasp of the picture seems to vanish: if what is a *unique* set of things *within a correct theory* may not be a unique set of things 'in reality', then the very heart of the picture is torn out.

WHY ALL THIS DOESN'T REFUTE INTERNAL REALISM

Suppose we try to stump the *internal* realist with the question, 'How do you know that "cow" refers to cows?' 'After all', we point out, 'there are other interpretations of your whole language – non-denumerably many interpretations (in the sense of satisfaction-relations), which would render true an *ideal* theory (in your language). Indeed, suppose God gave us the *set of all true sentences* in your language (pretend we have infinite memories, for this purpose). Call this set the *perfect* theory. Then there would *still* be infinitely many admissible interpretations of the *perfect* theory – interpretations which, as we saw, satisfied *all* the operational and theoretical constraints. Even the *sentence* "'Cow' refers to cows" is true in all of these interpretations. So how do you know that it is true in the sense of being true in a *unique* 'intended' interpretation? How do you know that "cow" refers to cows in the sense of referring to *one* determinate set of things, as opposed to referring to a determinate set of things *in each admissible interpretation*?' (This is, of course, just arguing against the internal realist exactly as we argued against the metaphysical realist.)

The internal realist should reply that '"Cow" refers to cows' follows immediately from the definition of 'refers'. In fact, '"cow" refers to cows' would be true even if internal realism were false: although we can revise '"Cow" refers to cows' by scrapping the theory itself (or at least scrapping or challenging the notion of a *cow*) – and this is how the fact that '"cow" refers to cows' is not *absolutely* unrevisable manifests itself – *relative to the theory*, '"Cow" refers to cows' is a logical truth.

The critic will now reply that his question hasn't been answered. '"Cow" refers to cows' is indeed analytic relative to the theory – but his question challenged *the way the theory is understood*. '"Cow" refers to cows' is true in all admissible interpretations of the theory – but that isn't at issue.

The internal realist should now reply that (1) 'the way the theory is understood' can't be discussed *within* the theory; and (2) the question whether the theory has a *unique* intended interpretation has *no* absolute sense. Viewed from within Story 1 (or a meta-language which contains the object language of Story 1), 'point' has a 'unique intended interpretation'. Viewed from within Story 2 (or a meta-language which contains the object language of Story 2), the term 'point' *as used in Story 1* has a plurality of admissible interpretations. The critic's 'how do you know?' question assumes a theory-independent fact of the matter as to what a term in a given theory corresponds to – i.e. assumes the picture of metaphysical realism; and this is a picture the *internal* realist need not (and better not) accept.

The critic now replies as follows: 'reference' (strictly speaking, 'satisfies') is defined so that

(1) 'Cow' refers to cows

just says (in the case of where 'cow' is a primitive expression of L) that the ordered pair < 'cow', {cows} > belongs to a certain *list* of ordered pairs. If anything, this *presupposes* that 'cow' refers (in some *other* sense of 'refers'); it doesn't *explicate* it.[9]

Answer: the *use* of 'cow' does presuppose that 'cow' is *understood*. And if my account of *understanding* was a *truth-conditional* (or *reference-conditional*) account, then the objection would be good. But I gave a verificationist account of understanding (in terms of degree of confirmation); thus *my use of the term 'cow' in the language* has already been explained, and I am free to use it – even to use it in explaining what 'cow' refers to.

What I am saying is that, in a certain 'contextual' sense, it is an *a priori* truth that 'cow' refers to a determinate class of things (or a more-or-less-determinate class of things – I neglect ordinary vagueness). Adopting 'cow talk' is adopting a 'version', in Nelson Goodman's phrase, from within which it is *a priori* that the word 'cow' refers (and, indeed, that it refers to cows).

One of the puzzling things about the metaphysical realist picture is that it makes it unintelligible how there can *be a priori* truths, even contextual ones, even as a (possibly unreachable) limit. An *a priori* truth would have to be the product of a kind of direct 'intuition' of the things themselves. Even verbal truth is hard to understand. Consider 'all bachelors are unmarried'. It can be 'verbal' that this is in some sense 'short for' 'all unmarried men are unmarried'. And this, in turn, is an instance of 'All AB are A'. But why is *this* true?

Suppose there *were* unrevisability – *absolute* unrevisability. And suppose we held 'All AB are A' (and even 'All unmarried men are unmarried') absolutely immune from revision. Why would this make it *true*?

Suppose, unimaginably, there are some AB that are not A. (After all, there are lots of things in modern science we can't imagine.) Then, on the metaphysical realist picture, our refusal to give up assenting to 'All AB are A' doesn't make *it* true – it just makes *us* stubborn.[10]

Once we abandon the metaphysical realist picture, the situation becomes quite different. Suppose we include a sentence S in the ideal theory T_1 just because it is a feature we *want* the ideal theory to have that it contain S. (Suppose we even hold S 'immune' from revision, as a behaviouristic fact about us.) Assuming S doesn't make T_1 inconsistent, T_1 *still* has a model. And since the model isn't fixed *independently* of the theory, T_1 will be *true* – true in *the* model (from the point of view of *meta*-T_1); true in all *admissible* models, from the point of view of a theory in which the terms of T_1 do not determinately refer to begin with. So S will be true! 'S' is 'analytic' – but it is an 'analyticity' that resembles Kant's account of the *synthetic a priori* more than it resembles his account of the analytic. For the 'analytic' sentence is, so to speak, part of 'the form of the representation' and not 'the content of the representation'. It can't be false of the world (as opposed to THE WORLD), because the world is not describable independently of our description.

Even if T_1 were *inconsistent* if we were consistently inconsistent (assigned 'truth' and 'falsity' to sentences in a stable way), this would not block this argument: for *stable* inconsistency can be viewed as *reinterpretation of the logical connectives*. When we give up the metaphysical realist picture we see for the first time how a truth can be 'about the world' ('All AB are A' *is* 'about the world' – it is about all *classes* A, B) and 'without content'.

In the foregoing, I used the idea of an absolutely 'unrevisable' truth as an idealization. Of course, I agree with Quine that this is an unattainable 'limit'. Any statement can be 'revised'. But what is often overlooked, although Quine stresses it again and again, is that the revisability of the laws of Euclid's geometry, or the laws of classical logic, does not make them mere 'empirical' statements. This is why I have called them *contextually a priori*.[11] Quine put the point very well when he said that 'the lore of our fathers' is black with fact and white with convention, and added that there are no *completely* white threads and no quite black ones. One might describe this as a soft (and demythologized) Kantianism. A trouble with the metaphysical realist picture is that one cannot see how there can be white at all – even greyish white.

Let me close with a last philosophical metaphor. Kant's image was of knowledge as a 'representation' – a kind of play. The author is me. But the author also appears as a character in the play (like a Pirandello play). The author in the play is not the 'real' author – it is the 'empirical me'. The 'real' author is the 'transcendental me'.

I would modify Kant's image in two ways. The authors (in the plural – my image of knowledge is social) don't write just *one* story: they write many versions. And the authors *in* the stories are the *real* authors. This would be 'crazy' if these stories were *fictions*. A fictitious character can't also be a real author. But these are true stories.

NOTES

1 This is spelled out in my 'What is "Realism"?', in *Proceedings of the Aristotelian Society* (1975–6), pp. 177–94. Part I, Lecture II of this volume.
2 Delivered in Jerusalem, May 1976, Part III of this volume.
3 If THE WORLD is finite, let the theory be compatible with there being only N individuals (where N is the cardinality of THE WORLD), and pick a model with N individuals instead of using the stronger model-theoretic theorem appealed to in the text.
4 Here, if SAT is a relation of the same logical type as 'satisfies', TRUE(SAT) is supposed to be defined in terms of SAT exactly as 'true' is defined in terms of 'satisfies' (by Tarski). Thus 'TRUE(SAT)' is the truth-property 'determined' by the relation SAT.
5 This is most completely spelled out in his (unpublished) William James Lectures. A partial account appears in his contribution to the conference on 'Language, Intentionality, and Translation Talk' reprinted in *Synthese,* vol. 27, nos 3/4 (July/August, 1974).
6 Cf. my 'The Meaning of "Meaning"' in my *Mind, Language, and Reality; Philosophical Papers,* vol. 2, Cambridge University Press, 1975.
7 Compare the alternative systems discussed in Nelson Goodman, *The Structure of Appearance*, Reidel, Dordrecht, 1951, 1977, ch. 1, sections 1–4.
8 The mathematical reader will note that in Story 3 there is no distinction between *open* and *closed* line segments – because there are no such things as *points*!
9 This objection to Tarskian definitions of reference is due to Hartry Field (cf. his 'Tarski's Theory of Truth', *Journal of Philosophy*, vol. 69, no. 13 (1972), pp. 347–75) and is discussed in my John Locke Lectures (Part I of this book).
10 The reader may be tempted to reply that even a metaphysical realist is entitled to the notion of a *verbal convention.* And why can't it be a verbal *convention* that no state of affairs is to be referred to as the *conjunction* of the states of affairs described by sentences *p, q* unless it entails both

p and *q* separately? This would mean that '*p*·*q* entails *p*' is 'true by convention'; and similarly, 'All AB are A' could be 'true by convention'.

The difficulty is that such a 'convention' only makes it analytic that the conjunction *p*·*q* entails *p if the state of affairs in question exists.* But how can the *existence* of a state of affairs with the properties of entailing *p* and entailing *q* and being entailed by every state of affairs which entails both *p* and *q* be *itself* a matter of *convention*, on the metaphysical realist picture?

To establish that this is not trifling, let me remark that there are logics (studied by David Finkelstein in connection with certain 'far out' physical theories – *not* standard quantum theory) in which (1) there are propositions *incompatible* with any given proposition; but (2) there is no such thing as *the* negation of a given proposition —i.e. no logically weakest proposition incompatible with a given proposition. (These logics correspond algebraically to lattices which are not orthocomplemented.) If 'the logic of the world' is one of *these* logics (as Finkelstein believes), then *the existence of a complement* to a given state of affairs is *false as a matter of fact* – and *no* linguistic convention *could* render it true!

It seems to me that a consistent metaphysical realist must *either* view logic as empirical, not just in the sense of thinking that logic is revisable (which I believe), but in the sense of having *no* conventional component at all (so that even our confidence that statements aren't *both* true and false becomes ultimately just *inductive* confidence), *or* he must believe that logic is *a priori* in a sense of *a priori* which is not explainable by the notion of convention at all.

11 In 'It Ain't Necessarily So', reprinted in my *Mind, Language and Reality; Philosophical Papers,* vol. 2, Cambridge University Press, 1975.

W. V. QUINE

ON EMPIRICALLY EQUIVALENT SYSTEMS OF THE WORLD

If all observable events can be accounted for in one comprehensive scientific theory – one system of the world, to echo Duhem's echo of Newton – then we may expect that they can all be accounted for equally in another, conflicting system of the world. We may expect this because of how scientists work. For they do not rest with mere inductive generalizations of their observations: mere extrapolation to observable events from similar observed events. Scientists invent hypotheses that talk of things beyond the reach of observation. The hypotheses are related to observation only by a kind of one-way implication; namely, the events we observe are what a belief in the hypotheses would have led us to expect. These observable consequences of the hypotheses do not, conversely, imply the hypotheses. Surely there are alternative hypothetical substructures that would surface in the same observable ways.

Such is the doctrine that natural science is empirically under-determined; under-determined not just by past observation but by all observable events. The doctrine is plausible insofar as it is intelligible, but it is less readily intelligible than it may seem. My main purpose in this paper is to explore its meaning and its limits.

This doctrine of empirical under-determination is not to be confused with holism. It is holism that has rightly been called the Duhem thesis and also, rather generously, the Duhem-Quine thesis. It says that scientific statements are not separately vulnerable to adverse observations, because it is only jointly as a theory that they imply their observable consequences. Any one of the statements can be adhered to in the face of adverse observations, by revising others of the statements. This holism thesis lends credence to the under-determination theses. If in the face of adverse observations we are free always to choose among various adequate modifications of our theory, then presumably all possible observations are insufficient to determine theory uniquely.

The holism thesis is less beset with obscurities than the under-determination thesis, and again it is a thesis that must command assent, with

Erkenntnis **9** (1975) 313–328. *All Rights Reserved*

reservations. One reservation has to do with the fact that some statements are closely linked to observation, by the process of language learning. These statements are indeed separately susceptible to tests of observation; and at the same time they do not stand free of theory, for they share much of the vocabulary of the more remotely theoretical statements. They are what link theory to observation, affording theory its empirical content. Now the Duhem thesis still holds, in a somewhat literalistic way, even for these observation statements. For the scientist does occasionally revoke even an observation statement, when it conflicts with a well attested body of theory and when he has tried in vain to reproduce the experiment. But the Duhem thesis would be wrong if understood as imposing an equal status on all the statements in a scientific theory and thus denying the strong presumption in favor of the observation statements. It is this bias that makes science empirical.

Another reservation regarding the Duhem thesis has to do with breadth. If it is only jointly as a theory that the scientific statements imply their observable consequences, how inclusive does that theory have to be? Does it have to be the whole of science, taken as a comprehensive theory of the world?

We should note that the sciences do link up more systematically than people are apt to realize who forget about logic and mathematics; for logic is shared by all branches of science, and much of mathematics is shared by many. People tend unduly to see the logical and mathematical components of science as different in kind from the rest, and hence fail to see these components as something common to all the branches. Ironically, this very neutrality, this fact of being shared by all branches of science, has encouraged people to think of the logical and mathematical components as different in kind from the rest, and hence to fail to recognize the unity that they confer. Thus I see science as a considerably integrated system of the world even now, though the explicit reduction of major branches to theoretical physics is incomplete.

But we can appreciate this degree of integration and still appreciate how unrealistic it would be to extend a Duhemian holism to the whole of science, taking all science as the unit that is responsible to observation. Science is neither discontinuous nor monolithic. It is variously jointed, and loose in the joints in varying degrees. In the face of a recalcitrant observation we are free to choose what statements to revise and what ones to hold fast, and these alternatives will disrupt various stretches of scientific theory in various ways, varying in severity. Little is gained by saying that the unit is in principle the whole of science, however defensible this claim may be in a legalistic way.

This is the end of my digression on the thesis of holism. I shall be concerned from here on with the empirical under-determination of natural science. It is a doctrine which, as I said, is plausible insofar as it is intelligible. My purpose will be to examine the meaning of this thesis more closely, and to consider its limits and its consequences.

A notion that is evidently central to the thesis is that of observation. This notion is subject to a curious internal tension. Observation affords the sensory evidence for scientific theory, and sensation is private. Yet observation must be shared if it is to provide the common ground where scientists can resolve their disagreements. The observation must be the distillate, somehow, of what is publicly relevant in the private sensations of present witnesses. This delicate process of distillation is already accomplished, happily, in our most rudimentary learning of language. One learns the word 'blue' from another speaker, in the presence of something blue. The other speaker has learned to associate the word with whatever inscrutable sensation it may be that such an object induces in him, and one now learns to associate the word with the sensation, same or different, that the object induces in oneself. All agree in calling the object blue, and even in calling their sensations blue.

We do well to recognize this crucial role of language, for in view of it we can spare ourselves the finicky task of defining the elusive notion of observation. We can speak rather of observation terms and observation sentences, thus cleaving to linguistic forms: 'blue', 'This is blue'. Observational expressions can be roughly distinguished from others by a behavioral criterion, involving no probing of sensations. For this is characteristic of them: witnesses will agree on the spot in applying an observation term, or in assenting to an observation sentence, if they are conversant with the language. Their verdicts do not vary with variations in their past experience.

It is sometimes objected that a specialist may recognize at a glance what the untrained observer cannot. The above behavioral criterion settles this discrepancy in favor of the untrained observer. Specialists rest content with the level of evidence that commands their expert agreement but in principle they usually could reduce this recondite evidence to observation terms at the layman's level.

Not always. There is expertise in tea tasting, in wine tasting, and in the recognition of tones, chords, and timbre, that resists conversion to the common coin.[1] We should like to be able to reckon the esoteric terms in these domains as observational for the experts, contrary to the proposed behavioral criterion. And indeed even a common observation term such as 'blue' has its penumbra of vagueness, where witnesses may disagree in their verdicts. The really distinctive trait of observation terms and sentences is to be sought not in concurrence of witnesses but in ways of learning. Observational expressions are expressions that can be learned ostensively. They are actually learned ostensively in some cases and discursively in others, but each of them could be learned by sufficiently persistent ostension. The behavioral manifestation of observationality, then, namely, the ready concurrence of witnesses, serves merely as a rough practical criterion.

Observation sentences are not incorrigible. A witness who has assented to an observation sentence on the spot is permitted to reconsider his verdict later in the face of conflicting theory. And of course the typical observation terms are not subjective in reference, but objective. 'Blue' was one; others are 'water', 'rabbit', 'ball', 'hard'. They recur in theoretical sentences.

Now the doctrine that is up for clarification is that scientific theory is under-determined by observable events. So we shall want to get clear on the relation of theories to observations. Or, now that we have taken to talking of observation sentences and terms, rather than of observations, let us look to the relation of theories to observation sentences.

An observation sentence is an occasion sentence: it commands assent on some occasions and not others, depending on what is happening where and when the sentence is queried. On the other hand the sentences of scientific theory are standing sentences. They are meant to be true or false independently of the occasion of utterance. The observation sentences cannot, as occasion sentences, be implied by theory; we must first change them into standing sentences, by incorporating specifications of place-times. Let us adopt, then, an arbitrary numerical system of spatio-temporal coordinates, and let us contemplate the infinite totality of what I shall call *pegged observation sentences*. Each observation sentence expressible in our language gets joined to each combination of spatio-temporal coordinates. The resulting sentences are standing sentences, some true and some false. The true ones do not depend for their truth on anyone's having made the observations; it matters only that the observable state or event in question occur, in fact, at the specified place-time. The time and place may be beyond the reach of all sentient life.[2]

Our move from occasional observation sentences to these pegged observation sentences is already an abrupt ascent from observation into theory. We need to know not only a bit of mathematics but also quite a lot about the physical world in order to establish a system of coordinates. I shall suppose this much achieved, in order to get on better with further questions; but let us not lose sight of the magnitude of our assumptions. The doctrine of under-determination says there is a certain slack between observation and theory; and we have already lost some of that slack by granting the system of coordinates. Just in order to define the slack, we are having to take some of it up.

We have moved from the occasional observation sentences to the pegged observation sentences because we want sentences that theories can imply. However, the goal is not yet. Typically a theory does not imply even a pegged observation sentence outright. Typically a theory traffics rather in generalities. Typically a theory will descend to particulars only conditionally upon other particulars, assumed as boundary conditions. I say 'typically' because I do not want to exclude particulars from theories altogether. I do not want to exclude such particular and unconditional conclusions as might concern the moon or Ur of the Chaldees or the cave painters of Altamira. But even such unconditional conclusions, though particular, will seldom be observational; they will seldom be pegged observation sentences. For a just view of the relation of scientific theory to the observations that support or refute it, we must look to the use of boundary conditions. The theory, plus some set of pegged observation sentences that have already been verified, implies some further pegged observation sentence that can now be checked; such is the test of a theory.

Instead of saying that the theory and the boundary conditions together imply the further pegged observation sentence, we could as well say that the theory implies, outright, a conditional sentence whose antecedent comprises the boundary conditions and whose consequent is the further pegged observation sentence. Such a conditional sentence I shall call an *observation conditional.* Its antecedent is a conjunction of pegged observation sentences and its consequent is a pegged observation sentence.

At last we can state, tentatively, the relation of theory to observation: the theory implies observation conditionals. However, we have still to tidy up the notion of a theory. I have spoken of a theory as implying sentences, as if the theory were itself a sentence or a set of sentences. It will be better to speak of a *theory formulation* as doing the implying. The theory formulation is simply a sentence – typically a conjunctive sentence comprising the so-called axioms of the theory. Currently the theory itself, then, is often identified with an infinite set of sentences, namely, the logical consequences of the theory formulation. Such has usually been my own usage. A single theory, in this sense, admits of many formulations; all that is required is that they be logically equivalent. But we shall find that even this requirement is too stringent to fit the more common and traditional usage of the term 'theory'; and we shall find also that some such broader notion of theory is wanted for making proper sense of the thesis of underdetermination.

Meanwhile a word about implication, or logical consequence, and logical equivalence. The notions are clear as long as the theory formulations and their consequences are couched in our regimented scientific language, with its explicit logical notation. If we try to accommodate formulations of theory in other languages, however, it is less clear when to say that one theory formulation implies another, or implies a given observation conditional. Problems of translation intrude. I prefer in the present study to set aside the whole question of formulating theories in other languages, and to think of theories only as formulated in our own language, with help of our own regimented logical notation. These parochial limitations will help us move forward to central issues.

Incidentally a yet more drastic limitation has been with us for some time, in the pegged observation sentences that go into our observation conditionals. For these are pegged to a coordinate system, and even to a single arbitrarily chosen coordinate system adhered to throughout the present theory of theories. When problems outside the scope of this paper call for lifting these arbitrary limitations, that will be soon enough to consider how best to do so.

For now, accordingly, we can say that theories are always formulated within our own language, of standard logical form. But even within these cozy limits I shall not want simply to identify a theory with the logical consequences of a theory formulation. The next consideration will show why.

Take some theory formulation and select two of its terms, say 'electron' and 'molecule'. I am supposing that these do not figure essentially in any observation sentences; they are purely theoretical. Now let us transform our theory formulation merely by switching these two terms throughout.[3] The new theory formulation will be logically incompatible with the old: it will affirm things about so-called electrons that the other denies. Yet their only difference, the man in the street would say, is terminological; the one theory formulation uses the technical terms 'molecule' and 'electron' to name what the other formulation calls 'electron' and 'molecule'. The two formulations express, he would say, the same theory. Someone else might urge, however perversely, that they express very different theories: both of them treat of molecules in the same sense but disagree sharply regarding the behavior of molecules, and correpondingly for electrons. Clearly, in any event, the two theory formulations are *empirically equivalent* – that is, they imply the same observation conditionals. I think, moreover, that we should individuate theories in such a way as to agree with the man in the street: the two formulations formulate the same theory, despite their overt logical incompatibility. This is why I do not want to identify a theory with the logical consequences of a formulation. I do not want to require that two formulations of a theory be logically equivalent, nor even logically compatible.

Certainly two formulations of a theory should be empirically equivalent, in the sense just defined, even if not logically equivalent. Still empirical equivalence must not be the only requirement, unless we are to repudiate the doctrine of underdetermination out of hand; for that doctrine says that empirically equivalent theories can conflict. What is required of two formulations of a theory must be, in short, some relation stronger than empirical equivalence and weaker than logical equivalence.

The unsatisfying example of 'molecule' and 'electron' consisted, intuitively speaking, in switching the meanings of the two words. When the man in the street protested that the conflict between the two theory formulations was merely terminological, his point was that it could be resolved by treating the one formulation as not quite English, and translating its respective words 'molecule' and 'electron' into the English words 'electron' and 'molecule'. I want to preserve this insight while avoiding problems raised by the terms 'meaning' and 'translation'. I can do so by appealing to little more than a permutation of vocabulary; in the present case, of course, a mere switching of the predicates 'molecule' and 'electron' throughout one of the theory formulations. I propose that we count two formulations as formulations of the same theory if, besides being empirically equivalent, the two formulations can be rendered identical by switching predicates in one of them.

This criterion needs a little broadening, in obvious respects. Since logically equivalent formulations were in any event to count as formulations of the same theory, we should not require that a switching of terms render formulations identical; we should only require it to render them logically equivalent. Further, we should not limit the permutation to a switching of two predicates; we should allow permutations of many. Finally, it would be arbitrary to require this transformation to carry predicates always into simple one-word predicates. The intuitive notion, after all, was a reconstruing of predicates; and the general way of reconstruing an n-place predicate is by supplying an open sentence in n variables, not caring whether there happens to be a word in our language with the same extension as that open sentence.

By a *reconstrual* of the predicates of our language, accordingly, let me mean any mapping of our lexicon of predicates into our open sentences (n-place predicates to n-variable sentences). Thus the predicate 'heavier than' might be mapped to the open sentence 'x is heavier than y', an identity mapping changing nothing, while the predicates 'molecule' and 'electron' might be mapped to the respective open sentences 'x is an electron' and 'x is a molecule', producing our example.

So I propose to individuate theories thus: two formulations express the same theory if they are empirically equivalent and there is a reconstrual of predicates that transforms the one theory into a logical equivalent of the other.[4]

I was able to define reconstrual more simply than otherwise only by assuming something about the form of our language that I now ought to make explicit. I am assuming the standard logical form of language at its most economical; there are just truth functions, quantification, and a finite lexicon of predicates. I made no provision for names or for functors, for there are well known ways of serving the purposes of these devices on the more austere basis.

We have now settled the individuation of theories, within our parochial confines. We have said when to count two formulations as expressing the same theory. Given this equivalence relation, it is a routine matter to say what a theory is. The method is artificial but familiar: theories are the equivalence classes of that equivalence relation. The theory expressed by a given formulation is the class of all the formulations that are empirically equivalent to that formulation and can be transformed into logical equivalents of it or vice versa by reconstrual of predicates.

It is usual in the literature to require of a theory that it be deductively closed. In our present terms, what this means is that if you change a formulation of a theory merely by annexing some logical consequences of that formulation, the result will still be a formulation of the same theory. We have insured this by requiring only that the reconstrual of predicates render the formulations logically equivalent, not identical. Thanks to the circumstance that any formulation is equivalent to itself plus any of its consequences, it is easily shown that theories as I have defined them are deductively closed.

So defined, theories are classes of theory formulations. But at this point we must liberalize the notion of a theory formulation, so as not to be limited to the few formulations that are physically available on paper. What are wanted rather are linguistic sequences in the abstract sense, in their infinite variety. Each single word or letter can still be explained as the class of all its tokens, a class of actual inscriptions, since we are assured that these classes all have members variously situated in space-time. Sentences, however, and longer expressions, are to be taken rather as mathematical sequences of their component words or letters. An expression in this sense is a function, or class of ordered pairs; the first word or letter of the expression is paired with the number 1, the second with 2, and so on. In this way we can assure the existence of all expressions however long, all theory formulations as yet unconceived, all texts as yet unwritten; all 'possible' expressions, as one might say. Theories, finally, are classes of formulations, hence classes of expressions in this abstract sense; classes of functions.

Because the question how to define a theory is interesting in itself, I have pursued it farther than required for what I want to say about the thesis of under-determination. For this purpose it is clearer to treat directly of the empirical equivalence of theory formulations, and of the reconstrual of predicates therein, and not to pause over the individuation of theories as such. In these terms, under-determination says that for any one theory formulation there is another that is empirically equivalent to it but logically incompatible with it, and cannot be rendered logically equivalent to it by any reconstrual of predicates.

We had one attempt at an example, and it failed: the trivial example of two theories that differed only in the switch of 'electron' and 'molecule'. A less trivial case, to which I now turn, is due to Poincaré. Here we have one formulation of cosmology that represents space as infinite, and another formulation that represents space as finite but depicts all objects as shrinking in proportion as they move away from center. The two formulations, again, are empirically equivalent. But again the example is disappointing as an example of under-determination, because again we can bring the two formulations into coincidence by reconstruing the predicates. The reconstrual called for here is less simple than the mere switch of 'electron' with 'molecule', but it presents no serious challenge. The two formulations are formulations, again, of a single theory.

Having disqualified these permutations as cases of under-determination, we might in passing consider how they stand as cases of indeterminacy of translation. If in the light of verbal behavior we translate two foreign words as 'molecule' and 'electron', what behavioral evidence could have obstructed the opposite choices? None, surely, except as we invoke what Neil Wilson called the principle of charity: maximize the agreement between the native and ourselves on questions of truth and falsity, other things being equal. Translation is not the recapturing of some determinate entity, a meaning, but only a balancing of various values. An observation sentence and its translation should command assent under similar stimulations; here is one value. Wide concomitance of assent to standing sentences is also a value. Good translation strikes some optimum combination of values, insofar as they can be compared.

Let us return now to the thesis of under-determination of natural science. We saw that what we need for illustration of this thesis are theory formulations that are empirically equivalent, logically incompatible, and irreconcilable by reconstrual of predicates. For substantiation of the thesis of under-determination we need more: we need to show not only that such branching alternatives exist, but that they are inevitable. Thus suppose we had an adequate theory of nature, and then we were to add to it some gratuitous further sentences that had no effect on its empirical content. By ringing changes on these excrescences we might get alternative theories, logically incompatible, yet always empirically equivalent. This gratuitous branching of theories would be of no interest to the thesis of under-determination, since the adequate original theory was itself logically compatible with each one of these gratuitous extensions; they were incompatible only with one another. What the thesis of under-determination calls for is unavoidable branching.[5] The adequate original theory, in our imaginary example, would itself have had to be one of several equivalent and logically incompatible theories if it was to illustrate the thesis of under-determination.

In its full generality, the thesis of under-determination thus interpreted is surely untenable. It must fail for weak theories, theories that imply no rich store of observation conditionals. If the implied observation conditionals (redundancies aside) are finite in number, we can simply take the conjunction of them, a single sentence, as our theory formulation. It contains its observation conditionals without remainder; they are all it is. It is implied by every empirically equivalent theory, and can conflict with none of them. Any that it conflicted with would have to be internally inconsistent, and so not empirically equivalent.

So we see that the thesis of under-determination must fail where only finitely many observation conditionals are implied. They, in conjunction, are their own theory formulation. But much the same thing can happen even where a theory irreducibly implies infinitely many observation conditionals; for it may happen that these can all be encompassed by a single universally quantified conditional, or by finitely many. Such a theory formulation, again, affords a tight fit. No theory formulation that implies just those same observation conditionals can conflict with it, unless indeed it is ω-inconsistent. And I think we may reject ω-inconsistent theories.

The empirical content of a theory formulation is summed up in the observation conditionals that the formulation implies. These are material conditionals; each is a truth function of pegged observation sentences. If we could check the truth values of all the pegged observation sentences, we could evaluate any observation conditional without consulting the theory formulation; theory could be dispensed with. But we cannot. Most of the pegged observation sentences are pegged to inaccessible place-times, and they are infinitely many. The theory formulation is thus a device for remote control and for mass coverage. The theory formulation serves to specify en masse the observation conditionals that we are rightly or wrongly taking to be true. It specifies them by implying them. Under-determination lurks where there are two irreconcilable formulations each of which implies exactly the desired set of observation conditionals plus extraneous theoretical matter, and where no formulation affords a tighter fit.

The only hope for a thesis of under-determination, evidently, is in application to theories that imply observation conditionals infinite in number and too ill-assorted to be exactly encompassed by any finite formulation; tightly encompassed, that is, without theoretical foreign matter. The thesis needs to be read as a thesis about the world. It needs to be read as saying, for one thing, that the observation conditionals that are in fact true in the world are thus ill-assorted. And it needs to be read as saying, further, that we can encompass more of these true observation conditionals in a loose formulation than in any tight one. And it needs to be read as saying, finally, that for any such loose formulation there will be others, empirically equivalent but logically incompatible with it and incapable of being rendered logically equivalent to it by any reconstrual of predicates.

Here, evidently, is the nature of under-determination. There is some infinite lot of observation conditionals that we want to capture in a finite formulation. Because of the complexity of the assortment, we cannot produce a finite formulation that would be equivalent merely to their infinite conjunction. Any finite formulation that will imply them is going to have to imply also some trumped-up matter, or stuffing, whose only service is to round out the formulation. There is some freedom of choice of stuffing, and such is the under-determination.

One may ask: why insist on a finite formulation? Why not just settle for the desired observation conditionals as they stand, in all their infinite variety? The answer is immediate, in the form of another question: how else are you going to specify this desired class of observation conditionals? But this response does not conclude the matter; for we must now notice a startling result due to William Craig.[6] Consider any formulation, and any desired class of consequences of it. For our purposes these consequences would be observation conditionals, but for Craig they can be any sentences. Then Craig shows how to specify a second or Craig class of sentences which are visibly equivalent, one by one, to the sentences of the desired first class; and the remarkable thing about this second class is that membership in it admits of a mechanical decision procedure.

In the cases that matter, these classes are infinite. Even so, the second or Craig class evidently makes the original finite formulation dispensable, by affording a different way of recognizing membership in the desired first class. Instead of showing that a sentence belongs to it by deducing it from the finite formulation, we show it by citing a visibly equivalent sentence that belongs, testably, to the Craig class.

This result does not belie under-determination, since the Craig class is not a finite formulation, but an infinite class of sentences. But it does challenge the interest of under-determination, by suggesting that the finite formulation is dispensable; and indeed the Craig class, for all its infinitude, is an exact fit, being a class of visible equivalents of the desired class. In fact I might say just how excessively visible these equivalences are. Each sentence in the Craig class is simply a repetitive self-conjunction, '*ppp*... *p*', of a sentence of the desired class.

Having said this much, I would do well to finish the Craig story. Why, when the desired class itself is undecidable, should this Craig class of its repetitive self-conjunctions be decidable? The trick is as follows. Each of the desired sentences (each of the desired observation conditionals, in our case) is deducible from the original finite formulation. Its proof can be coded numerically, Gödel fashion. Let the number be *n*. Then the corresponding sentence in the Craig class is the desired sentence repeated in self-conjunction *n* times. The resulting Craig class is decidable. To decide whether a given sentence belongs to it, count its internal repetitions; decode the proof, if any, that this number encodes; and see whether it is a proof of the repeated part of the given sentence.

Obviously there is no place for any of this in practice, as Craig was the first to emphasize. Just counting the repetitions and then decoding the proof from the Gödel number would require astronomical time, and each sentence in the Craig class would require astronomical space if it were to be written. And, after all that, the old original finite formulation still has to be consulted in checking the proof. Craig's point is of course strictly theoretical, and as such it is important.

As I said before, Craig's result does not refute the thesis of under-determination, since the Craig class, for all its tightness of fit, is not a finite formulation. However, this technicality is rather a frail reed at which to grasp. After all, one could reasonably extend the notion of theory formulations to apply not just to an expression but to a recursive set of expressions. So the thesis of under-determination would seem to be demoted to the status, at best, of a thesis affirming a certain contrast between expressions and recursive sets of expressions.

However, I see the importance of the thesis of under-determination as lying elsewhere. The more closely we examine the thesis, the less we seem to be able to claim for it as a theoretical thesis; but it retains significance in terms of what is practically feasible. A tempered version, the most favorable available, might run as follows. We, humanly, are capable of encompassing more true observation conditionals in a loose theory formulation than in any tight system that we might discover and formulate independently of any such loose formulation. And then the thesis would go on to say, as before, that for each such formulation there will be others, empirically equivalent but logically incompatible with it and incapable of being rendered logically equivalent to it by any reconstrual of predicates.

Even in this form the thesis is moot. It no longer stands to reason, as it seemed at first to do. The question now is whether we are underestimating the power of reconstrual of predicates. It does still stand to reason, overwhelmingly, that any theory formulation we may hope to devise as an adequate system of the world will be a loose one; that there will be others empirically equivalent to it and logically incompatible with it. This much is illustrated by the very trivial example where the words 'electron' and 'molecule' were switched, and by the half-trivial example from Poincaré; but these incompatibilities were reconciled by reconstrual of predicates that preserved empirical equivalence. What is moot is whether there are also bound to be cases not thus reconcilable.

The easy way to recognize empirical equivalence of two theory formulations is by seeing a reconstrual of predicates that will carry the one into the other. So it was with the examples just mentioned. But surely this is not the only way. We might study two incompatible theory formulations, trying in vain to imagine an observation that could decide between them, and we might conclude that they are empirically equivalent; we might conclude this without seeing a reconciling reconstrual of predicates. This we might; but there still could be a reconciling reconstrual of predicates, subtle and complex and forever undiscovered. The thesis of under-determination, even in my latest tempered version, asserts that our system of the world is bound to have empirically equivalent alternatives that are not reconcilable by reconstrual of predicates however devious. This, for me, is an open question.

Failing that, a last-ditch version of the thesis of under-determination would assert merely that our system of the world is bound to have empirically equivalent alternatives which, if we were to discover them, we would see no way of reconciling by reconstrual of predicates. This vague and modest thesis I do believe. For all its modesty and vagueness, moreover, I think it vitally important to one's attitude toward science. What it says in effect is just that there are undiscovered systematic alternatives much deeper and less transparent than, for instance, the Poincaré example.

It sets one to wondering about truth. Perhaps there are two best theories that imply all the true observation conditionals and no false ones. The two are equally simple, let us suppose, and logically incompatible. Suppose further, contrary to our last conjecture, that they are not reconcilable by reconstrual of predicates, however devious. Can we say that one, perhaps, is true, and the other therefore false, but that it is impossible in principle to know which? Or,taking a more positivistic line, should we say that truth reaches only to the observation conditionals at most, and, in Kronecker's words, that *alles übrige ist Menschenwerk*?

I incline to neither line. Whatever we affirm, after all, we affirm as a statement within our aggregate theory of nature as we now see it; and to call a statement true is just to reaffirm it. Perhaps it is not true, and perhaps we shall find that out; but in any event there is no extra-theoretic truth, no higher truth than the truth we are claiming or aspiring to as we continue to tinker with our system of the world from within. If ours were one of those two rival best theories that we imagined a moment ago, it would be our place to insist on the truth of our laws and the falsity of the other theory where it conflicts.

This has the ring of cultural relativism. That way, however, lies paradox. Truth, says the cultural relativist, is culture-bound. But if it were, then he, within his own culture, ought to see his own culture-bound truth as absolute. He cannot proclaim cultural relativism without rising above it, and he cannot rise above it without giving it up.

There is a final fantasy to contemplate. Suppose again two rival systems of the world, equally sustained by all experience, equally simple, and irreconcilable by reconstrual of predicates. Suppose further that we can appreciate their empirical equivalence. Must we still embrace one theory and oppose the other, in an irreducible existentialist act of irrational commitment? It seems an odd place for irrational commitment, and I think we can do better. It is the extreme situation where we would do well to settle for a frank dualism. Oscillation between rival theories is standard scientific procedure anyway, for it is thus that one explores and assesses alternative hypotheses. Where there is forever no basis for choosing, then, we may simply rest with both systems and discourse freely in both, using distinctive signs to indicate which game we are playing. This use of distinctive signs leaves us with two irreducible and unconflicting theories.

Harvard University
Cambridge, Mass.

NOTES

[1] I am indebted to Joseph Cowan here. For valuable criticism of the paper in general I am indebted to Burton Dreben.

[2] To dispel a misunderstanding in Harold Morick, 'Observation and Subjectivity in Quine', *Canadian Journal of Philosophy* **1** (1974), pp. 109–127, I must stress that a pegged observation sentence is not an observation sentence. It is a non-observational sentence obtained by pegging an observation sentence.

[3] There is substantially this idea in B. M. Humphries, 'Indeterminacy of Translation and Theory', *Journal of Philosophy* **67** (1970), pp. 167–178, particularly pp. 169f.

[4] Avishai Margalit has suggested to me that this amounts to equating theories that can be formulated by the same Ramsey sentence. (F. P. Ramsey, *The Foundations of Mathematics*, ed. by R. B. Braithwaite, Routledge and Kegan Paul, London, 1931, chapter ix(A), 'Theories'.)

[5] This requirement evidently disqualifies, for our purposes, an example of empirically equivalent and logically irreconcilable theories that is offered by Clark Glymour in his important paper 'Theoretical Realism and Theoretical Equivalence', *Boston Studies in the Philosophy of Science*, Vol. IIIV (1971), pp. 275–288. In his example the empirical evidence is covered by the statement that there are infinitely many objects. This content can be organized indifferently in a theory of dense order and a theory of discrete order, and these two theories are irreconcilable, sharing, as he says, no common model. But this is a case of avoidable branching.

[6] William Craig, 'Replacement of Auxiliary Expressions', *Philosophical Review* **65** (1956), pp. 38–55.

41. Equivalent theories

I defined empirical content in §7 only for testable theories, and I went on to point out that much solid experimental science fails of testability in the defined sense. This can happen, we saw, because of vague and uncalibrated probabilities in the backlog of theory. No doubt it happens also in more complex ways, not clearly understood. I have no definition of empirical content to offer for such theories, but it still seems to make reasonable intuitive sense to speak of empirical equivalence among them, since experimenta-

[5] See my *Set Theory and Its Logic,* 2d ed., pp. 234–238.

[6] See *Quiddities,* pp. 34–36.

tion is still brought to bear. The idea is that whatever observation would be counted for or against the one theory counts equally for or against the other. What I shall have to say about empirically equivalent theories applies indifferently to testable ones and to theories that are empirically equivalent in this ill-defined way.

Theories can differ utterly in their objects, over which their variables of quantification range, and still be empirically equivalent, as proxy functions show (§12). We hardly seem warranted in calling them two theories; they are two ways of expressing one and the same theory. It is interesting, then, that a theory can thus vary its ontology.

Effort and paper have been wasted, by me among others, over what to count as sameness of theory and what to count as mere equivalence. It is a question of words; we can stop speaking of theories and just speak of theory formulations. I shall still write simply 'theory', but you may understand it as 'theory formulation' if you will.

Theories (theory formulations) can be logically incompatible and still be empirically equivalent. A familiar example is Riemannian and Euclidean geometry as applied to the surface of a sphere. Riemannian geometry says that straight lines always meet. Euclidean geometry says that some do and some do not, and in particular that there are none on a sphere. The conflict is resolved by reinterpreting 'straight line' in the Riemannian glossary as 'great circle'.

The next example, due to Poincaré (ch. 4), is less trivial. We have on the one hand our commonsense conception of infinite space and rigid bodies that move freely without shrinking or stretching, and on the other hand the conception of a finite spherical space in which those bodies shrink uniformly as they move away from center. Both conceptions can be reconciled with all possible observations; they are empirically equivalent. Yet they differ, this time, more deeply than in the mere choice of words. The theory with the finite space makes crucial use of a theoretical term that admits of no counterpart in the theory with the infinite space—namely, 'center of space'.

Imagine now two theories, ours and another, such that we are persuaded of their empirical equivalence but we see no way of systematically converting one into the other by reinterpretation sentence by sentence, as we did in the example of the proxy function and that of the sphere. There are three cases to consider.

Case 1: The other theory is logically compatible with our own and is expressed directly in our own terms. It differs from ours in that it implies some theoretical sentences that ours leaves unsettled, and vice versa. Yet the theories are empirically equivalent. This case presents no problem. We would simply accept the other theory and incorporate it into our own as an enrichment, answering many theoretical questions that ours left open.

Case 2: Again the other theory is logically compatible with ours, but, like Poincaré's example, it hinges on some theoretical terms not reducible to ours.

Case 3: The two theories are logically incompatible. Donald Davidson showed me that this case can be reduced to Case 2 by the following maneuver. Take any sentence *S* that the one theory implies and the other denies. Since the theories are empirically equivalent, *S* must hinge on some theoretical term that is not firmly pinned down to observable criteria. We may then exploit its empirical slack by treating that term as two terms, distinctively spelled in the two theories. *S* thus gives way to two mutually independent sentences *S* and *S′*. Continuing thus, we can make the two theories logically compatible.

42. Irresoluble rivalry

So we may limit our attention to Case 2. Let us limit it further to global systems of the world, so that there is no question of fitting the rival theories into a broader context. So we are imagining a global system empirically equivalent to our own and logically compatible with ours but hinging on alien terms. It may seem that as staunch empiricists we should reckon both theories as true. Still, this line is unattractive if the other theory is less simple and natural than ours; and indeed there is no limit to how grotesquely cumbersome a theory might be and still be empirically equivalent to an elegant one. We do better, in such a case, to take advantage of the presence of irreducibly alien terms. We can simply bar them from our language as meaningless. After all, they are not adding to what our own theory can predict, any more than 'phlogiston' or 'entelechy' does, or indeed 'fate', 'grace', 'nirvana', 'mana'. We thus consign all contexts of the alien terms to the limbo of nonsentences.

We have here an encroachment of coherence considerations upon standards of truth. Simplicity and naturalness are making the difference between truth and meaninglessness.

We might still choose to enrich our original theory with any novel findings of the other theory that do not use the alien terms. It would be a matter of welcoming information from a presumed dependable outside source, much as supplementary truths of number theory are got by excursions through analysis, or as the four-color-map question was settled by an elaborate computer program.

But now suppose rather that the rival theory is as neat and natural as our own. Our empiricist scruples reawaken. Should we incorporate that theory into our own, as in Case 1? No, this would ill accord with the scientists' quest for simplicity and economy; for the irreducible new terms imported with the annexed theory have added no new coverage of observables. The two theories were already empirically equivalent to each other, and hence to their conjunction. The two theories were streamlined and neck-and-neck, but the tandem theory is loaded beyond necessity: loaded with all the sentences containing the new terms.

One possible attitude to adopt toward the two theories is a *sectarian* one, as I have called it:[7] treat the rival theory as in the preceding case, by rejecting all the contexts of its alien terms. We can no longer excuse this unequal treatment of the two theories on the ground that our own is more elegant, but still we can plead that we have no higher access to truth than our evolving theory, however fallible. Dagfinn Føllesdal and Roger Gibson abetted me in this sectarian attitude. The opposing attitude is the *ecumenical* one, which would count both theories true. Its appeal is empiricism: reluctance to discriminate invidiously between empirically equivalent and equally economical theories. The tandem theory, which we found prohibitively uneconomical, was one ecumenical line. But a different ecumenical line has been urged by Donald Davidson: that we account both theories separately true, the truth predicate being under-

[7]"Reply to Gibson."

stood now as disquotation in an inclusive and theory-neutral language in which both theories are couched. In recent years I have vacillated among these alternatives, which are now down to two: the sectarian line and the ecumenical line à la Davidson.

The latter alternative raises questions regarding the inclusive language. It would include all terms of both of the rival systems of the world, and its variables would range over both ontologies. Distinctive predicates would serve to delimit values of the variables to the one ontology or the other as needed. How much more widely should the variables range? And what of truth: will there be a hierarchy of truth predicates and a matching hierarchy of styles of variables, as at the end of §37? We must call a halt. We sought only an inclusive language, not a third theory.

What is to be gained is not evident, apart from the satisfaction of conferring the cachet of truth evenhandedly. The sectarian is no less capable than the ecumenist of appreciating the equal evidential claims of the two rival theories of the world. He can still be evenhanded with the cachet of warrantedness, if not of truth. Moreover he is as free as the ecumenist to oscillate between the two theories for the sake of added perspective from which to triangulate on problems. In his sectarian way he does deem the one theory true and the alien terms of the other theory meaningless, but only so long as he is entertaining the one theory rather than the other. He can readily shift the shoe to the other foot.

The fantasy of irresolubly rival systems of the world is a thought experiment out beyond where linguistic usage has been crystallized by use. No wonder the cosmic question whether to call two such world systems true should simmer down, bathetically, to a question of words. Hence also, meanwhile, my vacillation.

Fare these conventions as they may, the rival theories describe one and the same world. Limited to our human terms and devices, we grasp the world variously. I think of the disparate ways of getting at the diameter of an impenetrable sphere: we may pinion the sphere in calipers or we may girdle it with a tape measure and divide by pi, but there is no getting inside.

43. *Two indeterminacies*

There is an evident parallel between the empirical underdetermination of global science and the indeterminacy of translation. In both cases the totality of possible evidence is insufficient to clinch the system uniquely. But the indeterminacy of translation is additional to the other. If we settle upon one of the empirically equivalent systems of the world, however arbitrarily, we still have within it the indeterminacy of translation.

Another distinctive point about the indeterminacy of translation is that it clearly has nothing to do with inaccessible facts and human limitations. Dispositions to observable behavior are all there is for semantics to be right or wrong about (§14). In the case of systems of the world, on the other hand, one is prepared to believe that reality exceeds the scope of the human apparatus in unspecifiable ways.

Let us now look more closely to parallels. On the one hand we have the two incompatible but equally faithful systems of translation; each propounds some translations that the other rejects. On the other hand we have two incompatible but empirically equivalent systems of the world. We noted in §18 that we can reconcile the two systems of translation by recognizing them as defining different relations, $translation_1$ and $translation_2$. We noted in §41 that we can reconcile the two systems of the world by similarly splitting one or more theoretical terms.

What the indeterminacy of translation shows is that the notion of propositions as sentence meanings is untenable. What the empirical under-determination of global science shows is that there are various defensible ways of conceiving the world.

Acknowledgments

Maxwell, Grover. "The Ontological Status of Theoretical Entities." In *Minnesota Studies in the Philosophy of Science*, Vol. 3, *Scientific Explanation, Space, and Time*. Edited by Herbert Feigl and Grover Maxwell (Minneapolis: University of Minnesota Press, 1962): 3–27. Reprinted with the permission of the University of Minnesota Press.

Hanson, Norwood Russell. "Observation." In *Patterns of Discovery* (Cambridge: Cambridge University Press, 1961): 4–30. Reprinted with the permission of Cambridge University Press.

Hempel, Carl G. "Empiricist Criteria of Cognitive Significance." In *Aspects of Scientific Explanation* (New York: Free Press, 1965): 101–19. Reprinted with the permission of Revue Internationale de Philosophie.

Quine, Willard Van Orman. "Two Dogmas of Empiricism." In *From a Logical Point of View* (Cambridge: Harvard University Press, 1964): 20–46. Reprinted with the permission of Harvard University Press.

Hempel, Carl G. "The Theoretician's Dilemma: A Study in the Logic of Theory Construction." In *Minnesota Studies in the Philosophy of Science*, Vol. 2, *Concepts, Theories, and the Mind-Body Problem*. Edited by Herbert Feigel, Michael Scriven, and Grover Maxwell (Minneapolis: University of Minnesota Press, 1958): 37–98. Reprinted with the permission of the University of Minnesota Press.

Suppes, Patrick. "What Is a Scientific Theory?" In *Philosophy of Science Today*, edited by Sidney Morgenbesser (New York: Basic Books, 1967): 55–67. Reprinted with the permission of Basic Books, Division of Perseus Books Group.

van Fraassen, Bas C. "The Semantic Approach to Scientific Theories." In *The Process of Science: Contemporary Philosophical Approaches to Understanding Scientific Practice*, edited by Nancy J. Nersessian (Dordrecht: Martinus Nijhoff, 1987): 105–24. Reprinted with the permission of Martinus Nijhoff, Division of Wolters Kluwer.

van Fraassen, Bas C. "To Save the Phenomena." *Journal of Philosophy* 73 (1976): 623–32. Reprinted with the permission of the Journal of Philosophy, Inc., Columbia University, and the author.

Lewis, David. "How to Define Theoretical Terms." *Journal of Philosophy* 67 (1970): 427–46. Reprinted with the permission of the Journal of Philosophy, Inc., Columbia University, and the author.

Sklar, Lawrence. "Saving the Noumena." *Philosophical Topics* 13 (1982): 89–110. Reprinted with the permission of the University of Arkansas, Department of Philosophy.

Fine, Arthur. "The Natural Ontological Attitude." In *The Shaky Game: Einstein Realism and the Quantum Theory* (Chicago: University of Chicago Press, 1986): 112–135. Reprinted with the permission of the University of Chicago Press.

Cartwright, Nancy. "The Reality of Causes in a World of Instrumental Laws." *PSA 1980* 2 (1981): 38–48. Reprinted with the permission of the Philosophy of Science Association.

Sklar, Lawrence. "The Conventionalist Thesis and Its First Critics." In *Space, Time, and Spacetime* (Berkeley: University of California Press, 1974): 88–103. Reprinted with the permission of University of California Press.

Sklar, Lawrence. "The Framework of Philosophical Perplexity." In *Space, Time, and Spacetime* (Berkeley: University of California Press, 1974): 113–46. Reprinted with the permission of University of California Press.

Putnam, Hilary. "Realism and Reason." In *Meaning and the Moral Sciences* (London: Routledge and Kegan Paul, 1978): 123–38 and notes on pages 139–140. Reprinted with the permission of Routledge and Kegan Paul.

Quine, W.V. "On Empirically Equivalent Systems of the World." *Erkenntnis* 9 (1975): 313–28. Reprinted with the permission of Kluwer Academic Publishers.

Quine, W.V. "Equivalent Theories," "Irresoluble Rivalry," and "Two Indeterminacies." In *Pursuit of Truth,* Section V: 41–43 (Cambridge: Harvard University Press, 1990): 95–102. Reprinted with the permission of Harvard University Press.